Electromagnetic Signal and Information Theory

AF834376

Electromagnetic Signal and Information Theory

From Synergizing Three Realms to Evolution of Next Generation Communications

Edited by

Aryan Kaushik

OUF Innovative
London, United Kingdom

Marco Di Renzo

Paris-Saclay University
Paris, France

ELSEVIER

ACADEMIC PRESS
An imprint of Elsevier

Academic Press is an imprint of Elsevier
125 London Wall, London EC2Y 5AS, United Kingdom
50 Hampshire Street, 5th Floor, Cambridge, MA 02139, United States

Copyright © 2027 Elsevier Inc. All rights are reserved, including those for text and data mining, AI training, and similar technologies.

For accessibility purposes, images in electronic versions of this book are accompanied by alt text descriptions independently provided by Elsevier, generated with the aid of AI. For more information, see https://www.elsevier.com/about/accessibility or Radarweg 29, 1043 NX Amsterdam, Netherlands.

Books and Journals published by Elsevier comply with applicable product safety requirements. For any product safety concerns or queries, please contact our authorised representative, Elsevier B.V., at productsafety@elsevier.com.

Publisher's note: Elsevier takes a neutral position with respect to territorial disputes or jurisdictional claims in its published content, including in maps and institutional affiliations.

No part of this publication may be reproduced or transmitted in any form or by any means, electronic or mechanical, including photocopying, recording, or any information storage and retrieval system, without permission in writing from the publisher. Details on how to seek permission, further information about the Publisher's permissions policies and our arrangements with organizations such as the Copyright Clearance Center and the Copyright Licensing Agency, can be found at our website: www.elsevier.com/permissions.

This book and the individual contributions contained in it are protected under copyright by the Publisher (other than as may be noted herein).

Notices
Knowledge and best practice in this field are constantly changing. As new research and experience broaden our understanding, changes in research methods, professional practices, or medical treatment may become necessary.

Practitioners and researchers must always rely on their own experience and knowledge in evaluating and using any information, methods, compounds, or experiments described herein. In using such information or methods they should be mindful of their own safety and the safety of others, including parties for whom they have a professional responsibility.

To the fullest extent of the law, neither the Publisher nor the authors, contributors, or editors, assume any liability for any injury and/or damage to persons or property as a matter of products liability, negligence or otherwise, or from any use or operation of any methods, products, instructions, or ideas contained in the material herein.

ISBN: 978-0-443-36354-2

For information on all Academic Press publications
visit our website at https://www.elsevier.com/books-and-journals

Publisher: Matthew Deans
Acquisitions Editor: Ritika Manhas
Editorial Project Manager: Siddharth Anand
Production Project Manager: Anitha Sivaraj
Cover Designer: Raman Kumar

ELSEVIER Book Aid International
Working together to grow libraries in developing countries
www.elsevier.com • www.bookaid.org

Typeset by VTeX

Contents

Contributors

Qammer H. Abbasi James Watt School of Engineering, University of Glasgow, Glasgow, United Kingdom
School of Engineering, College of Science and Engineering, University of Glasgow, Glasgow, United Kingdom

Emre Arslan 6GEN. LAB, Turkcell, Istanbul, Turkiye

Mehmet Basaran 6GEN. LAB, Turkcell, Istanbul, Turkiye

Vimal Bhatia Department of Electronics & Communication Engineering, Indian Institute of Technology Indore, Prayagraj, India
Center for Wireless Communications, University of Oulu, Oulu, Finland

Frankie Chiu Department of Electronic and Computer Engineering, The Hong Kong University of Science and Technology, Hong Kong, Hong Kong

Bruno Clerckx University of Liverpool, Liverpool, United Kingdom

Fatih Cögen 6GEN. LAB, Turkcell, Istanbul, Turkiye

Ahmet Faruk Coskun 6GEN. LAB, Turkcell, Istanbul, Turkiye

Davide Dardari University of Bologna, Dipartimento di Ingegneria dell'Energia Elettrica e dell'Informazione "Guglielmo Marconi", Bologna, Italy

Özlem Tuğfe Demir Bilkent University, Ankara, Turkiye

Massimo Franceschetti Department of Electrical and Computer Engineering, University of California at San Diego, La Jolla, CA, United States

Chenxuan Gu National Mobile Communications Research Laboratory, School of Information Science and Engineering, Southeast University, Nanjing, China
Purple Mountain Laboratories, Nanjing, China

Saber Hassouna James Watt School of Engineering, University of Glasgow, Glasgow, United Kingdom
School of Engineering, College of Science and Engineering, University of Glasgow, Glasgow, United Kingdom

Jie Huang National Mobile Communications Research Laboratory, School of Information Science and Engineering, Southeast University, Nanjing, China
Purple Mountain Laboratories, Nanjing, China

Muhammad Imran James Watt School of Engineering, University of Glasgow, Glasgow, United Kingdom
School of Engineering, College of Science and Engineering, University of Glasgow, Glasgow, United Kingdom

Zheng-Rong Jin National Mobile Communications Research Laboratory, School of Information Science and Engineering, Southeast University, Nanjing, China
Purple Mountain Laboratories, Nanjing, China

Jalil Kazim James Watt School of Engineering, University of Glasgow, Glasgow, United Kingdom
School of Engineering, College of Science and Engineering, University of Glasgow, Glasgow, United Kingdom

Riichi Kudo NTT Network Innovation Laboratories, NTT Corporation, Yokosuka, Japan

Deepak Kumar Department of Electronics & Communication Engineering, Motilal Nehru National Institute of Technology Allahabad, Prayagraj, India

Xiazhi Lai School of Computer Science, Guangdong University of Education, Guangdong, China

Doohwan Lee NTT Network Innovation Laboratories, NTT Corporation, Yokosuka, Japan

Hongyu Li The Hong Kong University of Science and Technology (Guangzhou), Guangzhou, China

Junling Li National Mobile Communications Research Laboratory, School of Information Science and Engineering, Southeast University, Nanjing, China
Purple Mountain Laboratories, Nanjing, China

Qingyin Ma National Mobile Communications Research Laboratory, School of Information Science and Engineering, Southeast University, Nanjing, China
Purple Mountain Laboratories, Nanjing, China

Michail Matthaiou School of Electronics, Electrical Engineering and Computer Science, Queen's University Belfast, Belfast, United Kingdom

Marco Donald Migliore University of Cassino and Southern Lazio, Dipartimento di Ingegneria Elettrica e dell'Informazione "Maurizio Scarano", Cassino, Italy

Deepak Mishra School of Electrical Engineering and Telecommunications, University of New South Wales (UNSW) Sydney, Sydney, NSW, Australia

Mohammadali Mohammadi School of Electronics, Electrical Engineering and Computer Science, Queen's University Belfast, Belfast, United Kingdom

Ross Murch Department of Electronic and Computer Engineering, The Hong Kong University of Science and Technology, Hong Kong, Hong Kong

Matteo Nerini Imperial College London, London, United Kingdom

Hien Quoc Ngo School of Electronics, Electrical Engineering and Computer Science, Queen's University Belfast, Belfast, United Kingdom

Rui Ni Advanced Wireless Technology Lab, Huawei Technologies Canada Co., Ltd. Kanata, Ontario, Canada

Tianrun Qi National Mobile Communications Research Laboratory, School of Information Science and Engineering, Southeast University, Nanjing, China
Purple Mountain Laboratories, Nanjing, China

Hirofumi Sasaki NTT Network Innovation Laboratories, NTT Corporation, Yokosuka, Japan

Xingyao Shangguan National Mobile Communications Research Laboratory, School of Information Science and Engineering, Southeast University, Nanjing, China
Purple Mountain Laboratories, Nanjing, China

Shanpu Shen University of Macau, Taipa, Macau

Vidya Bhasker Shukla Center for Wireless Communications, University of Oulu, Oulu, Finland

Kosuke Suzuoki NTT Network Innovation Laboratories, NTT Corporation, Yokosuka, Japan

Jie Tang South China University of Technology, Guangdong, China

Wen Tong Advanced Wireless Technology Lab, Huawei Technologies Canada Co., Ltd. Kanata, Ontario, Canada

Cheng-Xiang Wang National Mobile Communications Research Laboratory, School of Information Science and Engineering, Southeast University, Nanjing, China
Purple Mountain Laboratories, Nanjing, China

Kai-Kit Wong Department of Electronic and Electrical Engineering, University College London, United Kingdom

Qun Wu National Mobile Communications Research Laboratory, School of Information Science and Engineering, Southeast University, Nanjing, China
Purple Mountain Laboratories, Nanjing, China

Tuo Wu South China University of Technology, Guangdong, China

Yasunori Yagi NTT Network Innovation Laboratories, NTT Corporation, Yokosuka, Japan

Runruo Yang National Mobile Communications Research Laboratory, School of Information Science and Engineering, Southeast University, Nanjing, China
Purple Mountain Laboratories, Nanjing, China

Yue Yang National Mobile Communications Research Laboratory, School of Information Science and Engineering, Southeast University, Nanjing, China
Purple Mountain Laboratories, Nanjing, China

Junteng Yao Faculty of Electrical Engineering and Computer Science, Ningbo University, Zhejiang, China

Zehra Yigit 6GEN. LAB, Turkcell, Istanbul, Turkiye

Kangda Zhi School of Electrical Engineering and Computer Science, Technical University of Berlin, Berlin, Germany

Qian Zhu Advanced Wireless Technology Lab, Huawei Technologies Co., Ltd. Shenzhen, Guangdong, China

About the editors

Prof. Aryan Kaushik is Director of OUF Innovative (UK). Previously he has been Associate Professor at Manchester Met (UK), Assistant Professor at University of Sussex (UK), and Research Fellow at UCL (UK). He has also held visiting positions at leading institutions across UK, Luxembourg, Italy, China, etc. Recognized among the Stanford/Elsevier World's Top 2% Scientists, he serves as an international assessor, PhD examiner, and funding evaluator. He chairs IEEE ComSoc TWG-ESIT and SIG-AITNTN, serves as one6G WG3 Vice-Chair, has edited five books, received seven IEEE Exemplary Editor Awards, and has delivered over 150 talks worldwide while chairing numerous flagship IEEE conferences.

Prof. Marco Di Renzo is Professor of Telecommunications Engineering, Director of the Centre for Telecommunications Research, and Head of the Telecommunications Group at King's College London, UK. He is also CNRS Research Director at IETR-CentraleSupélec, France. He is a Fellow of IEEE, IET, EURASIP, RSA, and AAIA; an Academician of AIIA; and an Ordinary Member of the European Academy of Sciences and Arts, Academia Europaea, and the Italian Academy of Technology and Engineering. He is an Ambassador of the European Association on Antennas and Propagation and a Highly Cited Researcher. He is Principal Investigator of an ERC Synergy Grant on metasurface-based information processing.

Acknowledgments

This book, being the third one, is dedicated to my parents, friends, and family.

Also, a big shout out to our collaborators and contributors of this book whose work has been phenomenal.

Thanks to Elsevier team for executing its publication from the beginning to the end.

And we are also grateful to the readers.

CHAPTER 1

ESIT fundamentals

Theoretical bounds and modeling

Marco Donald Migliore[a] **and Davide Dardari**[b]

[a]University of Cassino and Southern Lazio, Dipartimento di Ingegneria Elettrica e dell'Informazione "Maurizio Scarano", Cassino, Italy [b]University of Bologna, Dipartimento di Ingegneria dell'Energia Elettrica e dell'Informazione "Guglielmo Marconi", Bologna, Italy

1.1 Introduction

Over the past few decades, there has been a dramatic increase in the amount of information transmitted via wireless systems, a trend that shows no signs of slowing down. On the contrary, we expect an even more substantial increase in the near future, driven by an increasingly interconnected society. Addressing this exponential growth in data transmission demands will pose a significant technological challenge and require sophisticated use of the resources provided by radio channels. This situation raises two critical questions: what is the maximum amount of resources that the wireless communication channel can provide to the communication system, and what are the optimal strategies to exploit these resources? Electromagnetic signal and information theory (ESIT) integrates the fields of Electromagnetism, Information Theory, and Signal Processing to form a unified approach aimed at addressing these questions.

Figuratively speaking, ESIT delves deep into the OSI stack, right down to the physical laws that govern our universe. These laws set the ultimate limits on how much information can be reliably transmitted, and how to manipulate the physical world to maximize that capacity. From this perspective, ESIT extends information processing into the deepest area of the physical layer, the "deep physical layer" [1], operating directly on the electromagnetic (EM) field associated with information.

The study of the connections between electromagnetism and information traces its origins to the works of Toraldo di Francia [2], where he assessed the information contained in a coherent light image by introducing the concept of degrees-of-freedom (DoF) in optics. This concept was later extended to radiofrequency EM fields in the context of approximation theory [3], providing an estimate through band-limited approximation of the field [4]. With

Electromagnetic Signal and Information Theory
https://doi.org/10.1016/B978-0-44-336354-2.00008-3
Copyright © 2027 Elsevier Inc. All rights are reserved, including those for text and data mining, AI training, and similar technologies.

the introduction of multiple-input multiple-output (MIMO) systems, i.e., the exploitation of the spatial dimension, a systematic analysis of the information content of the electromagnetic field has begun. Studies on information processing techniques at the level of the EM field, although relatively recent, have experienced extremely rapid development, enabling the identification of a class of systems and algorithms that synergistically allow economical and effective processing directly at the field level [5–13]. Examples of such systems include reconfigurable intelligent surfaces (RISs), stacked intelligent metasurfaces (SIMs), extremely large-scale MIMO (XL-MIMO) arrays (also referred to as extremely large antenna arrays (ELAAs)) [14], large intelligent surfaces (LISs), also known as Holographic MIMO (HMIMO) surfaces [15,16], dynamic metasurface antennas (DMAs) [17], dynamic scattering arrays (DSAs) [18], fluid and reconfigurable antennas [19]. The availability of enabling technologies capable of implementing electrically large EM structures facilitates the exploitation of near-field propagation phenomena, thereby increasing the spatial DoF. This, in turn, enables beam focusing, adaptive interference shaping, and high-resolution sensing capabilities that are unattainable with conventional far-field operation [20–23].

The purpose of this chapter is to introduce some foundational concepts underlying ESIT. After a brief overview of electromagnetism and information theory, Section 1.4 identifies the limits of information that can be transmitted in various canonical geometries by exploring the subtle connection between the information associated with the electromagnetic field and the field's energy. These concepts are further applied in Section 1.5, demonstrating how the amount of information transmitted by wireless telecommunication systems can be maximized using signal processing techniques applied to physical structure allowing direct manipulation of the electromagnetic field. The same section shows how a generic reconfigurable scattering device can be seen as a mode converter and used to perform operations of the EM field whose constraints depend on the geometry and the technology adopted.

1.2 A quick look at Maxwell's equations in the harmonic domain

In this section, Maxwell's equations will be briefly recalled focusing on some aspects that are particularly relevant in ESIT. The discussion will be limited to time-varying sinusoidal fields. The time dependence $e^{j\omega t}$ (where ω is the angular frequency) will be understood and subsequently omitted. The position vector in three-dimensional space is denoted by $\mathbf{r}$. The fields are assumed to be square-integrable, and the MKSA measurement system is adopted. For simplicity, we will not consider magnetic sources in this chapter. While the physical existence of magnetic charges and currents has not been proven, they are introduced in certain electromagnetic problems for mathematical convenience, such as in the equivalence theorem. Extending the formulas to include magnetic sources is straightforward and can be found in books on applied electromagnetism.

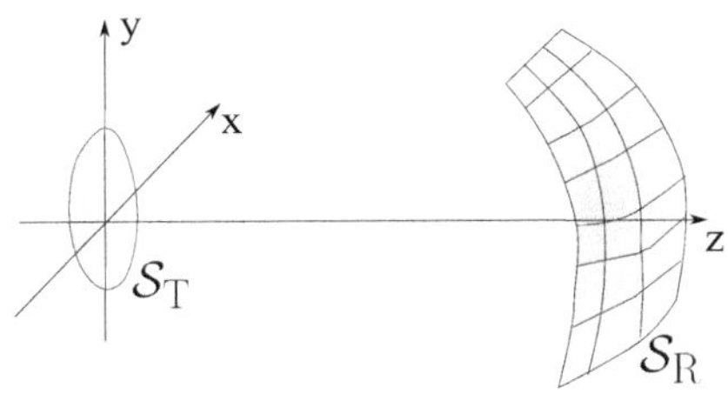

Figure 1.1:
Geometry of the electromagnetic problem.

A simple, but quite general, model used in electromagnetic problems consists of harmonic electromagnetic sources placed in a volume $\mathcal{S}_T$ radiating into free space (Fig. 1.1). An observation manifold $\mathcal{S}_R$ is positioned at a nonzero distance from the source $\mathcal{S}_T$. The electric current density distribution on $\mathcal{S}_T$ is denoted by $\mathbf{J}(\mathbf{r}')$, where $\mathbf{r}' \in \mathcal{S}_T$. We assume that both the electric field $\mathbf{E}(\mathbf{r})$ and the current density $\mathbf{J}(\mathbf{r}')$ belong to separable Hilbert spaces equipped with the $\mathbb{L}_2$ norm. The source current density is assumed to have finite energy, i.e., a finite $\mathbb{L}_2$ norm. Without loss of generality, we impose $\|\mathbf{J}(\mathbf{r}')\|^2 < 1$, where $\|\cdot\|$ denotes the norm. The problem is to evaluate the field radiated by the currents on the observation manifold.

The field radiated by a source placed in the domain $\mathcal{S}_T$ is [24]

$$\mathbf{E}(\mathbf{r}) = \int_{\mathcal{S}_T} \bar{\mathbf{G}}(\mathbf{r}, \mathbf{r}') \cdot \mathbf{J}(\mathbf{r}') \, dV \,, \tag{1.1}$$

where $\beta = \omega\sqrt{\epsilon_0 \mu_0}$ is the free space wavenumber, ϵ_0 is the permittivity of free space, μ_0 is the magnetic permeability of free space, $\bar{\mathbf{G}}$ is the dyadic Green function, and $\bar{\mathbf{I}}$ is the unit dyadic. In 3D space, the dyadic Green function is [24], [25, p. 38]:

$$\bar{\mathbf{G}}(\mathbf{r}, \mathbf{r}') = -j\omega\mu_0 \left[\bar{\mathbf{I}} + \frac{\nabla\nabla}{\beta^2} \right] G(\mathbf{r}, \mathbf{r}') \,, \tag{1.2}$$

$$G(\mathbf{r}, \mathbf{r}') = \frac{\exp(-j\beta|\mathbf{r} - \mathbf{r}'|)}{4\pi\,|\mathbf{r} - \mathbf{r}'|} \,. \tag{1.3}$$

where $\nabla\nabla f$ is the Hessian matrix, whose elements are the second-order partial derivatives of the function f. The integral operator (1.1), which maps the set of currents in the set of fields on $\mathcal{S}_R$, is called the "radiation operator." Using standard functional analysis tools, the radiation operator $\mathcal{A}$ maps the set $\mathcal{J}$, whose elements, denoted as x, are all the physically realizable currents on $\mathcal{S}_T$, into the set $\mathcal{E}$, whose elements, denoted as y, are the field distributions on $\mathcal{S}_R$ (Fig. 1.2).

Given the operator $\mathcal{A} : \mathcal{J} \to \mathcal{E}$, its adjoint $\mathcal{A}^* : \mathcal{E} \to \mathcal{J}$ is defined by the property $\langle \mathcal{A}\mathsf{x}, \mathsf{y} \rangle = \langle \mathsf{x}, \mathcal{A}^*\mathsf{y} \rangle$, $\forall \mathsf{x} \in \mathcal{J}, \mathsf{y} \in \mathcal{E}$, where $\langle \cdot, \cdot \rangle$ denotes the inner product. The kernel of the radiation

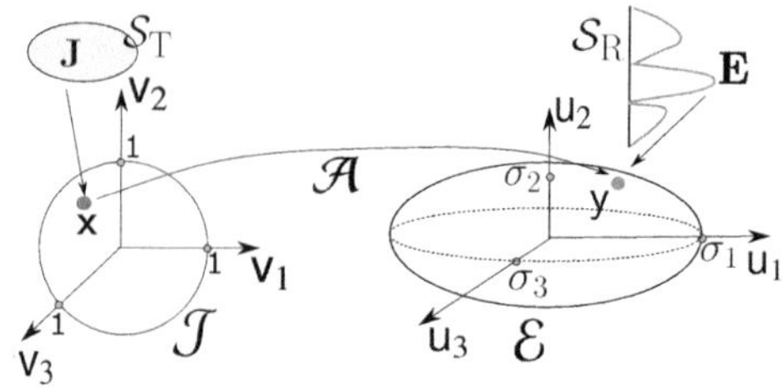

Figure 1.2:
Functional representation of the communication process.

operator is complex, and the operator is not self-adjoint. However, it is possible to consider the operators $\mathcal{A}^*\mathcal{A} : \mathcal{J} \to \mathcal{J}$ and $\mathcal{A}\mathcal{A}^* : \mathcal{E} \to \mathcal{E}$. These operators are nonnegative self-adjoint compact operators, and consequently, it is possible to apply the well-established literature on self-adjoint operators. In particular, both $\mathcal{A}^*\mathcal{A}$ and $\mathcal{A}\mathcal{A}^*$ admit a countably (possibly infinite) set of positive eigenvalues. Furthermore, $\mathcal{A}^*\mathcal{A}$ and $\mathcal{A}\mathcal{A}^*$ have exactly the same positive eigenvalues σ_k^2 with the same multiplicity.

Let u_k and v_k be the eigenfunctions of $\mathcal{A}\mathcal{A}^*$ and $\mathcal{A}^*\mathcal{A}$ associated with eigenvalues σ_k^2 of

$$\begin{aligned} \mathcal{A}\mathcal{A}^*\mathsf{u}_k &= \sigma_k^2\,\mathsf{u}_k, \\ \mathcal{A}^*\mathcal{A}\,\mathsf{v}_k &= \sigma_k^2\,\mathsf{v}_k. \end{aligned} \tag{1.4}$$

The set of eigenfunctions $\{\mathsf{u}_k\}$ forms an orthonormal basis in $\mathcal{N}(\mathcal{A})^\perp$ (wherein $\mathcal{N}(\mathcal{A})$ is the null space of $\mathcal{A}$, defined as $\mathcal{N}(\mathcal{A}) = \{\mathsf{x} \in \mathcal{J} : \mathcal{A}\mathsf{x} = 0\}$), while $\{\mathsf{v}_k\}$ forms an orthonormal basis in the $\mathcal{N}(\mathcal{A}^*)^\perp$, e.g., the closure of the subspace of the "exact" (i.e., "noiseless") observed data. They are also the solution of the shifted eigenvalue problem

$$\begin{aligned} \mathcal{A}\,\mathsf{v}_k &= \sigma_k\,\mathsf{u}_k, \\ \mathcal{A}^*\mathsf{u}_k &= \sigma_k\,\mathsf{v}_k\,. \end{aligned} \tag{1.5}$$

This allows diagonalizing the operator using the Hilbert–Schmidt decomposition (that reduces to the singular value decomposition (SVD) in finite-dimensional spaces) of $\mathcal{A}$ [25,26]

$$\mathsf{y} = \sum_{k=1}^{\infty} \sigma_k \langle \mathsf{x}, \mathsf{v}_k \rangle\, \mathsf{u}_k, \tag{1.6}$$

wherein the positive numbers σ_k are called *singular values* of the integral operator, and the functions u_k and v_k are the left and right singular functions. The triple $\{\mathsf{v}_k, \mathsf{u}_k, \sigma_k\}$ is called the *singular system* of the integral operator. The singular values are sorted in nonincreasing order, i.e., $\sigma_1 \geq \sigma_2 \geq \cdots \geq \sigma_k \geq \cdots$, and $\lim_{k\to\infty} \sigma_k = 0$. The asymptotic decay rate of the

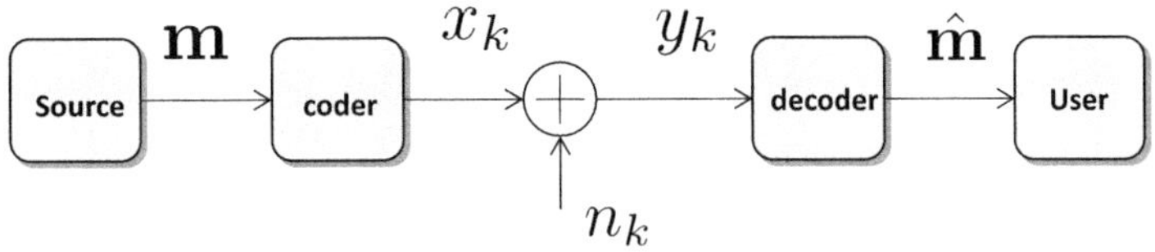

Figure 1.3:
The classical AWGN communication scheme.

singular values depends on the regularity characteristics of the operator kernel. In the case of an analytic function kernel, the decay is at least exponential, resulting in a sharp drop-off [27,28].

A further powerful method used in electromagnetic theory for the representation of the field is the modal expansion [29]. Starting from Maxwell's equations, and considering that the electric field $\mathbf{E}$ in free space is solenoidal, we obtain the homogeneous Helmholtz equation [24]

$$\nabla^2\mathbf{E}(\mathbf{r}) + \beta^2\mathbf{E}(\mathbf{r}) = 0, \tag{1.7}$$

with the condition $\nabla \cdot \mathbf{E} = 0$.

While (1.1) allows the evaluation of the field from the currents, (1.7) describes the propagation of the field in a free space portion of the space. In particular, the general field in free space is described by the linear superposition of solutions to (1.7). Each solution, also called a mode, represents an eigenfunction of the eigenproblem associated with the homogeneous wave equation. From these solutions, the modal representation of the field can be derived, consisting of a superposition of such solutions.

Analytic solutions to the wave equation are available only in a few coordinate systems where the equation is separable. They include rectangular, cylindrical, and spherical, coordinates [29].

1.3 A quick look at Shannon theory

The basic result of information theory is the channel coding theorem, introduced in the milestone paper by C. Shannon in 1948 [30], which proves that information can be sent reliably over a channel provided the data rate is less than or equal to the channel capacity.

The most important channel model is the time-discrete additive white Gaussian noise (AWGN) illustrated in Fig. 1.3, where symbol $x_k \in \mathbb{C}$ transmitted at discrete time instant k is corrupted by a zero mean complex Gaussian random variable $n_k \sim \mathcal{CN}\left(0, \sigma_{\mathrm{N}}^2\right)$ with variance

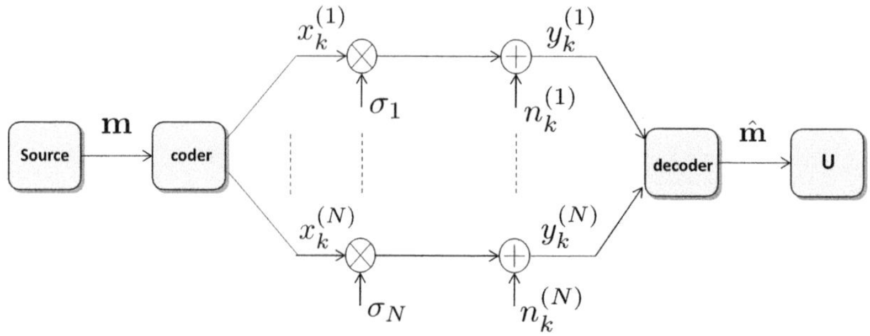

Figure 1.4:
Communication scheme with N parallel independent AWGN channels.

(power) σ_{N}^2, where $\mathcal{CN}(\cdot,\cdot)$ denotes a multivariate complex normal distribution. The received signal is given by $y_k = x_k + n_k$. The purpose of the coder block at the transmitter and the decoder block at the receiver is to make the transmission of the original message $\mathbf{m}$ of M bits as reliable as possible, that is, with the lowest probability of decoding error, by coding it using a suitable codeword $\mathbf{x} = [x_1, x_2, \ldots, x_K]$ under the constraint on the transmitted power $\frac{1}{K}\sum_k |x_k|^2 \leq P$. The channel capacity represents the maximum achievable data rate that can be sustained by the channel with an arbitrarily small decoding error probability. Specifically, the capacity of a complex time-discrete AWGN channel with a transmit power constraint P and noise variance σ_{N}^2 is [31]

$$C = \log_2\left(1 + \frac{P}{\sigma_{\mathrm{N}}^2}\right) = \log_2(1 + \mathrm{SNR}) \qquad \text{[bits/channel use]}, \tag{1.8}$$

where $\mathrm{SNR} = \frac{P}{\sigma_{\mathrm{N}}^2}$ denotes the signal-to-noise ratio (SNR). Looking at (1.8), it is evident that by increasing the SNR only relatively small capacity improvements can be obtained because of the log function. Moreover, constraints on the maximum available power are always present in practical applications (regulations, battery lifetime, etc.). Therefore, one might ask how it is possible to design high-speed communication systems capable of satisfying the exponential demand for wireless data transmission.

To answer this question, let us consider the following scenario in Fig. 1.4 where N parallel AWGN channels with different gains σ_n, for $n = 1, 2, \ldots, N$, and independent noise with power σ_{N}^2, are available to the communication system. For a fair comparison, suppose the constraint P on the total transmitted power is considered, but the transmitter can encode the message exploiting the parallel channels and assigning to each a specific transmitted power P_n, $n = 1, 2, \ldots, N$, such that $\sum_n P_n \leq P$. The receiver exploits the N (noisy) outputs to decode the original message. Due to (1.8) being channel independent, the total capacity is

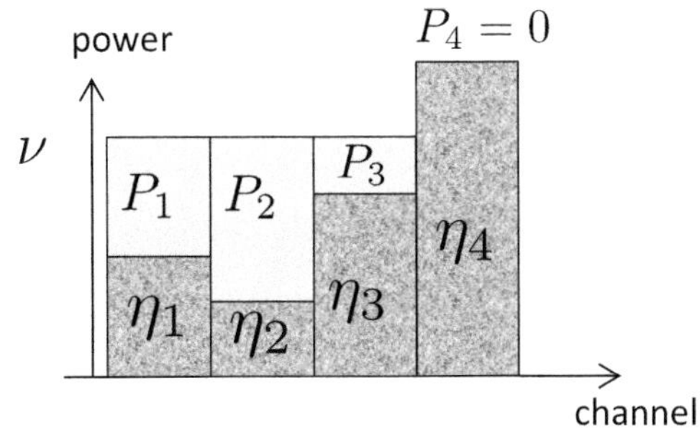

Figure 1.5:
Water-filling interpretation of the optimal power allocation scheme.

simply the sum of the channel capacities [31], that is,

$$C = \sum_{n=1}^{N} \log_2 \left(1 + \frac{P_n \sigma_n^2}{\sigma_{\mathrm{N}}^2}\right). \tag{1.9}$$

Differently from the single-channel case in (1.8), here we have at our disposal N DoF, i.e., the P_n's, minus one because of the power constraint, that we can use to maximize (1.9). This leads to the following constrained maximization problem:

$$C^{(\max)} = \max_{\{P_n\}:\sum_n P_n \le P} \sum_{n=1}^{N} \log_2 \left(1 + \frac{P_n \sigma_n^2}{\sigma_{\mathrm{N}}^2}\right). \tag{1.10}$$

The solution to such a maximization problem is the so-called *water-filling* solution according to which the capacity-maximizing power allocation is [31]

$$P_n = \left(\nu - \frac{\sigma_{\mathrm{N}}^2}{\sigma_n^2}\right)^+ \tag{1.11}$$

for $n = 1, 2, \ldots, N$, where ν is chosen to satisfy

$$\sum_{n=1}^{N} \left(\nu - \frac{\sigma_{\mathrm{N}}^2}{\sigma_n^2}\right)^+ = P \tag{1.12}$$

and $(x)^+ = \max(0, x)$.

A graphical interpretation of the water-filling solution is sketched in Fig. 1.5, where ν in (1.12) can be thought as the level of the water filling a valley whose orography is shaped by the levels $\eta_n = \sigma_{\mathrm{N}}^2/\sigma_n^2$. The level of the water at each channel, given by (1.11), determines

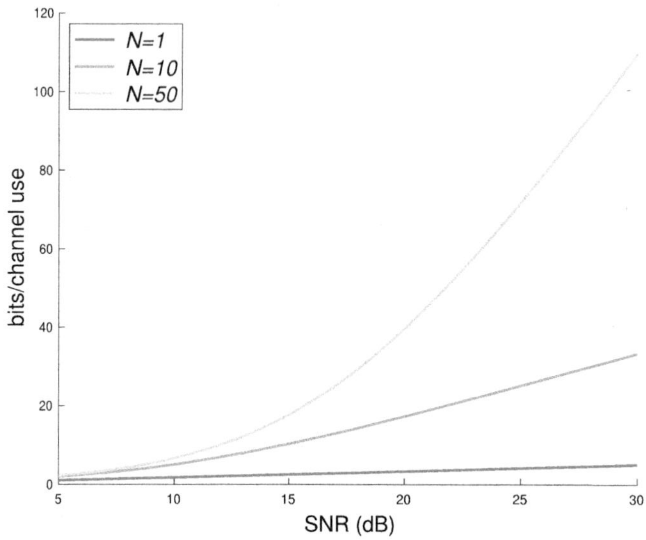

Figure 1.6:
Capacity of N parallel AWGN channels.

the power to be allocated to that channel. When the noise power of a channel is very high, it might happen that no power is allocated at all (channel 4 in the figure) and the actual number of channels used is less than N.

More generally, it can be shown that for large SNRs the optimal power allocation tends to be uniform. In contrast, for low SNRs the water-filling solution tends to allocate all the power to the strongest channel [32]. At first glance, the water-filling solution may appear counterintuitive because more power is allocated to the stronger channels, i.e., those less noisy, and less or no power is allocated to weak channels. It is like an anti-"Robin Hood" policy where money is taken from the poor and given to the rich. This solution derives from the particular structure of the capacity formula (1.9) as a sum of log functions.

To illustrate the potential advantage of considering multiple parallel channels, we consider the particular case of identical AWGN channels, i.e., where $\sigma_n^2 = \sigma^2$ for $n = 1, 2, \ldots, N$. Under this assumption, it can be easily seen that the water-filling solution leads to a uniform power allocation $P_n = P/N$, for $n = 1, 2, \ldots, N$, and the corresponding maximum capacity

$$C^{(\max)} = N \log_2\left(1 + \frac{P\sigma^2}{N\sigma_{\mathrm{N}}^2}\right) = N \log_2\left(1 + \frac{\mathsf{SNR}}{N}\right). \tag{1.13}$$

In Fig. 1.6, the maximum capacity in (1.13) as a function of the SNR is plotted for different numbers of identical AWGN channels. The single-channel classical capacity in (1.8) is obtained with $N = 1$ and reported for comparison. From the plots, it is evident the significant capacity boost that can be obtained when increasing the number of parallel channels. Therefore, the takeaway message is that to increase the overall capacity, it is better to allocate the available power among parallel independent AWGN channels instead of concentrating it on only one channel. The optimal power distribution maximizing the capacity is obtained through the water-filling algorithm.

Now, the main question is how to obtain a large number of parallel channels also dubbed as DoF or channel spatial dimensions. The classical way is to increase the available bandwidth, for instance, by exploring high-frequency bands such as millimeter wave and THz. However, this is not always possible due to the scarcity of the radio spectrum, propagation effects, and cost constraints. A breakthrough approach has been the introduction of the MIMO technology in the late 1990s. Here, a potentially large number of DoF is obtained by exploiting the spatial domain thanks to the adoption of multiple antennas both at the transmitter and the receiver – more generally, by considering electrically large antenna structures as will be discussed in Section 1.4.

1.4 Information carried by the electromagnetic field

In the previous section, it was demonstrated that the capacity of MIMO systems is dependent on the value of the DoF, representing the number of dimensions of the spatial channel, usually obtained by SVD of the channel matrix. This section aims to elucidate the relationship between the DoF and the electromagnetic problem, thereby describing the communication link from a physical perspective.

1.4.1 Degrees of freedom in spatial communication systems

In this section, the electromagnetic model shown in Fig. 1.1 will be analyzed from an information-theoretic perspective. From this viewpoint, the problem can be simplified as follows: there is a variety of possible different field configurations on $\mathcal{S}_\mathrm{R}$ that a receiver can distinguish. The transmitter selects a specific element x in the set $\mathcal{J}$ of the electric current distributions on the radiating source. The radiation operator then maps x to a specific element y in the set $\mathcal{E}$, which is observed by the receiver.

Loosely speaking, information gained by observing the field on $\mathcal{S}_\mathrm{R}$ is determined by the decrease in uncertainty about the set of possible field configurations on $\mathcal{S}_\mathrm{R}$: the larger the number of distinguishable elements in $\mathcal{E}$, the higher the amount of information that can be transmitted. The main goal of this section is to count the number of distinguishable elements in $\mathcal{E}$.

The first step is to identify a suitable representation of $\mathcal{E}$ in terms of linear superposition of orthogonal basis. Let $\mathcal{L}_n$ be an n-dimensional (linear) subspace of the set $\mathcal{E}$, and $\delta(\mathcal{E}, \mathcal{L}_n)$ the separation between $\mathcal{E}$ and $\mathcal{L}_n$, defined as $\delta(\mathcal{E}, \mathcal{L}_n) = \sup_{\mathcal{E}} \inf_{\mathcal{L}_n} \|\mathcal{E} - \mathcal{L}_n\|$. For a fixed n, let us find the infimum, $d_n(\mathcal{E})$, of the separations $\delta(\mathcal{E}, \mathcal{L}_n)$ of $\mathcal{E}$ from all the n-dimensional subspaces,

$$d_n(\mathcal{E}) = \inf_{\mathcal{L}_n} \delta(\mathcal{E}, \mathcal{L}_n). \tag{1.14}$$

This quantity is called the (Kolmogorov) n-width [33] (also known as n-diameter of $\mathcal{E}$). The minimum number of dimensions n (i.e., the number of basis functions) required to represent any element of $\mathcal{E}$ with an approximation error smaller than ϵ, is the n-width such that

$$d_n(\mathcal{E}) \leq \epsilon. \tag{1.15}$$

The number n represents the number of DoF of the electromagnetic field at the ϵ-level of accuracy (DoF_ϵ) [3]. In the asymptotic regime (wavelength $\lambda \to 0$), this number is scarcely dependent on ϵ and is simply called the number of degrees of freedom (DoF) of the field [3]. This "pure electromagnetic" definition has important connections to the effective number of spatial channels in MIMO communication systems. Indeed, the concept of degrees-of-freedom represents a bridge that connects electromagnetic theory to information theory, as clarified in the following discussion.

The elements of a set such that $d_n(\mathcal{E}) \leq \epsilon$ constitute an ϵ-net of $\mathcal{E}$. Since the elements of the ϵ-net of $\mathcal{E}$ have a distance smaller than ϵ, the set $\mathcal{E}$ can be substituted by any ϵ-net at ϵ level of accuracy. Let N_ϵ be the cardinality of the ϵ-net having the minimum number of elements. The Kolmogorov ϵ-entropy is defined as [34]

$$H_\epsilon = \log_2 N_\epsilon\,. \tag{1.16}$$

From a physical point of view, H_ϵ represents the minimum number of configurations of the electromagnetic field on $\mathcal{S}_\mathrm{R}$ that allow representing any field on $\mathcal{S}_\mathrm{R}$ with an approximation error smaller than ϵ, while DoF_ϵ denotes the minimum number of basis functions required to represent the field with a linear representation and an error smaller than ϵ (i.e., the effective dimension of the set $\mathcal{E}$).

The Kolmogorov n-width of $\mathcal{E}$ is related to the Hilbert–Schmidt expansion of the radiation operator (Eq. (1.6)) by the simple relationship [33]

$$d_k(\mathcal{E}) = \sigma_{k+1}, \tag{1.17}$$

wherein σ_k are the singular values of the radiation operator. Consequently, the essential dimensions of the subspace of the fields $\mathcal{E}$ approximated at ϵ is $n(\varepsilon) = \inf\{n : \sigma_n \leq \varepsilon\}$. Accordingly, it is possible to approximate the set $\mathcal{E}$ at an ϵ approximation level by the series in (1.6) truncated to the first DoF_ϵ terms since this assures that the distance between two field configurations on $\mathcal{S}_\text{R}$ is not larger than ϵ.

However, information transmission involves a *different* set of fields, each separated by at least 2ϵ to be distinguishable in the presence of an ϵ uncertainty. In this case, instead of solving an ϵ-covering problem, we must address a 2ϵ-packing problem. The challenge is to determine the least upper bound, denoted as $M_{2\epsilon}$, of the number of open and nonintersecting spheres with radius ϵ that can cover the set $\mathcal{E}$ of fields on $\mathcal{S}_\text{R}$ [35]. The Kolmogorov 2ϵ-capacity is defined as [34]

$$C_{2\epsilon} = \log_2 M_{2\epsilon} \,. \tag{1.18}$$

The amount of information that can be transmitted using the space resource, given a detection error not exceeding ϵ, is $C_{2\epsilon}$ bits.[1] The centers of the spheres represent the codewords of a codebook, whose elements are field configurations on $\mathcal{S}_\text{R}$. Note that, since $\mathcal{E}$ is a precompact set, both the ϵ-capacity and ϵ-entropy are *finite* for any $\epsilon > 0$ [26]. Consequently, the amount of information associated with the electromagnetic field radiated by finite-size, finite-energy sources is always finite when the measurements are affected by nonnull uncertainty [35].

Since H_ϵ involves an ϵ-covering problem, while $C_{2\epsilon}$ involves a 2ϵ-packing problem of the same set, it follows that $H_{2\epsilon} \leq C_{2\epsilon} \leq H_\epsilon$ [34,36]. Furthermore, the number of dimensions of the $\mathcal{L}_n$ subspace involved in the 2ϵ-packing problem is not larger than the number of effective dimensions of the subspace involved in the ϵ-covering problem, i.e., is not larger than the DoF_ϵ as defined in approximation theory.

Coming back to (1.6), it represents the field on $\mathcal{S}_\text{R}$ as a superposition of orthogonal field configurations on $\mathcal{S}_\text{R}$. Consequently, it plays the same role as the SVD expansion of the channel matrix in MIMO systems. Indeed the left and right singular functions of the radiation operator are special current configurations and field configurations that are related by a scalar, i.e., the singular value (see Eq. (1.5)). Each $(\mathsf{v}_k, \mathsf{u}_k, \sigma_k)$, also called "communication mode" [37], can carry independent information and represent the "physical" counterpart of an MIMO spatial channel [38]. The number of MIMO channels in the Kolmogorov approach is given by the effective dimension of the 2ϵ-packing set. Accordingly, the number of degrees of freedom of an electromagnetic field as defined in approximation theory is an upper bound for the number of spatial channels in MIMO systems [39].

[1] In a space-time communication system, $C_{2\epsilon}$ is the (Kolmogorov) channel capacity per channel use.

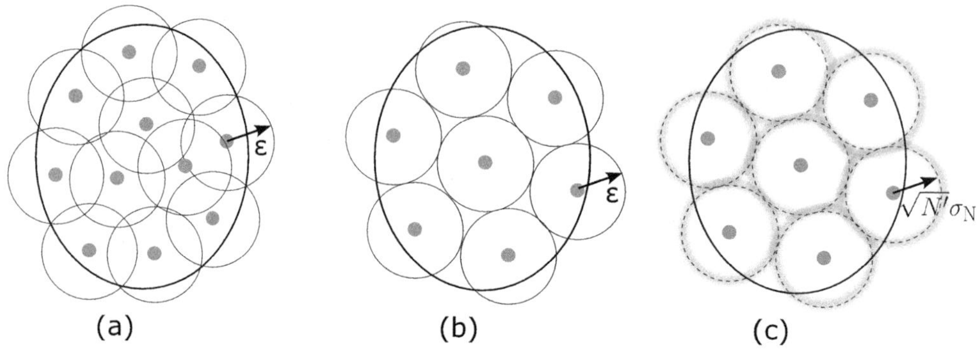

Figure 1.7:
(a) ϵ-covering involved in the definition of H_ϵ Kolmogorov ϵ-entropy; (b) ϵ-covering involved in the definition of $C_{2\epsilon}$ Kolmogorov 2ϵ-capacity; (c) covering in Shannon theory involving soft spheres; N' is the effective space dimensionality and σ_N the noise standard deviation; soft spheres tends to become hard spheres with $\sqrt{N'}\sigma_N$ radius for $N' \to \infty$.

Generally speaking, the radiation operator models the communication channel in an MIMO system. Consequently, the singular values involved in the definition of DoF are the same involved in MIMO communication Shannon theory [40]. A natural question regards the relationship between the Kolmogorov ϵ-capacity and Shannon capacity. Kolmogorov and Shannon consider two different noise models, bounded in the former and statistical in the latter. Loosely speaking, Kolmogorov considers "hard spheres" of uncertainty, while Shannon considers "soft spheres" of uncertainty (Fig. 1.7). Shannon capacity is an asymptotic result based on probabilistic concentration. Simply speaking, as the number of effective dimensions N' of the output space of the channel operator (e.g., the band-limiting Slepian operator in Shannon theory involving only time resources [41], and space-time channel operator in MIMO communication systems [35]) increases, the noise concentrates around its standard deviation $\sqrt{N'}\sigma_N$. This is achieved by increasing the observation time, i.e., considering arbitrarily long coding blocks, whose norm increases as $\sqrt{N'}$. However, in MIMO communication systems, it is not possible to increase the number of dimensions of the output space as it will use spatial resources since the DoF is limited by the dimensions of the source relative to the wavelength. Consequently, there is an asymmetry between space and time [10]; rigorously speaking, only time can be used to harden the noise sphere.

Kolmogorov, as discussed in this section, considers a single "channel use," and the temporal dimension is lost. To compare the two approaches, we imagine that the noise has been "hardened" in some way, and ϵ in the Kolmogorov model is equal to σ_N. Accordingly, $(\sigma_k/\epsilon)^2$ is equivalent to the SNR per channel normalized to unitary power.

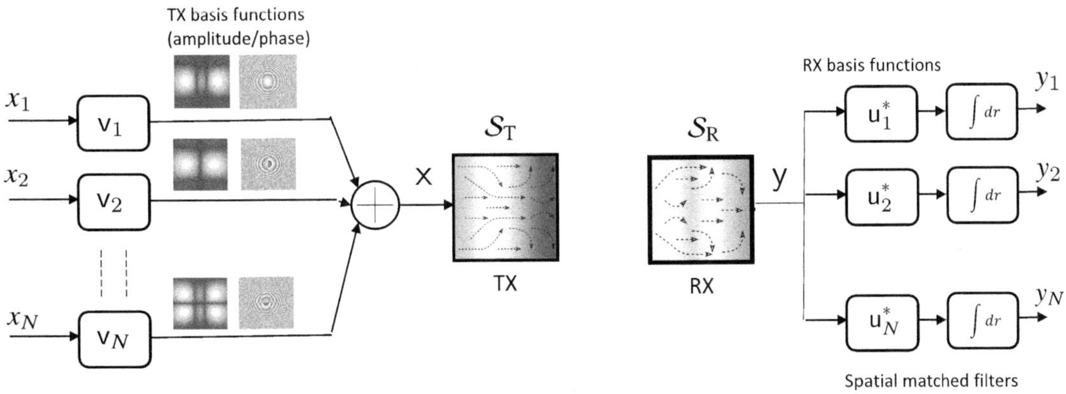

Figure 1.8:
Optimal communication scheme between spaces based on communication modes.

As an example, with reference to Section 1.3, let us use a channel operator having N effective singular values σ_k whose amplitude is such that $(\sigma_k/\epsilon)^2 = \mathsf{SNR}/N$ and negligible values for $k > N$. This condition is typical of many geometric configurations, for instance, that with two paraxial planar surfaces [37]. Considering the upper and lower bounds of the Kolmogorov capacity [42], and considering large $(\sigma_k/\epsilon)^2$ ratio, the capacity evaluated using Kolmogorov theory tends to the capacity in (1.13) obtained using Shannon theory. In fact, for uniform singular values σ_k, the water-filling solution leads to a uniform power allocation among the N dominant parallel channels, and the capacity is given by (1.13).

In terms of communication system representation, the Hilbert–Schmidt expansion can be associated with the communication scheme depicted in Fig. 1.8, where the N input data streams $\{x_n\}$ are associated with the TX basis functions $\{\mathsf{v}_k\}$ in $\mathcal{S}_\mathrm{T}$ to form the source current density $\mathsf{x} = \sum_{k=1}^{N} x_k \, \mathsf{v}_k$. At the receiver, they are recovered after the correlation of the received EM field with the corresponding RX basis functions $\{\mathsf{u}_k\}$ in $\mathcal{S}_\mathrm{R}$, i.e., $y_k = \langle \mathsf{y}, \mathsf{u}_k \rangle$, thus obtaining a set of matched filters working in the spatial domain. The corresponding input–output representation in terms of N parallel orthogonal channels is given by

$$y_k = \sigma_k \, x_k + n_k \, , \quad k = 1, 2, \ldots, N \, , \tag{1.19}$$

with n_k being the AWGN. The scheme resembles that explained in Section 1.3 and illustrated in Fig. 1.4, and then it is information-theoretical optimal when the power allocation among the channels is done according to the water-filling method. It is worthwhile to highlight that the capacity gain with respect to the case where $N = 1$, for a given SNR, could be significant. For example, supposing uniform singular values σ_k, such a gain is

$$G_C = \frac{N \log_2 \left(1 + \frac{\mathsf{SNR}}{N}\right)}{\log_2 \left(1 + \mathsf{SNR}\right)} \, , \tag{1.20}$$

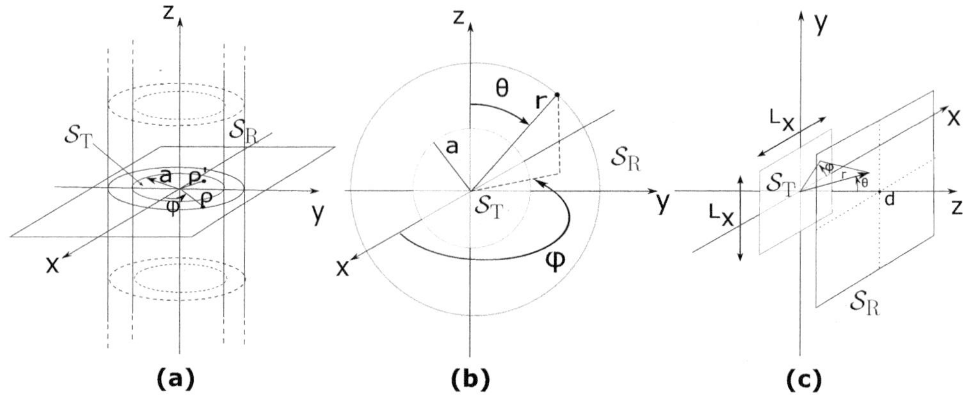

Figure 1.9:
Reference systems for cylindrical (a), spherical (b), and planar (c) geometry.

which increases with N.

1.4.2 Fundamental limits in cylindrical geometry

Let us consider an electromagnetic source consisting of an infinitely long cylinder with radius a, on which a current $I(\rho', \phi')\,\hat{\mathbf{z}}$ flows (Fig. 1.9(a)). The circular observation surface $\mathcal{S}_R$ is an infinitely long cylinder with radius ρ, concentric with the source cylinder $\mathcal{S}_T$ having radius a. The current distribution on the cylindrical source is directed along $\hat{\mathbf{z}}$ and is constant along z. Consequently, the fields vary along ρ and ϕ, but not along z, and the electric field radiated by the source can be represented as the superposition of cylindrical TM (transverse electric) modes [29] (see Appendix A).

With reference to the representation reported in Appendix A, the singular values of the electric field normalized to the first one are

$$\frac{\sigma_m^2}{\sigma_0^2} = f_m(\beta\rho)\, g_m(\beta a), \tag{1.21}$$

where

$$f_m(\beta\rho) = \left| \frac{H_m^{(2)}(\beta\rho)}{H_0^{(2)}(\beta\rho)|} \right|^2, \tag{1.22}$$

$$g_m(\beta a) = \frac{J_m^2(\beta a) - J_{m-1}(\beta a) J_{m+1}(\beta a)}{J_0^2(\beta a) + J_1^2(\beta a)}. \tag{1.23}$$

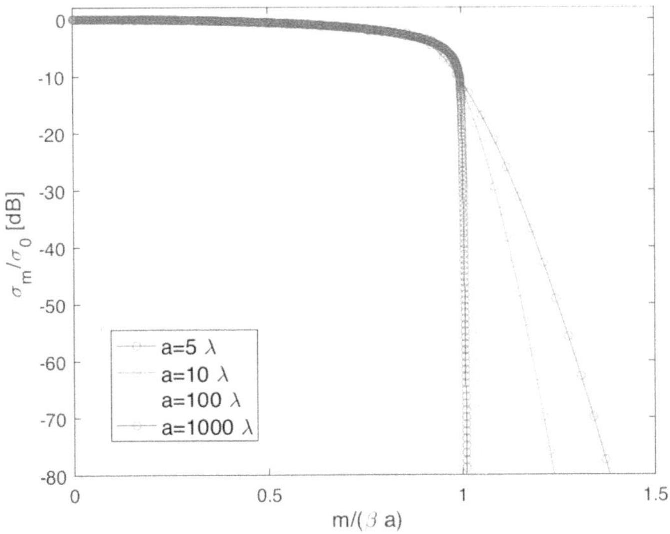

Figure 1.10:
Normalized singular values of the electric field in far-field.

Here $H_m^{(2)}(x)$ is the Hankel function of the second kind and of order m, and the index m ranges from $-\infty$ to $+\infty$ (Appendix A).

The field $E_z(\rho, \theta)$ is a (spatial) bandlimited function whose effective bandwidth depends on the behaviors of the singular values. Since the singular values associated with $\pm m$ are equal, we will restrict our analysis to $m \geq 0$.

From the asymptotic expansions of the Hankel function for large argument, we have that $f_m(\beta\rho) \to 1$ for $\rho \to \infty$. Consequently, under "far-field condition," the behavior of the singular values depends only on $g_m(\beta a)$. In Fig. 1.10 the normalized singular values in the case of sources whose radius ranges from 5λ to 1000 λ are plotted. For large electrical sources, a distinct knee emerges at $m = \beta a$ followed by an exponentially rapid decline. Considering the transition window around the knee, we have that DoF $= 2\beta a + o(\beta a)$, wherein $o(\cdot)$ is the Landau symbol [3]. Accordingly, the power carried by singular functions with index $|m|$ larger than βa tends to be so small that the received signal is essentially buried in noise. In order to utilize the "extra spatial channels" associated with indices $|m|$ significantly larger than βa, we need to synthesize radiating systems characterized by a large fraction of reactive power, obtaining a superdirective source [24]. Indeed, there is an amazing and conceptually intriguing parallelism between standard antenna synthesis and space-coding [10]: the possibility of using an unbounded number of singular functions would provide unbounded directivity in antenna synthesis and unbounded channel capacity in spatial communication systems.

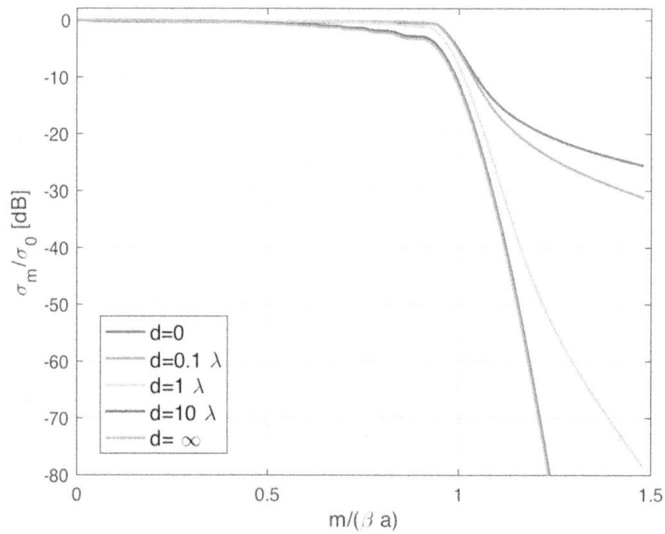

Figure 1.11:
Normalized singular values of the electric field varying the distance d between the source and the observation surface from 0 to 1000λ; $a = 10\lambda$.

A comprehensive understanding of the impact of the use of superdirective antennas on channel capacity requires an analysis of the channel in both temporal and spatial dimensions. In essence, superdirectivity leads to a narrowing of the antenna bandwidth [43]. Accordingly, from an informational standpoint, the use of additional spatial MIMO channels involves a trade-off, exchanging temporal degrees of freedom associated with singular values located *before* the Slepian operator's transition point, for spatial degrees of freedom related to singular functions occurring *after* the radiation operator's transition point. This typically results in a reduction of the channel capacity of the spatio-temporal communication system. This example also underscores the significance of in-band analysis for space-time systems [44]. Exceptional performance achieved through a single harmonic analysis may appear modest when considering the system's operation across the entire operational bandwidth. Limitation of the reactive power can be obtained in many ways, as introducing losses in the source [45], or also introducing an explicit constraint on the reactive energy in the evaluation of the channel capacity [46].

It is interesting to note that when the observation surface goes close to the source, the term $f_m(\beta\rho)$ becomes not negligible. In Fig. 1.11, the normalized singular values are plotted considering different distances of the observation domain in case of $a = 10\lambda$. The plot shows that in close electrical proximity (i.e., at short wavelength distance) to the source, the singular values exhibit a slow decay. Shortly, on the surface we have a huge amount of information that imposes using a number of basis functions much larger than $2\beta a$ to represent the field, in spite of the fact that the spatial pass-band filtering properties of the Green function reduces the relevant basis functions to only $\simeq 2\beta a$ in the far field.

In the above examples, we neglected the presence of fields associated with significant reactive power. The use of reactive fields for communication is not impossible, but requires an accurate electromagnetic analysis of the transmitter–receiver physical system. In fact, the analysis carried out in this paper considers a largely theoretical receiving system that is able to observe the field configuration over $\mathcal{S}_R$ without interfering with the field. In practice, there is always some interaction with the receiving probe. Generally, the distance between the source and the probe allows these interactions to be neglected. However, in close proximity, such interactions become nonnegligible and cause modifications to the currents on both the source and the probe. The effect is a modification of the field configuration. Loosely speaking, whenever the distances between objects are electrically short, mutual electromagnetic interactions must be considered, forcing us to solve the full electromagnetic problem. This effect has been studied in [47], confirming that that the number of significant spatial modes increases in the presence of objects placed at close distance, due to the interference effect of backscattered waves with the radiated field, which redistributes the reactive power to the modes associated with the active power flux. However, in the limit of electrically large systems, the environment can modify the amplitude of the singular value, but cannot increase their spectral support, and DoF is asymptotically limited to $2\beta a$ [48]. The use of controllable scattering does not change these results. However, even if the total number of spatial channels on a surface surrounding the source at a large distance does not change, scatterers allow control over how spatial information is distributed in space [49], allowing spatial processing. This topic will be the subject of Section 1.5.

As a final observation, the example considers a current directed along the z axis, which gives rise to the TM modes, i.e., field configurations whose magnetic field is transverse with respect to the axis of the cylinder. Currents flowing along ϕ give rise to the TE modes, i.e., field configurations whose electric field is transverse with respect to the axis of the cylinder. The analysis for TE modes is analogous to that carried out for TM modes. The excitation of both TE and TM modes doubles the number of degrees of freedom and the number of potential MIMO channels available for communication.

1.4.3 Fundamental limits in spherical geometry

Analysis in spherical geometry has been among the first approaches for evaluating the degrees of freedom using a physical model [50,51].

Spherical harmonics expansions [29] are solutions to Eq. (1.7) in a spherical coordinate system (r, θ, ϕ) (Fig. 1.9(b)). The field in a source-free space can be represented by a superposition of two sets of modes (see Appendix B), each having their electric and magnetic fields, respectively, transverse to the $\hat{\mathbf{r}}$ direction of a spherical system. These modes are called $TE_{m,n}$ and $TM_{m,n}$ modes and each mode is characterized by two indices, m and n.

The number of modes required to represent the field is theoretically infinite. However, for any spatially bounded radiating source, only a finite number of harmonics provides a practically relevant contribution to the field. To clarify the relationship between this property and the energy carried by the modes, we adopt the approach developed by Chu, as presented by Harrington [29].

Consider, for example, the TM modes. The local transverse wave impedance for an outgoing spherical TM wave of order n, defined as the ratio of the tangential electric and magnetic fields evaluated on a sphere of radius r, is

$$Z_n^{\mathrm{TM}}(r) = j\,\zeta_0\,\frac{\hat{H}_n^{(2)\prime}(\beta r)}{\hat{H}_n^{(2)}(\beta r)}, \tag{1.24}$$

where $\zeta_0 = \sqrt{\mu/\varepsilon}$ is the intrinsic impedance of the medium, $\hat{H}_n^{(2)}(x)$ is the Riccati–Hankel function of the second kind, defined as $\hat{H}_n^{(2)}(x) = x\,h_n^{(2)}(x)$ in terms of the spherical Hankel function $h_n^{(2)}$, and the prime denotes differentiation with respect to its argument x.

By using the recurrence formulas for the spherical Bessel function, the impedance can be represented by an electrical network of series capacitances and shunt inductances [29]. In particular, the network is the equivalent circuit of a high-pass filter with a cut-off frequency around $\beta r = n$. Consequently, we have a cut-off *distance* rather than a cut-off frequency: all the TM modes with an index $n > \beta r$ exhibit an impedance with a large imaginary part. These modes are associated with large reactive energy and provide a negligible contribution to the radiated field. Accordingly, terms associated with an index $n > \beta r$ can be neglected in the representation of the field on a sphere having radius r. Furthermore, m and n are related by the associated Legendre polynomials $P_n^m(x)$ of the first kind, order n, and degree m. Since $P_n^m(x) = 0$ for $m > n$, the number of TM modes on a sphere of radius r is approximately $(\beta r)^2$ and quadratically increases with the radius of the sphere.

Let us consider a spatially bounded radiating source, and let a be the radius of the smallest sphere containing the source. The analysis outlined above is valid in free space, i.e., outside this sphere. Consequently, the dimension of the smallest sphere containing the source limits the number of TM modes required to represent the fields outside the sphere to $\simeq (\beta a)^2$, i.e., $\simeq \frac{\pi A_{sp}}{\lambda^2}$ wherein A_{sp} is the area of the sphere. Each spherical mode can be associated with a MIMO channel. A number of MIMO spatial channels slightly larger than $(\beta a)^2$ can be utilized by taking advantage of the transition window of the singular values when there is sufficiently low noise. However, achieving a number of MIMO spatial channels significantly larger than $(\beta a)^2$ requires the use of modes associated with large reactive power, with all the pathological problems discussed in Section 1.4.2.

The discussion regarding the TE modes is similar. In particular, it can be shown that the number of TE modes on a sphere of radius r is approximately $(\beta r)^2$. Consequently, the DoF of the field radiated by any source contained within a sphere of radius a is effectively limited to $2(\beta a)^2$.

It is interesting to note that the same approach can be applied to evaluate the maximum directivity of a source bounded by a sphere of radius a, finding that the excitation of modes associated to singular values after the knee leads to superdirectivity [29], confirming the relationship between superdirectivity and extra-MIMO channels [35].

1.4.4 Fundamental limits in planar geometry

The solution of Eq. (1.7) in the Cartesian coordinate system (Fig. 1.9(c)) yields plane waves with spatial variations of the form $A\exp[j\mathbf{k}\cdot\mathbf{r}]$, where $\mathbf{k} = k_x\hat{\mathbf{x}} + k_y\hat{\mathbf{y}} + k_z\hat{\mathbf{z}}$ is the wave vector an $\mathbf{k}$, whose amplitude is $|\mathbf{k}| = \beta$ [29]. Consequently, it is possible to represent the field at any point in the free space $z > 0$ as a continuous superposition of plane waves,

$$\mathbf{E}(x, y, z) = \frac{1}{(2\pi)^2}\int_{-\infty}^{\infty}\int_{-\infty}^{\infty}\mathbf{f}(k_x, k_y)\exp[j\mathbf{k}\cdot\mathbf{r}]\,dk_x\,dk_y, \tag{1.25}$$

where $\mathbf{f}(k_x, k_y)$ is called the plane wave spectrum. The x and y components of the plane wave spectrum are the spatial Fourier transform of the x and y components of the electric field on the plane $z = 0$, while the z component of the plane spectrum, can be derived from the solenoidal condition of the electric field in free space (see Appendix C).

Coming back to Eq. (1.25), we can note that in the integrand function, the value of each plane wave in the $\mathbf{f}(k_x, k_y)$ spectrum is evaluated in the plane z where the observation point is located. During this process, waves with wavenumbers k_x and k_y such that $\beta^2 \geq k_x^2 + k_y^2$, have $k_z = \sqrt{\beta^2 - (k_x^2 + k_y^2)} \in \mathbb{R}$ and thus propagate. In contrast, waves with wavenumbers k_x and k_y such that $\beta^2 < k_x^2 + k_y^2$, have $k_z = -j\sqrt{(k_x^2 + k_y^2) - \beta^2} \in \mathbb{I}$. Solutions that possess an imaginary k_z wavenumber are called *evanescent waves*. They decay exponentially in the z direction and hence they contribute only very close to the $z = 0$ plane. Consequently, only those plane waves belonging to the part of the spectrum inside the circle of radius β in the k_x, k_y plane (referred to as the "visible domain") contribute to the radiation field. Meanwhile, the parts of the spectrum belonging to the “invisible domain” (i.e., outside the circle of radius β in the k_x, k_y plane) contribute only to a volume in close proximity to the $z = 0$ plane.

The plane wave spectrum allows us to directly obtain the far-field using the stationary phase method [24], which, in spherical coordinates (r, θ, ϕ), is expressed as

$$\mathbf{E}(\theta, \phi, r) = \frac{j\beta \cos\theta}{2\pi r} \exp[-j\beta r] \mathbf{f}(k_x, k_y), \tag{1.26}$$

$$k_x = \beta \sin\theta \cos\phi, \tag{1.27}$$

$$k_y = \beta \sin\theta \sin\phi. \tag{1.28}$$

The far-field in an observation direction (θ, ϕ) is given by the single contribution of the plane spectrum that propagates in the direction (θ, ϕ) of the observation point. Interestingly, the far field is the Fourier transform of the field on the aperture *restricted to the visible domain*, i.e., k_x, k_y values such that $k_x^2 + k_y^2 \leq \beta^2$, since only within this circle the plane waves are associated with real observation angles (see Eqs. (1.27) and (1.28)). Loosely speaking, propagation acts as a low pass filter for the spatial harmonics, and only the part of the spectrum inside the “visible range” range is “visible” in the far field, limiting the set $\mathcal{E}$ of far-field configurations.

The above expressions give a straightforward method to evaluate the DoF in the special case of the EM field radiated by a planar square surface area having $A_{\mathrm{T}} = L_{\mathrm{T}}^2$ and observed in the entire far-field half-sphere. The far field can be considered (apart from inessential constants) the Fourier transform of a spatial 2-dimensional function having support equal to the extension of the radiating source. The number of samples at Nyquist rate of a 2D function that is bandlimited with angular frequency bandwidth in the square support $\left(-\frac{L_x}{2}, \frac{L_x}{2}\right) \times \left(-\frac{L_x}{2}, \frac{L_x}{2}\right)$ and observed on the domain $(-\beta, \beta) \times (-\beta, \beta)$ is $N = \left(\frac{\beta L_x}{\pi}\right)^2 = \left(\frac{2L_x}{\lambda}\right)^2$. However, we must restrict the samples to the circle of the visible domain, obtaining a number of samples equal to $\pi A_{\mathrm{T}}/\lambda^2$.

Planar geometry naturally aligns with communication problems involving array antennas, which are typically planar. Consequently, significant attention has been directed towards analyzing the number of DoF in such cases. As an example of the relationship between system geometry and the number of DoF, consider two planar square surfaces with areas, respectively, $A_{\mathrm{T}} = L_{\mathrm{T}}^2$ and $A_{\mathrm{R}} = L_{\mathrm{R}}^2$, where $A_{\mathrm{T}} < A_{\mathrm{R}}$. These surfaces are located at a distance d in a paraxial configuration and outside the reactive region of the transmitting area. When $d > L_{\mathrm{R}}$, the classical expression for the available DoF (per polarization) is [37]

$$\mathsf{DoF} = \frac{A_{\mathrm{T}} A_{\mathrm{R}}}{\lambda^2 d^2}. \tag{1.29}$$

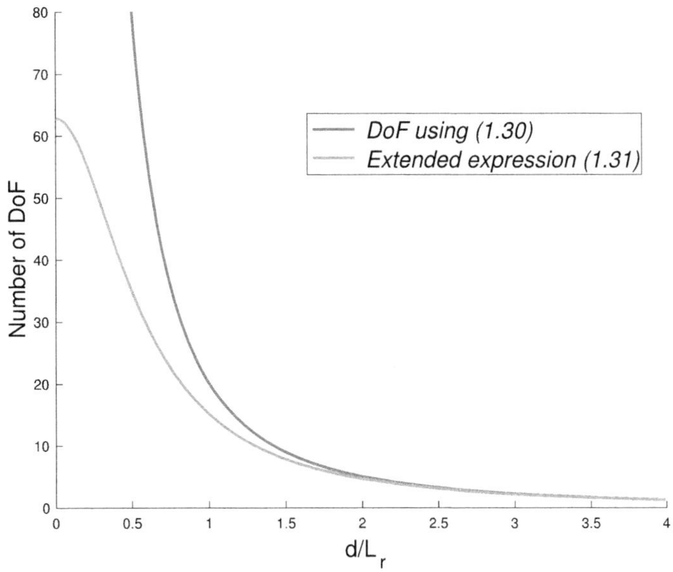

Figure 1.12:
The number of DoF for two parallel surfaces in the paraxial configuration as a function of the ratio between the distance and size of the receiving surface for $A_T/\lambda^2 = 20$.

In [15] the following extended expression is derived:

$$\mathsf{DoF} = \frac{4A_T}{\lambda^2} \frac{L_R \tan^{-1}\left(\frac{L_R}{\sqrt{4d^2+L_R^2}}\right)}{\sqrt{4d^2+L_R^2}}. \tag{1.30}$$

Differently from (1.29), Eq. (1.30) provides the actual number of DoF for planar paraxial surfaces of any size also showing a saturation as $d \ll L_R$, as evident in the numerical example provided in Fig. 1.12. In particular, when $L_R \to \infty$, we find the value $\mathsf{DoF} = \pi A_T/\lambda^2$ obtained in the previous discussion on the far field.

1.5 Spatial processing in reconfigurable environments

1.5.1 The reconfigurable scattering device as a mode converter

ESIT refers to the possibility of processing the EM wave "over-the-air" through the insertion of one or more reconfigurable scattering devices between a source and receiving spaces $\mathcal{S}_T$ and $\mathcal{S}_R$, respectively (see Fig. 1.13). Typically, the scattering device is linear and passive.

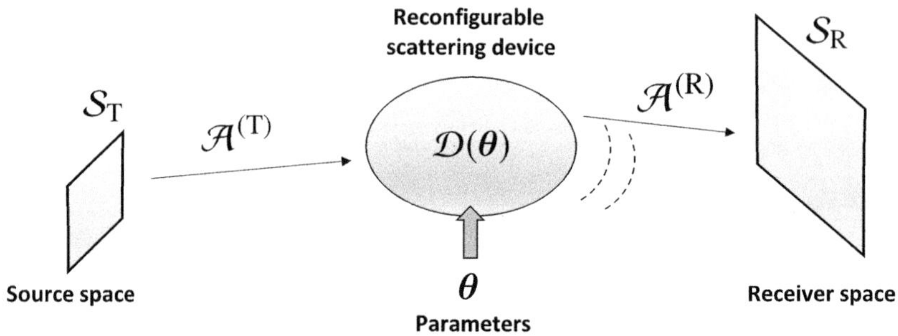

Figure 1.13:
Wave processing using a reconfigurable scattering device.

The scenario illustrated in Fig. 1.13 may involve various ESIT technologies. For example, an RIS functions as a scattering device situated in the radiating region between the source and receiving spaces [52,53]. RISs are primarily used to manipulate the propagation environment and are regarded as a cornerstone of the emerging *smart radio environments paradigm*, which aims to integrate the propagation environment into the system's design and optimization processes. Another example is provided by SIMs, which have recently been introduced as a key enabling technology for ESIT [54]. An SIM can be modeled as a scattering device positioned in the reactive region of either the source or the receiving spaces. It consists of a sealed vacuum structure incorporating multiple layers of metasurfaces. Each metasurface contains numerous reconfigurable metaatoms, small adaptable scattering elements—interconnected via a controller. Based on the well-established Huygens–Fresnel principle, when an EM wave interacts with a metaatom in any layer, the metaatom acts as a secondary point source, illuminating subsequent metaatoms and thereby increasing the DoF for realizing specific processing functionalities. A further example is the DSA [18,55], which can be viewed as an evolution of reactively controlled arrays, such as electrically steerable passive array radiators (ESPARs) [56].

The analysis carried out in Section 1.2 has shown that the operator $\mathcal{A}$, which maps the currents on the source space $\mathcal{S}_{\mathrm{T}}$ to the fields on $\mathcal{S}_{\mathrm{R}}$, is a Hilbert–Schmidt operator. Furthermore, up to N communication modes (i.e., orthogonal channels) can be identified for a given level of accuracy. Each mode is characterized by a singular system. According to (1.6), $(\mathsf{v}_k, \mathsf{u}_k, \sigma_k)$ represents the singular system of the kth mode, where v_k, u_k, and σ_k are, respectively, the kth singular functions of the current's expansion on $\mathcal{S}_{\mathrm{T}}$, the kth singular functions of the field expansion on $\mathcal{S}_{\mathrm{R}}$ space, and the kth singular value of the Hilbert–Schmidt decomposition.

Therefore, we can consider two separate coupling problems: from the source to the scattering device involving the operator $\mathcal{A}^{(\mathrm{T})}$, whose singular system is $\left(\mathsf{v}_n^{(\mathrm{T})}, \mathsf{u}_n^{(\mathrm{T})}, \sigma_n^{(\mathrm{T})}\right)$, for $n =$

$1, 2, \ldots, N$, and from the scattering device to the receiver, involving the operator $\mathcal{A}^{(R)}$, whose singular system in $\left(\mathsf{v}_m^{(R)}, \mathsf{u}_m^{(R)}, \sigma_m^{(R)}\right)$, for $m = 1, 2, \ldots, M$. The response of the scattering device is modeled by a linear operator $\mathcal{D} = \mathcal{D}(\boldsymbol{\theta})$ that maps the incident field y^S to the induced current x^S on the scatterers depending on a set of reconfigurable parameters $\boldsymbol{\theta}$. Denoting x the current on the source and y the field on the observation domain, in the absence of noise the whole communication system can be modeled by a linear (compact) operator $\mathcal{T}(\boldsymbol{\theta})$ such that

$$\mathsf{y} = \mathcal{A}^{(R)} \mathcal{D}(\boldsymbol{\theta}) \mathcal{A}^{(T)} \mathsf{x} = \mathcal{T}(\boldsymbol{\theta})\, \mathsf{x}. \tag{1.31}$$

It is understood that, if a direct connection is also present, the operator $\mathcal{T}(\boldsymbol{\theta})$ is modified as $\mathcal{T}(\boldsymbol{\theta}) = \mathcal{A}^{(R)} \mathcal{D}(\boldsymbol{\theta}) \mathcal{A}^{(T)} + \mathcal{A}$, wherein $\mathcal{A}$ is the free space (LOS) radiation operator mapping directly x to y.

Eq. (1.31) can also be represented in terms of the scattering operator $\mathcal{R}(\boldsymbol{\theta})$ that puts in relationship the incident field $\mathsf{y}^S = \mathcal{A}^{(T)} \mathsf{x}$ and the scattered field $\mathsf{z} = \mathcal{A}^{(s)} \mathcal{D}(\boldsymbol{\theta})\, \mathsf{y}^S = \mathcal{R}(\boldsymbol{\theta})\, \mathsf{y}^S$ at the device, where $\mathcal{A}^{(s)}$ is the operator mapping the induced current to the scattered field on the surface of the scattering object.

By choosing the singular functions as basis functions, we obtain:

$$\mathsf{x} = \sum_{n=1}^{N} x_n\, \mathsf{v}_n^{(T)}, \mathsf{y} = \sum_{m=1}^{M} y_m\, \mathsf{u}_m^{(R)}, \tag{1.32}$$

$$x_n = \left\langle \mathsf{x}, \mathsf{v}_m^{(T)} \right\rangle, y_m = \left\langle \mathsf{y}, \mathsf{u}_m^{(R)} \right\rangle. \tag{1.33}$$

Gathering the $\{x_n\}$'s in the vector $\mathbf{x} = [x_1, x_2, \ldots, x_N]^T \in \mathbb{C}^N$ and $\{y_m\}$'s in the vector $\mathbf{y} = [y_1, y_2, \ldots, y_M]^T \in \mathbb{C}^M$, (1.31) can be represented by a matrix equation (we have reintroduced the noise)

$$\mathbf{y} = \mathbf{A}^{(R)} \mathbf{D}(\boldsymbol{\theta})\, \mathbf{A}^{(T)}\, \mathbf{x} + \mathbf{n} = \mathbf{T}(\boldsymbol{\theta})\, \mathbf{x} + \mathbf{n}, \tag{1.34}$$

where $\mathbf{A}^{(T)} = \mathrm{diag}\left(\sigma_1^{(T)}, \sigma_2^{(T)}, \ldots, \sigma_N^{(T)}\right)$ and $\mathbf{A}^{(R)} = \mathrm{diag}\left(\sigma_1^{(R)}, \sigma_2^{(R)}, \ldots, \sigma_M^{(R)}\right)$ are the coupling matrices between, respectively, the source and device spaces, and the device and receiving spaces, and $\mathbf{n} \in \mathbb{C}^M$ is the AWGN, with $\mathbf{n} \sim \mathcal{CN}\left(\mathbf{0}, \sigma_N^2 \mathbf{I}_M\right)$. The generic element $[\mathbf{D}(\boldsymbol{\theta})]_{m,n}$ of matrix $\mathbf{D}(\boldsymbol{\theta}) \in \mathbb{C}^{M \times N}$ represents the coupling coefficient between the input communication mode n and the output mode m and it is given by $[\mathbf{D}(\boldsymbol{\theta})]_{m,n} = \left\langle \mathsf{v}_m^{(R)}, \mathcal{D}(\boldsymbol{\theta})\, \mathsf{u}_n^{(T)} \right\rangle$, for $n = 1, 2, \ldots, N$ and $m = 1, 2, \ldots, M$.

Matrix $\mathbf{D}(\boldsymbol{\theta})$ completely describes the controlled scattering object. It depends on the specific structure and technology of the scattering device and its configuration parameters $\boldsymbol{\theta}$ through which different end-to-end processing functionalities can be obtained. In essence, any linear scattering device can be seen as a *mode converter*. The maximum number of end-to-end communication modes is $K = \min(N, M)$.

As previously noted, Eq. (1.31) can be also represented using the scattering operator $\mathcal{R}(\boldsymbol{\theta})$. This allows us to define the widely-used reflection matrix (i.e., the "scattering matrix" in the electromagnetic community) of the scattering device[2] $\mathbf{R}(\boldsymbol{\theta}) = \mathbf{G}\,\mathbf{D}(\boldsymbol{\theta})$, where $[\mathbf{G}]_{m,i} = \left\langle \mathcal{A}^{(\mathrm{s})} \mathsf{v}_i^{(\mathrm{R})}, \mathsf{v}_m^{(\mathrm{R})} \right\rangle$, for $m, i = 1, 2, \ldots, M$. Then $\mathbf{T}(\boldsymbol{\theta})$ in (1.34) can be expressed as a function of $\mathbf{R}(\boldsymbol{\theta})$ as well, i.e., $\mathbf{T}(\boldsymbol{\theta}) = \mathbf{A}^{(\mathrm{R})}\,\mathbf{G}^{-1}\,\mathbf{R}(\boldsymbol{\theta})\,\mathbf{A}^{(\mathrm{T})}$.

In general, the EM processing problem can be formulated as finding the best configuration parameter set $\boldsymbol{\theta}$ such that a given target end-to-end processing functionality $\mathbf{T}_0$ is well approximated according to some criterium, for instance, by solving the following minimization problem:

$$\hat{\boldsymbol{\theta}} = \arg\min_{\boldsymbol{\theta}} \left\| \mathbf{T}_0 - \mathbf{A}^{(\mathrm{T})}\mathbf{D}(\boldsymbol{\theta})\,\mathbf{A}^{(\mathrm{R})} \right\|_{\mathrm{F}}, \tag{1.35}$$

constrained to a fixed radiated power and, possibly, to $\mathbf{D}(\boldsymbol{\theta})$ possessing some technology-dependent structure, where $\|\cdot\|_{\mathrm{F}}$ denotes the Frobenius norm.

Examples of functionalities and optimization algorithms regard beamforming, EM EM precoding, direction of arrival estimation, multiuser communication, and interference shaping [54,55,58,59]. More details on specific technologies and algorithms can be found in the other chapters of the book.

1.5.2 Maximum end-to-end capacity

In this section, we investigate the optimal matrix $\mathbf{D}(\boldsymbol{\theta})$ or, equivalently, the optimal reflection matrix $\mathbf{R}(\boldsymbol{\theta})$ in the following simply $\mathbf{D}$ and $\mathbf{R}$, respectively, that maximizes the end-to-end channel capacity of the communication system shown in Fig. 1.13, whose input–output relationship is given by (1.34), under the constraint that the reconfigurable scattering device is passive and lossless. Define $\tilde{\mathbf{R}}$ to be the $K \times K$ submatrix of $\mathbf{R}$ with $K = \min(N, M)$. Since the reconfigurable device is passive and lossless, it turns out that $\tilde{\mathbf{R}}$ must be a unitary matrix, i.e., $\tilde{\mathbf{R}}\tilde{\mathbf{R}}^{\mathrm{H}} = \mathbf{I}_K$, where $(\cdot)^{\mathrm{H}}$ is the Hermitian adjoint operator and $\mathbf{I}_K$ is the $K \times K$ identity matrix.

Let $\mathbf{Q}$ be the positive semidefinite (i.e., $\mathbf{Q} \succeq 0$) covariance matrix of the transmitted signal, i.e., $\mathbb{E}[\mathbf{x}\mathbf{x}^H] = \mathbf{Q}$, where $\mathbb{E}[\cdot]$ is the expectation operator. Then, the transmit power constraint

[2] According to the specific technological solutions, the scattering matrix can have some symmetries. For example, if we exclude the presence of nonreciprocal materials in the scatterer, the response of the scattering device must verify the reciprocity theorem. This results in a symmetric matrix [57].

is $\mathrm{tr}(\mathbf{Q}) \leq P$, where $\mathrm{tr}(\cdot)$ is the trace operator and P is the total transmit power. The authors in [60] proved that the end-to-end channel capacity of the scattering device-assisted communication is

$$C = \sum_{i=1}^{K} \log_2 \left(1 + \left(\sigma_i^{(\mathrm{T})}\right)^2 \left(\sigma_i^{(\mathrm{H})}\right)^2 \frac{P_i}{\sigma_N^2}\right), \tag{1.36}$$

where

$$P_i = \max\left(0, \mu - \frac{\sigma_N^2}{\left(\sigma_i^{(\mathrm{T})}\right)^2 \left(\sigma_i^{(\mathrm{H})}\right)^2}\right), \tag{1.37}$$

and $\sigma_i^{(\mathrm{H})}$ is the ith singular value of the SVD of the matrix $\mathbf{H} = \mathbf{A}^{(\mathrm{R})}\mathbf{G}^{-1}$, which represents the coupling between the scattered and the received fields. The parameter μ is obtained by fulfilling the identity $\sum_{i=1}^{K} P_i = P$. In addition, the capacity C is attained by setting

$$\mathbf{Q} = \mathrm{diag}\left(P_1, P_2, \ldots, P_K\right), \tag{1.38}$$

$$\tilde{\mathbf{R}} = \mathbf{V}^{(\mathrm{H})}, \tag{1.39}$$

where $\mathbf{V}^{(\mathrm{H})}$ is a unitary matrix containing the right eigenvectors of the SVD of $\mathbf{H}$.

From (1.39), it can be deduced that the end-to-end capacity is attained by designing the scattering matrix to be "matched" to the SVD of the two links, and by applying the water-filling power allocation to the ordered product of the singular values of the transmitter–device and device–receiver channels taken individually. The physical realizability of the resulting optimal device coupling matrix **R**, and hence **D**, depends on the specific technology under consideration.

To discuss the impact of the matrix **R** on the channel capacity, let us restrict our attention toward some specific RIS architectures. Loosely speaking, there are two main strategies for the synthesis of RIS. The first is to fully leverage the electromagnetic interactions among the cells, using metaatoms [61]. The second strategy involves using scattering cells about half the wavelength of the incident wave and includes resonant elements such as patches or dipoles printed on a dielectric substrate [52]. This approach is employed in reflectarrays. The use of resonant solutions imposes certain limitations on the reflectarrays compared to metasurface-based architectures, particularly regarding the usable bandwidth. However, from an analytical and practical standpoint, the resonance creates a configuration of the current distribution on the elements that weakly depends on the current distribution on other elements. Consequently, even though the amplitude and phase of the current distribution on an element of the reflectarray depend on the amplitude and phase of the currents on other elements, the "shape" of the current does not change. Excitations can be easily electronically controlled, for example using PIN diodes or varactors, making reflectarray configurations attractive for practical applications.

Restricting our attention toward reflectarray-base solutions, and simplifying the analysis, we represent the incident field and the scattered field of the reflectarray having L scattering elements as the superposition of proper functions $\Phi_n(\mathbf{r})$ and $\Psi_n(\mathbf{r})$:

$$\mathsf{x}^S(\mathbf{r}) = \sum_{n=1}^{L} x_n^S \, \Phi_n(\mathbf{r}), \tag{1.40}$$

$$\mathsf{y}^S(\mathbf{r}) = \sum_{n=1}^{L} y_n^S \, \Psi_n(\mathbf{r}). \tag{1.41}$$

In absence of nonlinear materials, the scattering process is linear, and the coefficients x_n^S and y_n^S are related by a matrix $\mathbf{C} \in \mathbb{C}^{L \times L}$, whose (m, n)th element represents the contribution of the incident field on the mth cell to the current on the nth cell. Since the scattering process is nonlocal, and generally depends on the entire incident field on $\mathcal{S}_\mathrm{S}$, $\mathbf{C}$ is a full matrix.

Supposing the mutual coupling between cells is negligible, we can approximate the matrix $\mathbf{C}$ with a diagonal matrix. This is the model typically adopted to characterize an RIS, also referred to as diagonal RIS, in which the argument of the coupling coefficient can be configured at each scattering element. The corresponding matrix $\mathbf{D}$ in (1.34) can be obtained by mapping $\mathbf{C}$ into the basis sets on which $\mathbf{D}$ is defined, that is,

$$\mathbf{D} = \mathbf{U}\,\mathbf{C}\,\mathbf{V}, \tag{1.42}$$

where $\mathbf{U}$ and $\mathbf{V}$ are matrices allowing the projection of $\mathbf{u}_m^{(\mathrm{T})}$ on $\Phi(\mathbf{r})$ and from $\Psi(\mathbf{r})$ to $\mathsf{v}_m^{(\mathrm{R})}$, respectively.

A diagonal RIS is, in general, not capacity-achieving because $\mathbf{D}$ in (1.42) cannot satisfy (1.39) in general because one has only L parameters to use being $\mathbf{C}$ diagonal. An RIS with a nondiagonal coupling matrix $\mathbf{C}$ might be, on the other hand, capacity-achieving because of the higher number of parameters at disposal. Nondiagonal RISs, where a programmable coupling between its elements is possible, have been recently investigated [62]. Nevertheless, in [60] it has been shown that for many geometrical configurations, a good approximation can be obtained even adopting lower cost non-diagonal RISs through a proper response design.

1.6 Conclusion

In this chapter, we revisited some foundational concepts underlying ESIT. We observed that EM laws set fundamental limits on the number of DoF that communication systems can exploit, depending on geometry and technological constraints. Recent advancements in metamaterials and EM technologies, such as RISs and SIMs, have enabled signal processing directly at the EM level using reconfigurable scattering structures. These solutions are expected to offer significant advantages over conventional fully digital and hybrid architectures, including reduced power consumption, lower latency, and more compact device designs [13]. While wave-based signal processing alone cannot address all these challenges, combining processing across the digital, analog, and EM domains can unlock substantial performance gains. This strategy, recently referred to as a trihybrid architecture, overcomes key limitations of traditional and hybrid MIMO systems by enhancing power efficiency and introducing an additional layer of signal processing [63].

However, despite these advantages, such systems face challenges, including limited processing flexibility and constraints imposed by EM laws and technology-specific characteristics. Overcoming these limitations requires the adoption of physically consistent models and EM-compliant signal processing methods. Another challenge is to determine how to optimally distribute processing tasks among different domains in hybrid architectures to achieve the desired trade-off between performance (e.g., achievable rate), energy efficiency, and latency, which is inherently application dependent.

Key research directions include developing optimization algorithms for high dimensional problems constrained by EM laws, hybrid schemes that balance digital and EM processing, and strategies for determining the optimal configuration of scatterers. Most of the existing literature has focused on linear, time-invariant ESIT solutions operating primarily in the spatial domain. Consequently, this chapter has largely addressed narrowband and time-invariant systems. Expanding ESIT to wideband and time-varying systems is a key priority, both from a theoretical perspective and in practice, with promising concepts such as space-time modulated metasurfaces leveraging frequency-dependent behaviors for tailored functionalities [64,65].

To fully realize the potential of ESIT, it will be essential to integrate advanced technologies such as artificial intelligence, cutting-edge materials, and principles from optics. This synergy could position ESIT as a cornerstone technology for future wireless networks.

Acknowledgment

This work was partially supported by the European Union under the Italian National Recovery and Resilience Plan (NRRP) of NextGenerationEU, partnership on "Telecommunications of the Future" (PE00000001 – program "RESTART").

Appendix A Hilbert–Schmidt decomposition in cylindrical geometry

With reference to the cylindrical geometry depicted in Fig. 1.9(a), and considering a current $I(\rho', \phi')\hat{\mathbf{z}}$, the electric field radiated by the source is [29, pp. 223–232]

$$E_z(\boldsymbol{\rho}) = -\frac{\omega\mu_0}{4}\int_{S_t} I(\boldsymbol{\rho}')H_0^{(2)}(\beta|\boldsymbol{\rho} - \boldsymbol{\rho}'|)\,dS, \tag{1.43}$$

where $\boldsymbol{\rho} = (\rho, \phi)$ and $\boldsymbol{\rho}' = (\rho', \phi')$ represent the vector positions of the observation point and the source point, respectively, and $H_0^{(2)}(x)$ is the Hankel function of the second kind, of order zero, and with argument x.

The magnetic field along ϕ, denoted by $H_\phi(\boldsymbol{\rho})$, can be derived directly from E_z using Maxwell's equations.

Applying the addition theorem for Hankel functions for $\rho > \rho'$ to $H_0^{(2)}(x)$, we have:

$$E_z(\rho, \phi) = -\frac{\omega\mu_0}{4}\int_0^a\int_0^{2\pi}\sum_{m=-\infty}^{\infty} I(\rho', \phi')J_m(\beta\rho')H_m^{(2)}(\beta\rho)\exp(jm(\phi - \phi'))\rho' d\rho' d\phi',$$

$$H_\phi(\rho, \phi) = \frac{j\beta}{4}\int_0^a\int_0^{2\pi}\sum_{m=-\infty}^{\infty} I(\rho', \phi')J_m(\beta\rho')H_m'^{(2)}(\beta\rho)\exp(jm(\phi - \phi'))\rho' d\rho' d\phi',$$

where $J_m(x)$ is the Bessel function of the first kind, of order m and argument x, $H_m^{(2)}(x)$ is the Hankel function of the second kind, of order m and argument x, $H_m'^{(2)}(x)$ is the derivative of $H_m^{(2)}(x)$ with respect to x, and μ_0 is the magnetic permeability of the vacuum.

From the above formulae, we directly obtain the singular system for TM polarization:

$$E_z(\rho, \phi) = \sum_{m=-\infty}^{\infty} \sigma_m \langle I, v_m\rangle u_m, \tag{1.44}$$

$$H_\phi(\rho, \phi) = \sum_{m=-\infty}^{\infty} \sigma_m' \langle I, v_m\rangle u_m', \tag{1.45}$$

$$u_m = -\frac{1}{\sqrt{2\pi}}\frac{H_m^{(2)}(\beta\rho)}{|H_m^{(2)}(\beta\rho)|}\exp(jm\phi), \tag{1.46}$$

$$\sigma_m = \frac{\pi\beta\zeta_0}{2}|H_m^{(2)}(\beta\rho)|\sqrt{\int_0^a J_m^2(\beta\rho')\rho' d\rho'}, \tag{1.47}$$

$$v_m = \frac{1}{\sqrt{2\pi}}\frac{J_m(\beta\rho')}{\sqrt{\int_0^a J_m^2(\beta\rho')\rho' d\rho'}}\exp(jm\phi') \tag{1.48}$$

$$u'_m = \frac{j}{\sqrt{2\pi}} \frac{H'^{(2)}_m(\beta\rho)}{|H'^{(2)}_m(\beta\rho)|} \exp(jm\phi), \tag{1.49}$$

$$\sigma'_m = \frac{1}{\zeta_0} \frac{|H'^{(2)}_m(\beta\rho)|}{|H^{(2)}_m(\beta\rho)|} \sigma_m, \tag{1.50}$$

where ζ_0 is the free space impedance.

Appendix B Spherical harmonics expansion

Spherical harmonics arise in the solution of the homogeneous Helmholtz equation in spherical coordinates (r, θ, ϕ) (Fig. 1.9(b)). The field in a source-free region is represented by a superposition of TE modes, whose electric fields are transverse to $\hat{\mathbf{r}}$, and TM modes, whose magnetic fields are transverse to $\hat{\mathbf{r}}$ [29]:

$$\mathbf{E}(r, \theta, \phi) = \sum_{n,m} \left(a_{mn}\, \mathbf{E}^{TM}_{m,n}(r, \theta, \phi) + b_{mn}\, \mathbf{E}^{TE}_{m,n}(r, \theta, \phi) \right), \tag{1.51}$$

$$\mathbf{H}(r, \theta, \phi) = \sum_{n,m} \left(a_{mn}\, \mathbf{H}^{TM}_{m,n}(r, \theta, \phi) + b_{mn}\, \mathbf{H}^{TE}_{m,n}(r, \theta, \phi) \right), \tag{1.52}$$

with the modal fields:

$$\mathbf{H}^{TM}_{m,n} = \nabla \times \left(\hat{\mathbf{r}}\, A_{m,n} \right), \tag{1.53}$$

$$\mathbf{E}^{TM}_{m,n} = \frac{1}{j\omega\epsilon} \nabla \times \mathbf{H}^{TM}_{m,n}, \tag{1.54}$$

$$\mathbf{E}^{TE}_{m,n} = -\nabla \times \left(\hat{\mathbf{r}}\, F_{m,n} \right), \tag{1.55}$$

$$\mathbf{H}^{TE}_{m,n} = -\frac{1}{j\omega\mu} \nabla \times \mathbf{E}^{TE}_{m,n}, \tag{1.56}$$

$$A_{m,n}(r, \theta, \phi) = h^{(2)}_n(kr)\, P^m_n(\cos\theta)\, e^{im\phi}, \tag{1.57}$$

$$F_{m,n}(r, \theta, \phi) = h^{(2)}_n(kr)\, P^m_n(\cos\theta)\, e^{im\phi}, \tag{1.58}$$

where $P^m_n(x)$ is the associated Legendre function of the degree n and $h^{(2)}_n$ the spherical Hankel function of the second kind.

Appendix C Plane wave expansion

Supposing that all the sources are in the $z < 0$ half-plane ((Fig. 1.9(c)), the electric field in any point of the $z > 0$ half-space is given by [24]

$$\mathbf{E}(x, y, z) = \frac{1}{(2\pi)^2} \int_{-\infty}^{\infty} \int_{-\infty}^{\infty} \hat{\mathbf{f}}(k_x, k_y, k_z, z) \exp[-j(k_x x + k_y y)]\, dk_x\, dk_y, \tag{1.59}$$

where

$$\hat{\mathbf{f}}(k_x, k_y, k_z, z) = \mathbf{f}(k_x, k_y) \exp[-jk_z z], \tag{1.60}$$

$$\mathbf{f}(k_x, k_y) = f_x(k_x, k_y)\hat{\mathbf{x}} + f_y(k_x, k_y)\hat{\mathbf{y}} + f_z(k_x, k_y)\hat{\mathbf{z}}, \tag{1.61}$$

$$f_x(k_x, k_y) = \int_{-\infty}^{\infty} \mathbf{E}_x(x, y, z = 0) \exp[j(k_x x + k_y y)] dx dy, \tag{1.62}$$

$$f_y(k_x, k_y) = \int_{-\infty}^{\infty} \mathbf{E}_y(x, y, z = 0) \exp[j(k_x x + k_y y)] dx dy, \tag{1.63}$$

$$f_z(k_x, k_y) = -\frac{k_x f_x(k_x, k_y) + k_y f_y(k_x, k_y)}{k_z(k_x, k_y)}, \tag{1.64}$$

$$k_z = \begin{cases} +\sqrt{\beta^2 - (k_x^2 + k_y^2)} & \text{when } \beta^2 \geq k_x^2 + k_y^2, \\ -j\sqrt{(k_x^2 + k_y^2) - \beta^2} & \text{when } \beta^2 < k_x^2 + k_y^2. \end{cases} \tag{1.65}$$

Eq. (1.59) can be interpreted as a representation of the electric field at any point in the half-space $z > 0$ as a superposition of plane waves $\mathbf{f}(k_x, k_y) \exp[-j\mathbf{k} \cdot \mathbf{r}]$, where $\mathbf{k} = k_x\hat{\mathbf{x}} + k_y\hat{\mathbf{y}} + k_z\hat{\mathbf{z}}$ is the wave vector, whose amplitude is $|\mathbf{k}| = \beta$.

The knowledge of the two tangential components of the electric field on the plane at $z = 0$ allows for the determination of the tangential components of the plane spectrum, $f_x(k_x, k_y)$ and $f_y(k_x, k_y)$. The third component, $f_z(k_x, k_y)$, is derived from the condition $\nabla \cdot \mathbf{E}(x, y, z) = 0$, which in the Fourier spatial domain becomes $\mathbf{k} \cdot \mathbf{f}(k_x, k_y) = 0$ (Eq. (1.64)). Finally, the magnetic field can be obtained directly by $\mathbf{H} = -\frac{1}{j\omega\mu}\nabla \times \mathbf{E}$.

Consequently, the six scalar components of the fields are not independent, according to the uniqueness theorem, and the field in the entire $z > 0$ half-space is uniquely determined by the two tangent components of the electric or magnetic fields on the plane $z = 0$. A popular application of this behavior is holographic imaging, where the three-dimensional field is reconstructed from the interference patterns captured on a two-dimensional planar photographic plate.

References

[1] Marco Donald Migliore, The world beneath the physical layer: an introduction to the deep physical layer, IEEE Access 9 (2021) 77106–77126.

[2] G. Toraldo di Francia, Resolving power and information, Journal of the Optical Society of America 45 (7) (1955) 497–501.

[3] Ovidio M. Bucci, Giorgio Franceschetti, On the degrees of freedom of scattered fields, IEEE Transactions on Antennas and Propagation 37 (7) (1989) 918–926.

[4] Ovidio M. Bucci, Giorgio Franceschetti, On the spatial bandwidth of scattered fields, IEEE Transactions on Antennas and Propagation 35 (12) (1987) 1445–1455.
[5] Jieao Zhu, Zhongzhichao Wan, Linglong Dai, Mérouane Debbah, H. Vincent Poor, Electromagnetic information theory: fundamentals, modeling, applications, and open problems, IEEE Wireless Communications 31 (3) (2024) 156–162, https://doi.org/10.1109/MWC.019.2200602.
[6] Marco Di Renzo, Marco Donald Migliore, Electromagnetic signal and information theory, IEEE BITS Information Theory Magazine (2024) 1–13, https://doi.org/10.1109/MBITS.2024.3359523.
[7] Emil Björnson, Chan-Byoung Chae, Robert W. Heath Jr., Thomas L. Marzetta, Amine Mezghani, Luca Sanguinetti, Fredrik Rusek, Miguel R. Castellanos, Dongsoo Jun, Özlem Tugfe Demir, Towards 6G MIMO: massive spatial multiplexing, dense arrays, and interplay between electromagnetics and processing, arXiv:2401.02844, https://doi.org/10.48550/arXiv.2401.02844, 2024.
[8] Li Wei, Tierui Gong, Chongwen Huang, Zhaoyang Zhang, Wei E.I. Sha, Zhi Ning Chen, Linglong Dai, Merouane Debbah, Chau Yuen, Electromagnetic information theory for holographic MIMO communications, arXiv:2405.10496, https://doi.org/10.48550/arXiv.2405.10496, 2024.
[9] Christophe Caloz, Shulabh Gupta, Qingfeng Zhang, Babak nikfal, Analog signal processing: a possible alternative or complement to dominantly digital radio schemes, IEEE Microwave Magazine 14 (6) (2013) 87–103, https://doi.org/10.1109/MMM.2013.2269862.
[10] Marco Donald Migliore, Horse (electromagnetics) is more important than horseman (information) for wireless transmission, IEEE Transactions on Antennas and Propagation 67 (4) (2018) 2046–2055.
[11] Davide Dardari, Reconfigurable electromagnetic environments: a general framework, IEEE Journal on Selected Areas in Communications (ISSN 1558-0008) 42 (6) (2024) 1479–1493, https://doi.org/10.1109/JSAC.2024.3389117.
[12] Davide Dardari, Giulia Torcolacci, Gianni Pasolini, Nicolo Decarli, Over-the-Air Electromagnetic Signal Processing: From fundamentals to emerging paradigms, IEEE Signal Processing Magazine 43 (1) (2026) 6–28, https://doi.org/10.1109/MSP.2025.3622001.
[13] Rohit Singh, Jiancheng An, Aryan Kaushik, et al., Redefining Electromagnetic Signal and Information Theory with Programmable Surfaces: Open Critical Questions and Future Horizon, TechRxiv, 2025.
[14] Zhe Wang, Jiayi Zhang, Hongyang Du, Wei E.I. Sha, Bo Ai, Dusit Niyato, Merouane Debbah, Extremely large-scale MIMO: fundamentals, challenges, solutions, and future directions, IEEE Wireless Communications (2023) 1–9, https://doi.org/10.1109/MWC.132.2200443.
[15] Davide Dardari, Communicating with large intelligent surfaces: fundamental limits and models, IEEE Journal on Selected Areas in Communications (ISSN 1558-0008) 38 (11) (2020) 2526–2537, https://doi.org/10.1109/JSAC.2020.3007036.
[16] Tierui Gong, Panagiotis Gavriilidis, Ran Ji, Chongwen Huang, George C. Alexandropoulos, Li Wei, Zhaoyang Zhang, Mérouane Debbah, H. Vincent Poor, Chau Yuen, Holographic MIMO communications: theoretical foundations, enabling technologies, and future directions, IEEE Communications Surveys and Tutorials 26 (1) (2024) 196–257, https://doi.org/10.1109/COMST.2023.3309529.
[17] Haiyang Zhang, Nir Shlezinger, Francesco Guidi, Davide Dardari, Yonina C. Eldar, 6G wireless communications: from far-field beam steering to near-field beam focusing, IEEE Communications Magazine (ISSN 1558-1896) 61 (4) (2023) 72–77, https://doi.org/10.1109/MCOM.001.2200259.
[18] Davide Dardari, Over-the-air multifunctional wideband electromagnetic signal processing using dynamic scattering arrays, IEEE Transactions on Wireless Communications (2025) 1, https://doi.org/10.1109/TWC.2025.3615730, early access.
[19] Kai-Kit Wong, Wee Kiat New, Xu Hao, Kin-Fai Tong, Chan-Byoung Chae, Fluid antenna system – part I: preliminaries, IEEE Communications Letters 27 (8) (2023) 1919–1923, https://doi.org/10.1109/LCOMM.2023.3284320.
[20] Haiyang Zhang, Nir Shlezinger, Francesco Guidi, Davide Dardari, Mohammadreza F. Imani, Yonina C. Eldar, Beam focusing for near-field multiuser MIMO communications, IEEE Transactions on Wireless Communications (ISSN 1558-2248) 21 (9) (2022) 7476–7490, https://doi.org/10.1109/TWC.2022.3158894.

[21] Zicheng Lin, Shuhao Zeng, Aryan Kaushik, Hongliang Zhang, Beamforming design for wideband near-field communications with reconfigurable refractive surfaces, arXiv preprint, arXiv:2501.01012, 2025.
[22] Nuria González-Prelcic, Musa Furkan Keskin, Ossi Kaltiokallio, Mikko Valkama, Davide Dardari, Xiao Shen, Yuan Shen, Murat Bayraktar, Henk Wymeersch, The integrated sensing and communication revolution for 6G: vision, techniques, and applications, Proceedings of the IEEE (ISSN 1558-2256) 112 (7) (2024) 676–723, https://doi.org/10.1109/JPROC.2024.3397609.
[23] Shupei Zhang, Boya Di, Aryan Kaushik, Yonina C. Eldar, Holographic-pattern based multi-user beam training in RHS-aided hybrid near-field and far-field communications, IEEE Transactions on Wireless Communications (2025).
[24] Robert E. Collin, Antennas and Radiowave Propagation, McGraw-Hill Book Company, 1985.
[25] George W. Hanson, Alexander B. Yakovlev, Operator Theory for Electromagnetics: An Introduction, Springer Science & Business Media, 2013.
[26] A.N. Kolmogorov, S.V. Fomin, Elements of the Theory of Functions and Functional Analysis, Dover Puplications Inc, 1999.
[27] Hermann Weyl, Über die asymptotische Verteilung der Eigenwerte, Nachrichten Von der Gesellschaft der Wissenschaften Zu Göttingen, Mathematisch-Physikalische Klasse 1911 (1911) 110–117.
[28] H.J. Landau, On Szegö's eingenvalue distribution theorem and non-Hermitian kernels, Journal d'Analyse Mathématique 28 (1) (1975) 335–357.
[29] Roger F. Harrington, Time Harmonic Electromagnetic Fields, McGraw-Hill, New York City, New York, 1961.
[30] C.E. Shannon, A mathematical theory of communication, The Bell System Technical Journal 27 (3) (1948) 379–423, https://doi.org/10.1002/j.1538-7305.1948.tb01338.x.
[31] Thomas A. Cover, Joy A. Thomas, Elements of Information Theory, vol. 10158, first ed., John Wiley & Sons, Inc., New York, NY, 1991.
[32] David Tse, Pramod Viswanath, Fundamentals of Wireless Communication, Cambridge University Press, New York, NY, ISBN 0521845270, 2005.
[33] Allan Pinkus, N-widths in Approximation Theory, vol. 7, Springer Science & Business Media, 2012.
[34] Vladimir M. Tikhomirov, ε-Entropy and ε-Capacity of Sets in Functional Spaces, Selected Works of AN Kolmogorov: Volume III: Information Theory and the Theory of Algorithms, 1993, pp. 86–170.
[35] Marco Donald Migliore, On electromagnetics and information theory, IEEE Transactions on Antennas and Propagation 56 (10) (2008) 3188–3200.
[36] Reese T. Prosser, William L. Root, The ε-entropy and ε-capacity of certain time-invariant channels, Journal of Mathematical Analysis and Applications 21 (2) (1968) 233–241.
[37] A.B. David Miller, Waves, modes, communications, and optics: a tutorial, Advances in Optics and Photonics 11 (3) (2019) 679–825.
[38] Marco Donald Migliore, Information flows at the deep physical layer level, in: A Glimpse Beyond 5G in Wireless Networks, Springer, 2022, pp. 59–87.
[39] Marco Donald Migliore, On the role of the number of degrees of freedom of the field in MIMO channels, IEEE Transactions on Antennas and Propagation 54 (2) (2006) 620–628.
[40] Marco Donald Migliore, Notes on deterministic and stochastic approaches in electromagnetic information theory, arXiv preprint, arXiv:2508.16601, 2025.
[41] David Slepian, Henry O. Pollak, Prolate spheroidal wave functions, Fourier analysis and uncertainty—I, The Bell System Technical Journal 40 (1) (1961) 43–63.
[42] Taehyung J. Lim, Massimo Franceschetti, Information without rolling dice, IEEE Transactions on Information Theory 63 (3) (2017) 1349–1363, https://doi.org/10.1109/TIT.2016.2647602.
[43] Robert C. Hansen, Fundamental limitations in antennas, Proceedings of the IEEE 69 (2) (1981) 170–182.
[44] Casimir Ehrenborg, Mats Gustafsson, Miloslav Capek, Capacity bounds and degrees of freedom for MIMO antennas constrained by Q-factor, IEEE Transactions on Antennas and Propagation 69 (9) (2021) 5388–5400, https://doi.org/10.1109/TAP.2021.3069432.

[45] Matthew L. Morris, Michael A. Jensen, Jon W. Wallace, Superdirectivity in MIMO systems, IEEE Transactions on Antennas and Propagation 53 (9) (2005) 2850–2857.

[46] Fred K. Gruber, Edwin A. Marengo, New aspects of electromagnetic information theory for wireless and antenna systems, IEEE Transactions on Antennas and Propagation 56 (11) (2008) 3470–3484.

[47] Ramakrishna Janaswamy, On the EM degrees of freedom in scattering environments, IEEE Transactions on Antennas and Propagation 59 (10) (2011) 3872–3881.

[48] Massimo Franceschetti, Marco Donald Migliore, Paolo Minero, Fulvio Schettino, The information carried by scattered waves: near-field and nonasymptotic regimes, IEEE Transactions on Antennas and Propagation 63 (7) (2015) 3144–3157.

[49] Daniele Pinchera, Jon W. Wallace, Marco Donald Migliore, Michael A. Jensen, Experimental analysis of a wideband adaptive-MIMO antenna, IEEE Transactions on Antennas and Propagation 56 (3) (2008) 908–913.

[50] Ada S.Y. Poon, Robert W. Brodersen, David N.C. Tse, Degrees of freedom in multiple-antenna channels: a signal space approach, IEEE Transactions on Information Theory 51 (2) (2005) 523–536.

[51] Mats Gustafsson, Sven Nordebo, Characterization of MIMO antennas using spherical vector waves, IEEE Transactions on Antennas and Propagation 54 (9) (2006) 2679–2682.

[52] Marco Di Renzo, Alessio Zappone, Merouane Debbah, Mohamed-Slim Alouini, Chau Yuen, Julien de Rosny, Sergei Tretyakov, Smart radio environments empowered by reconfigurable intelligent surfaces: how it works, state of research, and the road ahead, IEEE Journal on Selected Areas in Communications 38 (11) (2020) 2450–2525.

[53] Marco Di Renzo, Fadil H. Danufane, Sergei Tretyakov, Communication models for reconfigurable intelligent surfaces: from surface electromagnetics to wireless networks optimization, Proceedings of the IEEE 110 (9) (2022) 1164–1209, https://doi.org/10.1109/JPROC.2022.3195536.

[54] Jiancheng An, Chao Xu, Derrick Wing Kwan Ng, George C. Alexandropoulos, Chongwen Huang, Chau Yuen, Lajos Hanzo, Stacked Intelligent Metasurfaces for Efficient Holographic MIMO Communications in 6G, 2023.

[55] Davide Dardari, 3D electromagnetic signal processing, in: Proc. Asilomar Conf. on Signals, Systems and Computers, 2024.

[56] Juan Carlos Bucheli Garcia, Mohamed Kamoun, Alain Sibille, Low-complexity adaptive spatial processing of ESPAR antenna systems, IEEE Transactions on Wireless Communications 19 (6) (2020) 3700–3711, https://doi.org/10.1109/TWC.2020.2975800.

[57] O.M. Bucci, T. Isernia, Electromagnetic inverse scattering: retrievable information and measurement strategies, Radio Science 32 (6) (1997) 2123–2137.

[58] Anastasios Papazafeiropoulos, Jiancheng An, Pandelis Kourtessis, Tharmalingam Ratnarajah, Symeon Chatzinotas, Achievable rate optimization for stacked intelligent metasurface-assisted holographic MIMO communications, IEEE Transactions on Wireless Communications (2024).

[59] Jiancheng An, Chau Yuen, Yong Liang Guan, Marco Di Renzo, Mérouane Debbah, H. Vincent Poor, Lajos Hanzo, Two-dimensional direction-of-arrival estimation using stacked intelligent metasurfaces, IEEE Journal on Selected Areas in Communications (2024).

[60] Giulio Bartoli, Andrea Abrardo, Nicolo Decarli, Davide Dardari, Marco Di Renzo, Spatial multiplexing in near field MIMO channels with reconfigurable intelligent surfaces, IET Signal Processing 17 (3) (2023) e12195, https://doi.org/10.1049/sil2.12195, https://ietresearch.onlinelibrary.wiley.com/doi/abs/10.1049/sil2.12195.

[61] E. Martini, S. Maci, Theory, analysis, and design of metasurfaces for smart radio environments, Proceedings of the IEEE 110 (9) (2022) 1227–1243, https://doi.org/10.1109/JPROC.2022.3171921.

[62] Hongyu Li, Shanpu Shen, Matteo Nerini, Bruno Clerckx, Reconfigurable intelligent surfaces 2.0: beyond diagonal phase shift matrices, IEEE Communications Magazine 62 (3) (2024) 102–108, https://doi.org/10.1109/MCOM.001.2300019.

[63] Robert W. Heath Jr., Joseph Carlson, Nitish Vikas Deshpande, Miguel Rodrigo Castellanos, Mohamed Akrout, Chan-Byoung Chae, The tri-hybrid MIMO architecture, arXiv:2505.21971, https://doi.org/10.48550/arXiv.2505.21971, 2025.

[64] Mayur V. Katwe, Aryan Kaushik, Lina Mohjazi, Mohammad Abualhayja'a, Davide Dardari, Keshav Singh, Muhammad Ali Imran, M. Majid Butt, Octavia A. Dobre, An overview of intelligent meta-surfaces for 6G and beyond: opportunities, trends, and challenges, IEEE Communications Standards Magazine 8 (4) (2024) 62–69.

[65] Rohit Singh, Aryan Kaushik, Wonjae Shin, George C. Alexandropoulos, Mesut Toka, Marco Di Renzo, Indexed multiple access with reconfigurable intelligent surfaces: the reflection tuning potential, IEEE Communications Magazine 62 (4) (2023) 120–126.

CHAPTER 2

Electromagnetic information: a signal-space, nonstochastic approach

Massimo Franceschetti

Department of Electrical and Computer Engineering, University of California at San Diego, La Jolla, CA, United States

2.1 Introduction

Since Shannon's early work [1], information theory has been developed along rigorous mathematical lines. By abstracting out the physics of the channel, this theory yields results within the context of a given mathematical model. This abstraction is certainly useful, and sometime necessary to handle the complexity of the design, but it also comes at the price of hiding physical insights. Especially in the context of wireless communications, simple channels models may be inadequate to describe the most recent technological advancements and architectures. In this framework, the growing field of electromagnetic information theory (EIT) attempts to merge information theory with Maxwell's theory of electromagnetism [2]. This is useful, for example, in the context holographic multiple antenna systems (HMIMO) [3], where the main novelty is the availability of an essentially continuous antenna surface for transmission. In this case, the pointwise discretization of transmitters and receivers used in classical information-theoretic models, predicting an infinite number of degrees of freedom for continuous arrays, is not well suited to fully capture the limitations these systems. Furthermore, in communication theory, the channel's input to output relationship is typically modeled using a stochastic model that tries to mimic the real propagation conditions occurring between transmitters and receivers. When using such models, mutual coupling between antennas, near-field effects, and the crucial role of scattering in enabling signal diversity can easily be overlooked. For example, this has led to overestimating the correct capacity scaling law of wireless networks that is obtained using electromagnetic arguments [4]. Recently, EIT has expanded into a field of growing interest [5,6]. Some key early contributions date back to the 1980s and include the determination of the spatial bandwidth and degrees of freedom of scattered fields [7,8]. These results have been generalized from single-frequency transmission to bandlimited transmission [9]. A comprehensive theoretical treatment, along with extensions and applications appears in the book [10]. Additional applications of electromagnetic degrees of freedom in the context of MIMO systems appear in [11–14] and in the context of sampling and recon-

Electromagnetic Signal and Information Theory
https://doi.org/10.1016/B978-0-44-336354-2.00009-5
Copyright © 2027 Elsevier Inc. All rights are reserved, including those for text and data mining, AI training, and similar technologies.

struction in [15–20]. Near field effects on degrees of freedom are considered in [21–23]. Information-theoretic limits of bandlimited radiation from a quantum perspective are considered in [24]. Recent developments in the context of holographic MIMO systems appear in [23,25–27]. Another promising application of EIT is in the design of antennas optimized for spatial multiplexing, and that exploit reconfigurable elements [28–32].

In this chapter, we wish to show how information theory can be used to investigate the fundamental question of how much information can be encoded using electromagnetic waves and how much information can be transported by the propagating field. A novel aspect of our treatment is that we explore information-theoretic quantities in a nonstochastic setting, following an approach originally developed by Kolmogorov, who followed with interest Shannon's original development of the field of information theory [33]. This approach allows determining the amount of information carried by arbitrary electromagnetic fields, independent of stochastic assumptions on channel conditions. Some additional details on the relationship between stochastic and nonstochastic approaches can be found in [34] and [35]. In the next section we discuss the dimensionality of the space of electromagnetic waveforms and we argue that this is limited due to the inherent bandlimitation of the propagating waves following from Maxwell's equations. In Section 2.3 we introduce a resolution constraint that is consistent with the quantized form of any physically observable field quantity, and show how this constrain can be combined with the dimensionality constraint, leading to the notion of entropy of the propagating field. In Section 2.4 we argue that the entropy is only an upper bound on the amount of information that the field can carry and introduce the related notion of capacity that can be used to characterize the amount of information carried by the field. We show that the capacity can be tied to the notion of communication error, and to the notion of mutual information between the transmitted and received waveforms. Finally, we provide a summary discussion in Section 2.5.

2.2 Dimensionality of the signal space

The physics of propagation dictate that any observed electromagnetic field is an essentially bandlimited function. This basic property allows to define the dimensionality of the electromagnetic signals' space in terms of number of degrees of freedom. This gives a first-order characterization of the amount of information carried by the field in terms of number of real coefficients that need to be specified to identify any propagating waveform. We first review results for temporal signals, and then extend the treatment to space-time electromagnetic waveforms.

2.2.1 Temporal signals

Consider a one-dimensional, real, scalar waveform f of a single scalar variable t that is bandlimited to Ω, namely has angular frequency support over the interval $[-\Omega, \Omega]$. We assume that f is square-integrable, and

$$\int_{-\infty}^{\infty} f^2(t)dt \leq E. \tag{2.1}$$

This ensures that the waveform can be expanded in a series of, possibly complex, orthonormal basis functions $\{\psi_n\}$,

$$f(t) = \sum_{n=1}^{\infty} a_n \psi_n(t), \tag{2.2}$$

where

$$a_n = \int_{-\infty}^{\infty} f(t)\psi_n^*(t)dt. \tag{2.3}$$

The waveform can then be viewed as a single point in an infinite-dimensional space of coordinates given by the coefficients $\{a_n\}$ in (2.3). By varying the values of these coefficients, we can create distinct waveforms and use them to communicate information. If the orthonormal set of basis functions $\{\psi_n\}$ is complete, then using (2.2) we can construct any element in the space of signals defined by (2.1). One may reasonably expect that only a finite number of coefficients is in practice needed to specify the waveform up to any given accuracy, while using a larger number does not significantly improve resolution. We are then interested in determining the smallest N beyond which varying higher order coefficients does not change the form of the waveform significantly. Determining this number, over all possible choices of basis functions, is a question first posed by Kolmogorov [36] and corresponds to determining the *number of degrees of freedom* of the waveform. Thus, the number of degrees of freedom corresponds to the effective dimensionality of the space of waveforms, and we can think each waveform as encoding an amount of information proportional to the number of coefficients that need to be specified to identify it.

To make the above observations mathematically precise, consider an observation interval $[-T/2, T/2]$, and introduce the norm

$$\|f\| = \left(\int_{-T/2}^{T/2} f^2(t)dt\right)^{1/2}. \tag{2.4}$$

We want to determine the minimum value of N for which the normalized energy difference between any signal f and its approximation f_N obtained by N-dimensional truncation of (2.2) is

$$\frac{\|e_N\|^2}{E} = \frac{\|f(t) - f_N(t)\|^2}{E} \leq \epsilon. \tag{2.5}$$

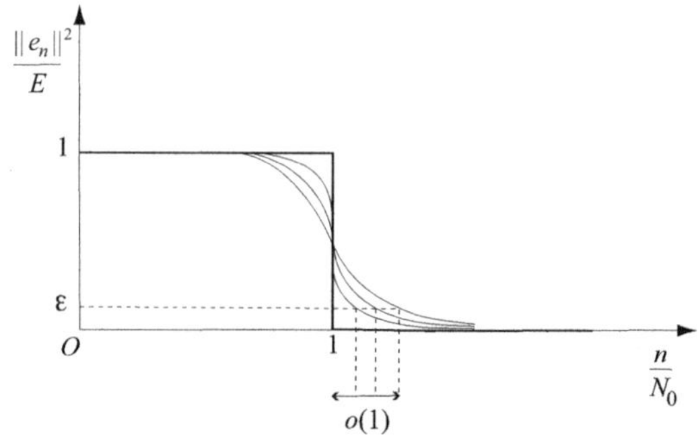

Figure 2.1:
Phase transition of the approximation error. The transition becomes sharper and its width, viewed at the scale of N_0, shrinks to zero as $N_0 \to \infty$.

It turns out that letting

$$N_0 = \frac{\Omega T}{\pi}, \tag{2.6}$$

we have that such a minimum value is

$$N_\epsilon = N_0 + o(N_0) \quad \text{as } N_0 \to \infty. \tag{2.7}$$

It follows that in the asymptotic regime of wide observation interval $T \to \infty$ or wide band $\Omega \to \infty$, the number of degrees of freedom is well identified by the number $N_0 = \Omega T/\pi$, indicating the scale at which the amount of information carried by the field concentrates. This result was a crowning achievement of Landau, Pollak, and Slepian [37,38]. It shows that the number of degrees of freedom is, up to first order, independent of the level of approximation ϵ that appears only as a preconstant of the second order term $o(N_0)$. Hence, varying the approximation level ϵ does not affect the first-order scaling of the result. Fig. 2.1 shows the transition of the approximation error, when viewed at the scale of N_0, for the optimal choice of basis functions. The transition appears sharper and the transition window tends to zero as $o(1)$ as $N_0 \to \infty$. The optimal interpolating functions $\{\psi_n\}$ in (2.2), called prolate spheroidal wave functions, that achieve the smallest approximation error are obtained by solving an eigenvalue problem [38], while sampling representations using $\text{sinc}(\cdot)$ functions are suboptimal approximations widely used for computational convenience.

2.2.2 Space-time signals

The electromagnetic field is in general a function of four scalar variables, three spatial and one temporal. It follows that in order to appreciate the total field's informational content in terms of degrees of freedom, the treatment above should be extended to multiple variables.

Landau's theorem

The extension to multivariable signals is possible thanks to Landau's theorem [39], which states that for waveforms of d variables whose spectral support is over a set Q and that are observed over a set P, letting

$$N_0 = \frac{m(P)m(Q)}{(2\pi)^d}, \tag{2.8}$$

where $m(\cdot)$ indicates the Lebesgue measure, we have that (2.7) holds, provided that one takes an appropriate scaling of either of the sets P or Q, so that $N_0 \to \infty$. Namely, letting $(x_1, x_2, \ldots, x_d)$ be the space-time coordinates of the waveform and $(w_1, w_2, \ldots, w_d)$ the corresponding coordinates in the transformed spectral domain, we can let P and Q be two fixed sets and for a given $s > 0$ consider the scaling of the observation set P, namely

$$P_s = \{sx_1, sx_2, \ldots, sx_d : (x_1, x_2, \ldots, x_d) \in P\}, \tag{2.9}$$

or the scaling of the spectral support,

$$Q_s = \{sw_1, sw_2, \ldots, sw_d : (w_1, w_2, \ldots, w_d) \in Q\}. \tag{2.10}$$

In either of these two cases, the number of degrees of freedom concentrates around the value N_0 as $s \to \infty$, in the sense of (2.7). In practice, the number of degrees of freedom can be identified with the number N_0 provided that one enlarges the observation set P or the spectral support set Q by scaling all of its coordinates by a factor $s \to \infty$. The case $d = 1$ of temporal signals is recovered by noting that (2.8) recovers (2.6), since in this case $m(Q) = 2\Omega$ and $m(P) = T$, and we can let either $T \to \infty$ or $\Omega \to \infty$ to visualize the transition of the approximation error.

Landau's theorem extension

A more general statement of Landau's theorem can also be obtained by considering a hybrid scaling of the domains P and Q [9]. Namely, let x_1 correspond to the time coordinate of the field and $x_2, \ldots, x_d$ correspond to the space coordinates, with $d = 4$ in the case of electromagnetic fields. In this case, we can scale the time and wavenumber domains simultaneously letting, for given τ and $\rho > 0$,

$$P_\tau = \{\tau x_1, x_2, \ldots, x_d \ : (x_1, x_2, \ldots, x_d) \in P\}, \tag{2.11}$$

and

$$Q_\rho = \{w_1, \rho w_2, \ldots, \rho w_d \ : (x_1, x_2, \ldots, x_d) \in Q\}, \tag{2.12}$$

and letting both $\tau, \rho \to \infty$ to achieve concentration of the number of degrees of freedom around the value N_0. In this case, the temporal observation set is scaled, as $\tau \to \infty$, along with the spectral wavenumber domain, as $\rho \to \infty$, while maintaining the spatial domain and frequency domains constant.

The most general statement of the result is given in [9, Theorem 3] for arbitrary $d \times d$ matrices $\mathrm{A} = \mathrm{A}(\tau)$ and $\mathrm{B} = \mathrm{B}(\rho)$ such that

$$\lim_{(\tau,\rho)\to\infty} \mathrm{B}^T \mathrm{A} U = \mathbb{R}^d, \tag{2.13}$$

where U is the d-dimensional unit ball. We can let A_P be the scaling set of points of the form $\mathrm{A}x$ where x is the column vector composed of the elements of $\mathbf{x} \in P$, and B_Q be the scaling set of points of the form $\mathrm{B}w$ where w is the column vector composed of the elements of $\mathbf{w} \in Q$, and obtain the number of degrees of freedom in (2.8) as $\tau, \rho \to \infty$.

Landau's theorem in electromagnetics

In order to apply the results above in an electromagnetic setting, we need to specify a given space-time set P where the field is observed and determine the total spectral support Q of the field over this set using Maxwell equations. In practice, we need to argue that given bounded sources radiating in a given geometric configuration and observing the corresponding field in a certain spatial domain P, the frequency and spatial (i.e., wavenumber) bandwidths are contained in a given set Q. This is where the theory of electromagnetism comes into the picture, as it can be done by studying the spectral support of Green's propagation operator given by Maxwell's equations in some canonical geometries where the spatial bandwidth can be computed [7,8].

For example, consider a spherical geometry where arbitrary sources located inside a sphere of radius r radiate a time-harmonic (single-frequency) field that is observed over a spherical surface concentric with the radiating sphere and described by two angular coordinates and whose total angular aperture is of measure 4π, and let us compute the number of space-wavenumber degrees of freedom. In this case, analysis of the field's propagation operator yields that the spectral support set Q in the wavenumber domain is given by [7,10]

$$Q = [-W, W] \times [-W, W], \tag{2.14}$$

where $W = 2\pi r/\lambda$ is the spatial (i.e., wavenumber) bandwidth for each angular coordinate of the radiated field and λ is the wavelength of radiation. It follows that $m(Q) = 4W^2$. On the other hand, the observation set P corresponds to the solid angle of measure $m(P) = 4\pi$. It now follows by applying Landau's theorem that

$$N_0 = \frac{m(P)m(Q)}{(2\pi)^2} = \frac{4\pi\, 4(2\pi r)^2}{(2\pi\lambda)^2} = \frac{4\pi r^2}{(\lambda/2)^2}, \tag{2.15}$$

and letting $r \to \infty$, so that $W \to \infty$, we obtain the desired scaling of the support set Q, leading to the concentration of the number of degrees of freedom at the scale of N_0. This result has the nice physical interpretation of spatial degrees of freedom in terms of surface area through which the information must flow, normalized by patches of area half a wavelength squared. The same formula has been derived in the literature using different approaches, including in optics [40] and using sampling arguments [15]. Similar formulas also arise in other geometric configurations, for example, $N_0 = \pi A/\lambda^2$, for rectangular apertures of area A [18,41]. It follows that each degree of freedom of the field, representing an independent information channel, occupies a spatial resource of the order of the wavelength squared on the cut-set surface through which the information flows.

Above results can be extended from single frequency to a radiating a waveform of angular frequency bandwidth Ω, for which we have

$$N_0 = \frac{\Omega T}{\pi} \frac{4\pi r^2 \Omega^2}{3c^2\pi^2} = \frac{\Omega T}{\pi} \frac{4\pi r^2}{3(\Lambda/2)^2}, \tag{2.16}$$

where c is the speed of light and Λ is the interval of wavelengths corresponding to the bandwidth Ω. This result is obtained in [9] by applying the extended version of Landau's theorem using the hybrid scaling technique. In this case, in order to obtain the number of degrees of freedom to concentrate around the value of N_0, the temporal observation set is scaled, as $T \to \infty$, along with the spectral wavenumber domain, as $r \to \infty$, while maintaining the angular observation domain 4π and frequency domain Ω constant. It follows that in this case the number of degrees of freedom is the product of two factors, each viewed in an appropriate asymptotic regime: one accounting for the number of degrees of freedom in the time–frequency domain, $\Omega T/\pi$, as $T \to \infty$, and another accounting for the number of degrees of freedom in the space–wavenumber domain, that is now given by $4\pi r^2\Omega^2/(3c^2\pi^2)$, as $r \to \infty$. This second factor also corresponds to the surface area of the sphere of radius r through which the information must flow, normalized by 3/4 times the square of the interval of radiated wavelengths. Similar results can be obtained in more complex geometries, and the result can always be written, at least in integral form [9,42], as the number of time–frequency degrees of freedom times the wavelength-normalized surface area through which the waveform propagates.

A less rigorous, but immediate, derivation of (2.16) can be obtained by integrating the single-frequency result (2.15) over the bandwidth Ω and multiplying by T/π to account for the degrees of freedom in the frequency domain. In this case, we have

$$N_0 = \frac{T}{\pi} \int_0^{\Omega} \frac{4\pi r^2 \omega^2}{(c\pi)^2} d\omega = \frac{\Omega T}{\pi} \frac{4\pi r^2 \Omega^2}{3c^2\pi^2}, \tag{2.17}$$

which also explains the appearance of the factor 3 at the denominator when going from the spatial degrees of freedom formula (2.15) to the space–frequency degrees of freedom formula (2.16).

Finally, we point out that these results do not tell us what are the optimal interpolation basis functions and parallel research has concentrated in finding good representations that can approximate the field with a number of functions close to the number of degrees of freedom. This entails solving an eigenvalue problem in the given geometric configuration [10,43] and corresponds to performing the Hilbert–Schmidt (singular-value) decomposition of the propagation operator in the given environment, which often needs to be performed numerically; see, for example, [23,25,27]. Sampling representations can also be employed, yielding good approximations [16–18].

2.3 Entropy

The informational characterization of EM waveforms in terms degrees of freedom is a basic consequence of the laws of propagation. Maxwell's equations impose a bandlimitation property, which in turn leads to a number N_0 of real coefficients essentially describing the waveform up to arbitrary precision ϵ in a well-defined asymptotic sense. In other words, the space of electromagnetic signals is inherently finite-dimensional. However, in our discussion each coordinate of the space, representing a coefficient in the interpolation (2.2), can still be described with arbitrary accuracy and hence carry an infinite amount of information. It follows that some kind of indeterminacy must be introduced to limit the resolution, in addition to the dimensionality, at which the waveform can be observed.

Once again, we can proceed on physical grounds and assert that waveforms that are distinct as mathematical objects cannot be resolved as distinct physical quantities at a scale smaller than the noise affecting the measurement process. For this reason, we may consider two waveforms *distinguishable* only if, for a given $\epsilon > 0$,

$$\|f_1 - f_2\| > \epsilon. \tag{2.18}$$

This point of view yields a discretization of the space, which allows an information-theoretic description of electromagnetic waveforms in terms of the universal unit of information – the *bit*.

Kolmogorov's ϵ-entropy

Kolmogorov's ϵ-entropy H_ϵ is the *minimum* number of bits that can represent any waveform within accuracy ϵ. Viewing electromagnetic waveforms as points in a space of N_0 dimensions

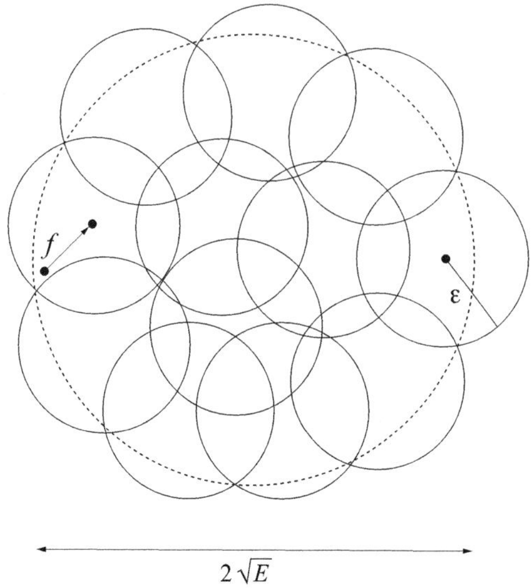

Figure 2.2:
Every signal in the space is within ϵ of the center of a covering ball.

and of radius

$$\left(\int_{\mathbb{R}^3}\int_{-\infty}^{\infty} f^2(t)dtd\mathbf{x}\right)^{1/2} = \left(\sum_{n=1}^{N_0} a_n^2\right)^{1/2} \leq \sqrt{E}, \tag{2.19}$$

we have that the ϵ-entropy geometrically corresponds to the logarithm base two of the *covering number* $Q_\epsilon(E)$, that is, the minimum number of balls of radius ϵ with their centers located inside the space and such that their union contains the whole space, and we have

$$H_\epsilon = \log Q_\epsilon(E) \quad \text{bits}. \tag{2.20}$$

This follows by noticing that performing a covering of the space by ϵ-balls, any point in the space is indistinguishable from the center of a corresponding ϵ-ball covering it, see Fig. 2.2. It follows that signals in the covering set can approximate all signals in the space, with error at most ϵ, and these can be identified using only H_ϵ bits. For example, an encoding procedure can be to associate any waveform with the closest center of a ball covering it. In this way, any waveform is uniquely associated to a point specified by H_ϵ bits that is indistinguishable from it.

The ϵ-entropy is related to the number of degrees of freedom. Given the interpretation of the number of degrees of freedom as the effective dimensionality of an infinite-dimensional functional space, and since the covering number of this space is roughly of the order of the volume of the space divided by the volume of the ϵ-ball, we can expect the entropy to be roughly

$$H_\epsilon = \log(\sqrt{E}/\epsilon)^{N_0} = \frac{N_0}{2}\log(E/\epsilon^2) \text{ bits}. \tag{2.21}$$

This is shown rigorously in [35] and should be intended in an asymptotic sense as $N_0 \to \infty$, showing that the entropy scales linearly with the number of degrees of freedom. The preconstant of the scaling law is given by half the logarithm of the ratio of the maximum energy of the signal to the energy of the measurement uncertainty. This can be interpreted as the signal-to-noise ratio, and can be geometrically viewed as imposing a quantization of the space at level ϵ due to the noise inherent in the measurement process.

2.4 Capacity

The ϵ-entropy gives only an upper bound on the number of bits that can be used for communication using electromagnetic waveforms. In a communication setting, we not only need to be able to construct a nonredundant codebook that represents all signals in the space, but in order to use this codebook for communication, we also need to ensure that any pair of waveforms in the codebook are distinguishable in the sense of (2.18). In this way, when we transmit a given waveform from the codebook, we can be sure that this cannot be confused with any other waveform in the codebook. Hence, rather than looking for a *minimum* covering set of points such that any point in the space is within ϵ from a point in the covering, we may wish to look for the *maximum* set of points that are distinguishable, namely such that any two points in the set are at distance at least ϵ from each other. This leads to the maximum number of distinguishable signal configurations that the waveform can attain and to the notion of Kolmogorov ϵ-capacity.

Kolmogorov's ϵ-capacity

The ϵ-capacity corresponds to the logarithm base two of the *packing number* $M_\epsilon(E)$, that is, the maximum number of disjoint balls of radius $\epsilon/2$ with their centers situated inside the signals' space, see Fig. 2.3. Since any two signals in the packing are at least ϵ apart, they are distinguishable and they can be identified by

$$C_\epsilon = \log M_\epsilon(E) \quad \text{bits}. \tag{2.22}$$

An upper bound on the ϵ-capacity can be computed in terms of ϵ-entropy using Kolmogorov's inequality

$$C_\epsilon \le H_{\epsilon/2}, \tag{2.23}$$

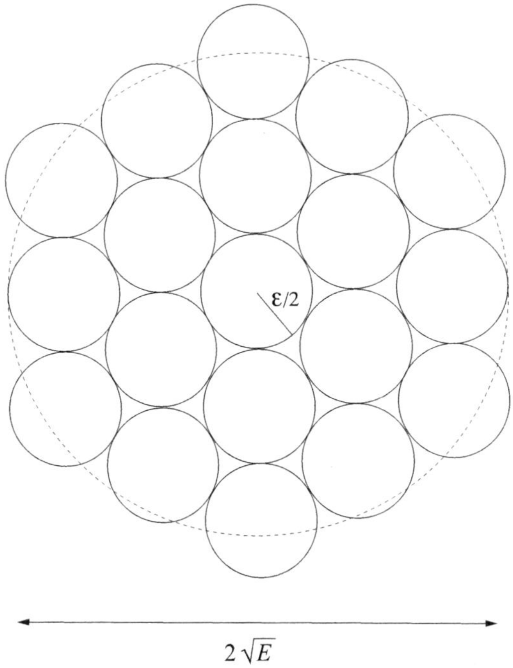

Figure 2.3:
Any two signals identified by the centers of the packing balls are at a distance at least ϵ of each other.

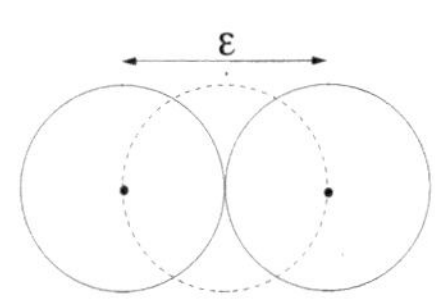

Figure 2.4:
Illustration of the Kolmogorov inequality $C_\epsilon \leq H_{\epsilon/2}$.

and substituting the value of the ϵ-entropy in terms of degrees of freedom (2.21), we have

$$C_\epsilon \leq N_0 \log(2\sqrt{E}/\epsilon) \tag{2.24}$$

Kolmogorov's inequality (2.23) follows from the observation that if by contradiction $M_\epsilon > Q_{\epsilon/2}$, then there must be one ϵ-ball that covers at least two points that are more than ϵ apart, which is impossible, see Fig. 2.4. Tighter upper and lower bounds on the ϵ-capacity are given in [35,44].

It now follows that the problem of least redundant communication using electromagnetic waves is reduced to the problem of designing the largest distinguishable codebook, namely

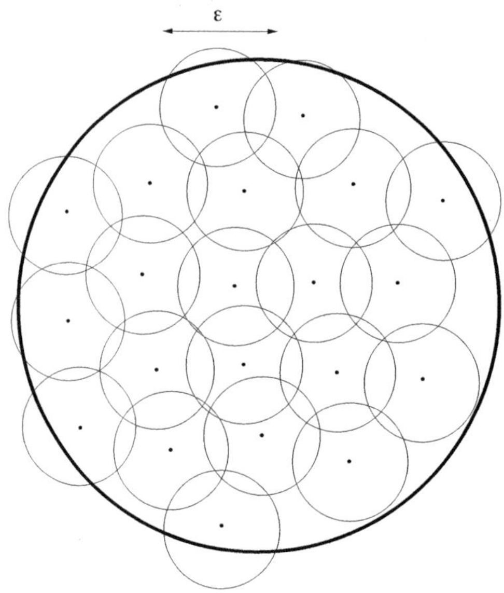

Figure 2.5:
Illustration of the (ϵ, δ)-capacity.

a maximal packing of the space with balls of radius $\epsilon/2$, and this achieves a capacity proportional to the number of degrees of freedom.

Relaxing Kolmogorov's ϵ-capacity

Another interpretation of the ϵ-capacity can be given by assuming that any transmitted signal can only be observed at the receiver subject to a perturbation at most $\epsilon/2$. It then immediately follows that any signal in the packing set can be communicated without error because, since the noise perturbation balls of all signals in the codebook do not overlap, the received signal cannot fall into the uncertainty region of any other transmitted signal. We can then think about relaxing the notion of ϵ-capacity by allowing a certain amount of overlap among the ϵ-balls. This would allow fitting more balls in the packing, hence increasing the capacity in terms of number of bits that can be communicated using a waveform, but it would also open to the possibility of potentially having some errors in the communication process. Nevertheless, if the amount of overlap among the ϵ-balls is kept small, we might be able to tolerate an occasional possibility of the perturbed signal falling into the overlap regions, where it can be confused with some other signal. Letting δ be the measure of the maximum allowable overlap region among the balls, we consider the maximum number M_ϵ^δ of balls of radius $\epsilon/2$ that overlap at most δ with their centers situated inside the signal space, see Fig. 2.5. The (ϵ, δ)-capacity is then defined as [35]

$$C_\epsilon^\delta = \log M_\epsilon^\delta \text{ bits,} \tag{2.25}$$

which provides another way of characterizing the information associated to the field when one allows for a small possibility of error δ. Like the Kolmogorov ϵ-capacity, the (ϵ, δ)-capacity is, up to first order as $N_0 \to \infty$, proportional to the number of degrees of freedom of the field times the logarithm of the signal-to-noise ratio [35] and independent on both ϵ and δ. Furthermore, measuring capacity in bits per unit time, we can also define the (ϵ, δ)-capacity as the largest bit-rate having vanishing error $\delta \to 0$ as $T \to \infty$, which provides a direct comparison withe the classical (probabilistic) definition in information theory [46].

Capacity and mutual information

In information theory, the capacity is also expressed as the maximum mutual information between a transmitter and a receiver. The mutual information quantifies the amount of information obtained about the unknown transmitted signal by observing the received signal. While mutual information is traditionally defined in a probabilistic setting, this notion can also be introduced in a nonstochastic setting and related to the ϵ-packing or (ϵ, δ)-packing of the space. Specifically, the work in [45] shows that for ϵ-distinguishable signals and error $\delta = 0$, the capacity can be expressed as the supremum of a mutual information functional between transmitted and received signals. On the other hand, the work in [46] generalizes this functional introducing a confidence level $(1 - \delta)$ for the mutual information and shows that the largest mutual information defined in this context corresponds to the (ϵ, δ)-capacity. Namely, accounting for the possibility of having an error in the communication process, quantified by the δ-overlap, translates in a corresponding limited confidence $(1 - \delta)$ in the information acquired by the receiver about the transmitter. In a more traditional stochastic setting, the work [47] considers communication using as random fields and, following the traditional probabilistic information theory approach, derives the mutual information between transmitter and receiver in this setting and provides a numerical scheme for its computation.

2.5 Conclusion

In this chapter, we have characterized the amount of information carried by electromagnetic waves using a signal space approach. First, we have argued that the dimensionality of the space is limited by the number of degrees of freedom of the propagating wave, thus posing a limit on the number of real coefficients that can be used to specify the waveform up to arbitrary accuracy. This is because propagation ensures that the observed field is a filtered version of the transmitted one through a propagation operator that projects it onto the space of effectively bandlimited signals of essentially N_0 dimensions. Once this result is established using Maxwell's theory, the number of degrees of freedom can be computed using functional analysis tools like Landau's theorem and its extensions. Then, introducing a resolution limit on the signals' space, which is consistent with the physical quantized nature of observable fields, leads to two mathematical notions that quantify information in terms of bits. The Kolmogorov ϵ-entropy represents the smallest number of bits needed to identify any signal in the space and is geometrically represented in terms of sphere covering at resolution ϵ, while the Kolmogorov ϵ-capacity represents the largest number of bits carried by any signal in the space and is geometrically represented in terms of sphere packing at resolution $\epsilon/2$. Both entropy and capacity are linearly related to the number of degrees of freedom and logarithmically related to the ratio of the maximum norm of the signal to the maximum norm of the perturbation with which the signal is observed. The notion of capacity can also be relaxed using a packing with partial overlap among the balls and this leads to the notion of (ϵ, δ)-capacity corresponding to the largest bitrate achieved in communication with waves with the possibility of error.

Kolmogorov's ϵ-entropy and capacity are also closely related to the probabilistic notions of entropy and capacity used in classical information theory. Shannon's classical entropy notion represents the smallest number of bits needed to identify any typical realization of a continuous stochastic process, quantized at level ϵ. On the other hand, Shannon's capacity represents the largest number of bits that can be transported by any signal in the space, when this is subject to a stochastic perturbation of standard deviation ϵ. In this setting, a geometric picture completely analogous to the one described in the nonstochastic case arises, where the value of ϵ is replaced by the statistical dispersion of the noise associated to the measurement process, that probabilistically concentrates around its typical value. Furthermore, the Shannon capacity can be expressed as the supremum of the mutual information between transmitted and received signals. This notion can also be constructed in a nonstochastic setting by defining a mutual information functional. In this case, the δ-overlap in the (ϵ, δ)-capacity translates in a corresponding limited confidence $(1 - \delta)$ in the information acquired by the receiver about the transmitter.

References

[1] C.E. Shannon, A mathematical theory of communication, The Bell System Technical Journal 27 (3) (July 1948) 379–423.

[2] M.D. Migliore, On electromagnetics and information theory, IEEE Transactions on Antennas and Propagation 56 (10) (Oct. 2008) 3188–3200.

[3] T. Gong, et al., Holographic MIMO communications: theoretical foundations, enabling technologies, and future directions, IEEE Communications Surveys and Tutorials 26 (1) (2023) 196–257.

[4] M. Franceschetti, M.D. Migliore, P. Minero, The capacity of wireless networks: information-theoretic and physical limits, IEEE Transactions on Information Theory 55 (8) (Aug. 2009) 3413–3424.

[5] Special Issue on Electromagnetic Signal and Information Theory, IEEE Journal on Selected Areas in Communications (June 2024).

[6] Special issue on electromagnetic information theory, Journal of Selected Areas in Information Theory (2025), in press.

[7] O. Bucci, G. Franceschetti, On the spatial bandwidth of scattered fields, IEEE Transactions on Antennas and Propagation 35 (12) (December 1987) 1445–1455.
[8] O.M. Bucci, G. Franceschetti, On the degrees of freedom of scattered fields, IEEE Transactions on Antennas and Propagation 37 (7) (July 1989) 918–926.
[9] M. Franceschetti, On Landau's eigenvalue theorem and information cut-sets, IEEE Transactions on Information Theory 61 (9) (Sept. 2015) 5042–5051.
[10] M. Franceschetti, Wave Theory of Information, Cambridge University Press, 2017.
[11] M.D. Migliore, On the role of the number of degrees of freedom of the field in MIMO channels, IEEE Transactions on Antennas and Propagation 54 (2) (Feb. 2006) 620–628.
[12] K. Gruber, E.A. Marengo, New aspects of electromagnetic information theory for wireless and antenna systems, IEEE Transactions on Antennas and Propagation 56 (11) (2008) 3470–3484.
[13] M.A. Jensen, J.W. Wallace, Capacity of the continuous-space electromagnetic channel, IEEE Transactions on Antennas and Propagation 56 (2) (Feb. 2008) 524–531.
[14] M. Franceschetti, K. Chakraborty, Space-time duality in multiple antenna channels, IEEE Transactions on Wireless Communications 8 (4) (April 2009) 1733–1743.
[15] O.M. Bucci, C. Gennarelli, C. Savarese, Representation of electromagnetic fields over arbitrary surfaces by a finite and nonredundant number of samples, IEEE Transactions on Antennas and Propagation 46 (3) (March 1998) 351–359.
[16] O.M. Bucci, C. Gennarelli, G. Riccio, C. Savarese, Sampling representation of electromagnetic fields over three-dimensional domains, Radio Science 34 (3) (May 1999) 763–774.
[17] M.D. Migliore, On the sampling of the electromagnetic field radiated by sparse sources, IEEE Transactions on Antennas and Propagation 63 (2) (Feb. 2015) 553–564.
[18] A. Pizzo, A.d.J. Torres, L. Sanguinetti, T.L. Marzetta, Nyquist sampling and degrees of freedom of electromagnetic fields, IEEE Transactions on Signal Processing 70 (2022) 3935–3947.
[19] O.M. Bucci, G. Di Massa, The truncation error in the application of sampling series to electromagnetic problems, IEEE Transactions on Antennas and Propagation 36 (7) (1988) 941–949.
[20] O.M. Bucci, C. Gennarelli, C. Savarese, Optimal interpolation of radiated fields over a sphere, IEEE Transactions on Antennas and Propagation 39 (11) (Nov. 1991) 1633–1643.
[21] M. Franceschetti, M.D. Migliore, P. Minero, F. Schettino, The information carried by scattered waves: near-field and nonasymptotic regimes, IEEE Transactions on Antennas and Propagation 63 (7) (July 2015) 3144–3157.
[22] J. Xu, R. Janaswamy, Electromagnetic degrees of freedom in 2-D scattering environments, IEEE Transactions on Antennas and Propagation 54 (12) (Dec. 2006) 3882–3894.
[23] N. Decarli, G.L. Giorgi, M. Di Renzo, S.M. Giordano, M.Z. Shakir, Communication modes with large intelligent surfaces in the near field, IEEE Access 9 (Nov. 2020) 16564–16580.
[24] M. Franceschetti, Quantum limits of bandlimited radiation, Journal of Statistical Physics 127 (2017) 813–823.
[25] D. Dardari, Communicating with large intelligent surfaces: fundamental limits and models, IEEE Journal on Selected Areas in Communications 38 (11) (Nov. 2020) 2526–2537.
[26] A. Pizzo, T.L. Marzetta, L. Sanguinetti, Spatially-stationary model for holographic MIMO small-scale fading, IEEE Journal on Selected Areas in Communications 38 (9) (Sept. 2020) 1964–1979.
[27] J.C. Ruiz-Sicilia, M. Di Renzo, M.D. Migliore, M. Debbah, H.V. Poor, On the degrees of freedom and eigenfunctions of line-of-sight holographic MIMO communications, arXiv preprint, arXiv:2308.08009v2, 2023.
[28] D. Pinchera, M.D. Migliore, F. Schettino, Optimizing antenna arrays for spatial multiplexing: towards 6G systems, IEEE Access 9 (2021) 53276–53291.
[29] M. Di Renzo, F.H. Danufane, S. Tretyakov, Communication models for reconfigurable intelligent surfaces: from surface electromagnetics to wireless networks optimization, Proceedings of the IEEE 110 (9) (Sept. 2022) 1164–1209.

[30] M.D. Migliore, D. Pinchera, F. Schettino, A simple and robust adaptive parasitic antenna, IEEE Transactions on Antennas and Propagation 53 (10) (Oct. 2005) 3262–3272.
[31] H. Li, S. Shen, B. Clerckx, Beyond diagonal reconfigurable intelligent surfaces: from transmitting and reflecting modes to single-, group- and fully-connected architectures, IEEE Transactions on Wireless Communications 22 (4) (April 2023) 2311–2324.
[32] Y. Zhao, H. Li, M. Franceschetti, B. Clerckx, Channel shaping using beyond diagonal reconfigurable intelligent surface: analysis, optimization, and enhanced flexibility, arXiv:2407.15196, 2024.
[33] A. Kolmogorov, On the Shannon theory of information transmission in the case of continuous signals, I.R.E. Transactions on Information Theory 2 (4) (Dec. 1956) 102–108.
[34] D.L. Donoho, Wald lecture I: Counting bits with Shannon and Kolmogorov's, Tech. Rep. Dept. Statist., Stanford Univ., Stanford, CA, USA, 2000.
[35] T.J. Lim, M. Franceschetti, Information without rolling dice, IEEE Transactions on Information Theory 63 (3) (March 2017) 1349–1363.
[36] A. Pinkus, n-Widths in Approximation Theory, Springer, 1985.
[37] D. Slepian, On bandwidth, Proceedings of the IEEE 64 (3) (1976) 292–300.
[38] D. Slepian, Some comments on Fourier analysis, uncertainty and modeling, SIAM Review 25 (3) (1983) 379–393.
[39] H.J. Landau, On Szegö's eigenvalue distribution theorem and non-hermitian kernels, Journal d'Analyse Mathématique 28 (1975) 335–357.
[40] D.A.B. Miller, Waves, Modes, Communications, and Optics: A Tutorial, Advances in Optics and Photonics, vol. 11(3), Optical Society of America, 2019, pp. 679–706, https://doi.org/10.1364/AOP.11.000679.
[41] M. Gustafsson, Degrees of freedom for radiating systems, IEEE Transactions on Antennas and Propagation (2026), in press.
[42] M. Franceschetti, M.D. Migliore, P. Minero, F. Schettino, The degrees of freedom of wireless networks via cut-set integrals, IEEE Transactions on Information Theory 63 (7) (2011) 3144–3157.
[43] D. Miller, Communicating with waves between volumes: evaluating orthogonal spatial channels and limits on coupling strengths, Applied Optics 39 (11) (May 2006) 1681–1691.
[44] A.D. Wyner, A bound on the number of distinguishable functions which are time-limited and approximately band-limited, SIAM Journal on Applied Mathematics 24 (3) (May 1973) 289–297.
[45] G.N. Nair, A nonstochastic information theory for communication and state estimation, IEEE Transactions on Automatic Control 58 (6) (June 2013) 1497–1510.
[46] A. Rangi, M. Franceschetti, Non-stochastic information theory, arXiv:1904.11632 [cs.IT], May 2019.
[47] Z. Wan, J. Zhu, Z. Zhang, L. Dai, C.-B. Chae, Mutual information for electromagnetic information theory based on random fields, IEEE Transactions on Communications 71 (4) (April 2023) 1982–1996.

CHAPTER 3

3D continuous-space radio channel research based on electromagnetic information theory

Yue Yang[a,b], Cheng-Xiang Wang[a,b], Jie Huang[a,b], Junling Li[a,b], Chenxuan Gu[a,b], Zheng-Rong Jin[a,b], Qingyin Ma[a,b], Tianrun Qi[a,b], Xingyao Shangguan[a,b], Qun Wu[a,b], and Runruo Yang[a,b]

[a]*National Mobile Communications Research Laboratory, School of Information Science and Engineering, Southeast University, Nanjing, China* [b]*Purple Mountain Laboratories, Nanjing, China*

3.1 Introduction

During the era of rapid advancement in wireless communication technologies, the investigation into channel models has surpassed the scope of traditional discrete models and integrated the continuous features of electromagnetic fields. Wireless channels are oriented to be three-dimensional (3D) and realize space–air–ground–sea global coverage in future communication systems. Base stations and users tend to move in 3D continuous space, while antennas are deeply coupled with wireless channels. These trends bring challenges to channel modeling, antenna design, and transmission system optimization. A 3D continuous-space perspective is essential for understanding the spatial, temporal, and frequency-dependent behaviors of wireless channels, especially in scenarios involving high mobility systems, complex environments, and wide frequency bands. The combination of electromagnetic theory and information theory lays a solid groundwork for the analysis and modeling of these channels, facilitating more precise forecasts of system performance and a more profound understanding of the fundamental boundaries of communication.

This chapter is set to tackle the challenges pertaining to channel modeling in 3D continuous space. By incorporating electromagnetic information theory (EIT), it establishes a comprehensive framework that bridges the physical properties of wireless propagation with their implications for system design and capacity analysis. The chapter also introduces innovative modeling approaches, measurement techniques, and design methodologies tailored for continuous-space scenarios, offering a unified perspective that goes beyond conventional

Electromagnetic Signal and Information Theory
https://doi.org/10.1016/B978-0-44-336354-2.00010-1
Copyright © 2027 Elsevier Inc. All rights are reserved, including those for text and data mining, AI training, and similar technologies.

methods. These advancements aim to meet the demands of next generation communication systems, including sixth-generation (6G) and beyond 6G (B6G), where coverage, adaptability, and efficiency are important.

Serving as both a theoretical reference and a practical guidance, this chapter synthesizes fundamental principles, detailed modeling strategies, and forward-looking applications. It provides researchers, engineers, and system designers with the tools needed to explore the potential of 3D continuous-space radio channels. Subsequent sections will delve into the theoretical underpinnings, propose novel models, analyze channel capacities, and offer design guidelines for antennas and systems, culminating in discussions on future research directions and applications.

The remaining sections are structured as follows: Section 3.2 covers the essential theories, including electromagnetic theory, information theory, and antenna theory, establishing the foundation for the rest of the book. Section 3.3 introduces various measurement techniques and models for 3D continuous-space radio channels, offering a robust basis for channel characterization. Section 3.4 explores capacity analysis for these models, highlighting their performance under different conditions. In Section 3.5, general antenna design is discussed, focusing on its role in optimizing continuous-space communication systems. Section 3.6 examines the channel map-based system design, offering a strategy for adapting systems to channel characteristics. Finally, Section 3.7 expounds on future research directions and potential applications, thereby laying the foundation for additional innovation within this topic. This structure ensures a comprehensive exploration of the subject, empowering readers to advance the field of wireless communications.

3.2 Fundamental theories

In this section, four essential theories, namely electromagnetic theory, information theory, wireless propagation channel modeling theory, and antenna theory, are presented. Then, the relationships and gaps of these fundamental theories are given. Finally, the definition and necessity of EIT are illustrated.

In the 6G/B6G wireless communication systems, the 3D continuous-space radio channels will be distributed across large-scale spaces, including space, air, ground, and sea, and deeply coupled with the environment. The 3D continuous-space radio channel demonstrates multiscale correlation features within the space–time–frequency domains, thereby presenting a multitude of challenges for the design of wireless communication systems. The foundational pillars of wireless communications mainly comprise four theories: electromagnetic theory, information theory, wireless propagation channel modeling theory, and antenna theory. The four theories have been developed separately and independently to meet the requirements of wireless communication systems up to the fifth-generation (5G). However, future applications will require new technologies, which give rise to new characteristics in the 3D continuous-space radio channel. This makes it challenging to achieve the designed requirements based on a single theory.

3.2.1 Electromagnetic theory

Electromagnetic theory describes the emergence and transmission of electromagnetic waves and investigates the correlations among assorted physical quantities within electromagnetic fields. As a cornerstone of physics, electromagnetic theory deals with the inherent relationship between electric and magnetic fields, providing profound insights into the underlying principles governing our universe. It is based on the three experimental laws of electromagnetism, including Coulomb's law, Ampere's law, and Faraday's law of electromagnetic induction, which describe the relationship between electric and magnetic fields, providing a set of differential and integral equations that articulate how field quantities are related to their sources [1].

In the context of wireless communications, where the transmission of information is dependent upon manipulating electromagnetic waves, electromagnetic theory proves indispensable for channel modeling [2], antenna design [3], and radio positioning [4]. Maxwell's equations, formulated in the 19th century, summarize this theory, offering a unified framework that elegantly explains the propagation of electromagnetic waves. Electric and magnetic fields interact to produce electromagnetic waves. Using Maxwell's equations, researchers can analyze signal propagation, uncovering the origins of multipath fading. This approach is essential for understanding near-field channels and modeling frequency-dependent phenomena, supporting efficient wireless signal transmission and reception [5]. The theoretical framework of electromagnetic theory underpins our ability to detect signals and enhance reliability, offering a lens through which researchers can address contemporary challenges and pioneer innovative solutions. In the evolution of communication technologies, electromagnetic theory contributes significantly to advancements like holographic multiple-input multiple-output (MIMO), reconfigurable intelligent surfaces (RIS), integrated sensing and communication (ISAC), and ultramassive MIMO.

However, electromagnetic theory faces challenges in 6G/B6G wireless communications, particularly in statistical channel modeling and full-space channel capacity estimation. Real-world communication channels are dynamic and time-varying, making them difficult to model with electromagnetic theory, which relies on predefined initial conditions, boundary conditions, and constitutive relations for accurate predictions. Key statistical variations, such as fading and shadowing, often remain unknown and require advanced statistical models for thorough analysis. Estimating continuous-space channel capacity necessitates accounting for all propagation paths and environmental factors, a computationally demanding process. While electromagnetic theory offers critical insights into wave propagation, it falls short as a comprehensive and efficient tool for capacity estimation in complex environments with multipath effects and scattering.

3.2.2 Information theory

Information theory pertains to the conveyance and manipulation of information. It centers on the proficient and precise transmission of information within the confines of restricted bandwidth and power. Information theory constitutes a scientific examination of information metrics, channel capacity, as well as the encoding and decoding operations of communication systems. Established by Claude Shannon [6], information theory is primarily comprised of three theorems. Shannon's first theorem is known as the variable length lossless coding theorem, which primarily describes the transformation of original source symbols into new encoded symbols to maximize the information carried by each encoded symbol. Shannon's second theorem is referred to as the noisy channel coding theorem, which measures the upper limit of the error-free data transmission rate, considering limitations in both bandwidth and power, specifically in an additive white Gaussian noise (AWGN) channel. This theorem signifies that if the information transmission rate of a channel remains below its capacity, implementing suitable channel coding methods can lead to exceptionally high levels of transmission reliability. However, reliable transmission becomes impossible if the information transmission rate surpasses the channel capacity. Shannon's third theorem is named the rate-distortion theorem. This theorem indicates that, given a sufficiently long code length, it is invariably feasible to discover a source coding approach that attains a marginally superior information transmission rate in comparison to the rate-distortion function, while the average distortion of the code does not surpass the specified allowable distortion.

Information theory is fundamental to communication systems, providing essential tools to evaluate performance and optimize signals for highly efficient information transmission [7]. By quantifying the uncertainty in information transmission through AWGN channels, it enables the development of effective encoding methods to reduce transmission errors. In system design, information theory helps researchers determine channel capacity and degree-of-freedom (DoF) in wide-sense stationary fading channels [8].

However, as wireless communication technologies evolve, information theory faces challenges in requiring accurate abstract models that align with real-world conditions, especially for continuous-space channels in the 6G/B6G era [9]. These channels introduce complexities beyond traditional information theory, such as calculating the capacity of nonstationary fading channels and addressing multiuser and multinetwork scenarios. Effectively describing information transmission in the continuous space of 6G/B6G necessitates integrating existing theoretical frameworks with innovative approaches tailored to the unique characteristics of these channels.

3.2.3 Wireless propagation channel modeling theory

Wireless propagation channel modeling theory offers a systematic framework for depicting the channels between transmitter (Tx) and receiver (Rx) antennas [10]. While radio channel modeling theory integrates the wireless propagation channel modeling theory and antenna theory. Wireless propagation channel modeling theory employs various models to analyze signal propagation in wireless channels with large bandwidth and complex environments, including deterministic, stochastic, and artificial intelligence (AI)-assisted models [11]. Deterministic models primarily include ray-tracing (RT) channel models and channel models based on channel measurements, which focus on precisely defining the characteristics of wireless channels based on specific environmental conditions and parameters [12]. Stochastic models, in contrast, encompass geometry-based stochastic models (GBSMs), correlation-based stochastic models (CBSMs), and beam domain channel models (BDCMs), and they employ a statistical methodology, accounting for the inherent variability in wireless channels. Integrating AI into channel modeling introduces adaptive and prediction capabilities, allowing more dynamic and accurate models to adapt to changing channel conditions [13].

Enhancing channel model accuracy involves integrating theories from various fields into wireless propagation channel modeling. Incorporating elements such as antenna patterns, electromagnetic waves, channel fading, and information transmission, the wireless propagation channel modeling theory serves as a link between electromagnetic theory, information theory, and antenna theory, offering valuable insights for optimizing spectrum utilization.

3.2.4 Antenna theory

Antenna theory focuses on the design, examination, and description of antennas. The principal design methodologies depend on the analysis of radiation fields in order to fulfill the demands of communication systems. The antenna serves as a guiding apparatus for transmitting and receiving radio waves [14]. Antenna theory stands as a comprehensive scientific discipline delving into the intricacies of antenna design, performance, and operational principles. Radiation patterns, indicative of how antennas distribute emitted energy in space, are crucial parameters in antenna theory [15]. Gain, which measures the effectiveness of an antenna in concentrating radiation in a particular direction, is another vital factor influencing communication system performance [16]. Polarization, representing the orientation of the electric field, and directionality, indicating the antenna's sensitivity to signals from different directions, further contributes to the antenna theory [17]. Antenna theory spans a multitude of antenna types, shapes, and operating frequency bands, exploring their interactions within the wireless channel. By employing key parameters of antennas, researchers can more accurately model the spatial propagation of signals, especially for new antenna arrays in 6G/B6G, such as holographic MIMO, RIS, fluid and movable antennas, and ultramassive MIMO.

However, while antenna theory offers valuable insights into individual antennas, directly deriving the channel capacity considering the Tx and Rx antenna parameters remains challenging. The intricate nature of wireless channels, influenced by environmental factors and the dynamic behavior of communication systems, extends beyond the capabilities of conventional antenna theory. Moreover, the comprehension of double-ended antenna design, which assesses both Tx and Rx performance for possible improvements, is still restricted [18].

3.2.5 EIT

EIT is the integration of the aforementioned four theories [19]. The evolutionary trend of 6G/B6G wireless systems poses challenges to wireless channel modeling theory and methods, channel capacity limits, antenna design, and transmission design and optimization. To reach the objective of enhanced channel capacity, it is essential to take into account the combination of electromagnetic theory, information theory, wireless propagation channel modeling theory, and antenna theory. The key research directions, including 3D continuous-space radio channel measurements and modeling, channel capacity analysis, general antenna design, and system design, need further exploration in wireless communication systems. Through the examination of the 3D continuous-space radio channel model, the integration of electromagnetic theory, antenna theory, and wireless propagation channel modeling theory can be realized. Additionally, by analyzing the achievable rate of the 3D continuous-space radio channel model, devising the antenna with regard to the channel characteristics, and formulating the optimal transmission approach, the four theories can be merged to establish the EIT.

3.3 3D continuous-space radio channel measurements and modeling

In this section, several 3D continuous-space radio channel measurements are shown. The statistical properties of channel are analyzed. Then, 3D continuous-space electromagnetic channel models, circuit-based radio channel models are proposed. The paradigm of 3D continuous-space radio channel models is envisioned.

3.3.1 3D continuous-space radio channel measurements

With the growth in the number of antenna elements and the reduction in the spacing among them, antenna arrays become highly intertwined with the communication environment, essentially turning into an integral part of it. Technologies in 6G like ISAC, RIS, ultramassive

MIMO, and holographic MIMO will be extensively utilized. Thus, it is not only requisite to gauge the wireless propagation channel and analyze the characteristics of the wireless propagation channel within the 3D continuous-space, but also to carry out measurements on the radio channel by taking into account the transmitting and receiving antenna arrays and analyze the properties of the radio channel in the 3D continuous-space.

Channel measurement is the most direct and effective method for exploring channel characteristics. Conducting research on 3D continuous-space radio channel measurement, parameter extraction, and characteristics analysis can provide significant support for the accurate modeling of continuous-space radio channels. By equipping the unmanned aerial vehicle at the Tx with a single antenna or arranging antenna arrays with different apertures, element types, topological structures, and circuit parameters on the ground, and by taking into account the mutual coupling effect among antenna elements and rationally governing the movement paths of the Tx and Rx within the wireless propagation channel setting, experiments regarding radio channel measurement in the 3D continuous-space can be conducted. According to the channel parameter extraction algorithm, the channel state information in the 3D continuous-space can be obtained, the characteristics of the radio channel in the 3D continuous-space can be analyzed, and the influence of the wireless propagation channel environment and significant antenna parameters such as antenna impedance, gain, and radiation efficiency in relation to 3D continuous-space radio channel properties like space-time-frequency correlation functions, power spectral density, and level crossing rate can be investigated.

Radio channel measurement experiments need to consider the radio channel measurement including the transmitting and receiving antennas. Consequently, during the process of performing radio channel measurements, it is essential to measure the impedance matrix of the antenna array, the antenna radiation pattern, and the antenna efficiency. However, the following technical challenges are also faced:

1. The impedance matching network at the Tx and Rx serves as the transitional segment between the antenna array at the Tx and Rx and its subsequent radio frequency (RF) devices. Measuring the matching network requires a multiport vector network analyzer with high cost. In practical measurements, calibration of the antennas and circuits is a necessity, and attaining precise calibration and impedance matching proves to be a challenging task, resulting in the difficulty in fairly comparing the integrated response of each part with the total radio channel response.
2. The nonlinear effect of the power amplifier is difficult to eliminate through calibration, which affects the accurate acquisition of the radio channel.
3. The phase noise of the local oscillator presents difficulties in being modeled and eliminated via calibration.

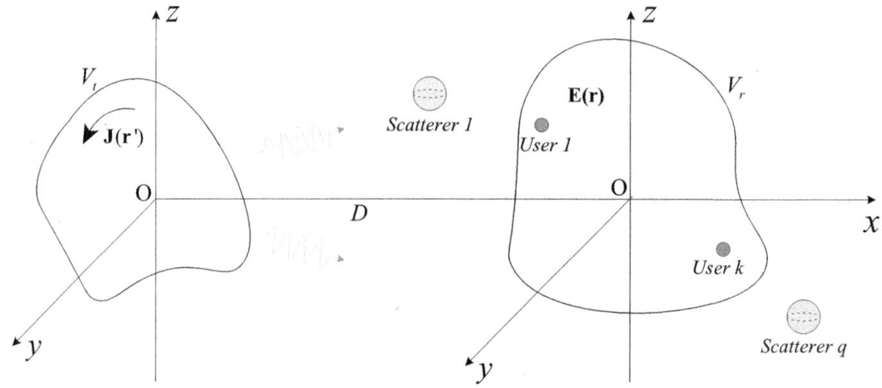

Figure 3.1:
A 3D continuous-space electromagnetic channel model applicable to 6G multiuser communications.

To deal with these difficulties, the following approaches can be utilized:

1. Use a vector network analyzer to calibrate the responses of various parts such as the baseband at the Tx and Rx, the up/down conversion modules, the power amplifier/low noise amplifier, and the transmitting and receiving antennas.
2. Adopt the power back-off method and the predistortion method to overcome the nonlinear effect of the power amplifier.
3. Use components with better frequency stability and lower phase noise, or conduct experiments at lower frequencies to minimize the impact of phase noise as much as possible. A phase tracking reference signal can also be introduced to estimate the common phase error and then perform compensation.

3.3.2 A 3D continuous-space electromagnetic channel model

A novel 3D continuous-space electromagnetic channel model jointly considering dyadic Green's function, scattering environment, and spherical wave expansion is proposed [20]. To construct the continuous-space electromagnetic channel model, the radiation process is introduced. In Fig. 3.1, a 3D electromagnetic channel model in the continuous-space is shown. The current source is dispersed within the volume at the Tx, and users are situated in the volume at the Rx. Scatterers are distributed in the vicinity of the users to reflect the impact of physical communication settings. The separation between the Tx and Rx is denoted as D. The volumes of the Tx and Rx volume are V_t and V_r, respectively.

In the free space, the radiation process of the source currents and electric fields can be expressed as [21]

$$\mathbf{E}(\mathbf{r}) = i\omega\mu \int_{V_t} \overline{\mathbf{G}}(\mathbf{r}, \mathbf{r}')\mathbf{J}(\mathbf{r}')d^3\mathbf{r}' \tag{3.1}$$

where $\mathbf{r}$ and $\mathbf{r}'$ stand for the locations at the Rx and Tx, respectively. The symbol ω represents the angular frequency, μ is the permeability in free space, $\mathbf{J}(\mathbf{r}')$ is the current density at position $\mathbf{r}'$, and $\overline{\mathbf{G}}(\mathbf{r}, \mathbf{r}')$ is the dyadic Green's function, which is defined as [22]

$$\begin{aligned}\overline{\mathbf{G}}(\mathbf{r}, \mathbf{r}') &= \left(\overline{\mathbf{I}} + \frac{\nabla\nabla}{k^2}\right) g(\mathbf{r}, \mathbf{r}') \\ &= \left(1 + \frac{i}{kr} - \frac{1}{k^2r^2}\right)\overline{\mathbf{I}}g(\mathbf{r}, \mathbf{r}') + \left(-1 - \frac{3i}{kr} + \frac{3}{k^2r^2}\right)\hat{\mathbf{r}}\hat{\mathbf{r}}g(\mathbf{r}, \mathbf{r}'),\end{aligned} \tag{3.2}$$

where $\overline{\mathbf{I}}$ is the unit dyad, ∇ is the differential operator, $k = \omega\sqrt{\mu\varepsilon}$ is the wave number in free space, in which ε is the permittivity. The distance and unit vector of direction between the Tx and Rx point are denoted as $r = |\mathbf{r} - \mathbf{r}'|$ and $\hat{\mathbf{r}}$, respectively. The scalar Green's function $g(\mathbf{r}, \mathbf{r}')$ in free space can be given by

$$g(\mathbf{r}, \mathbf{r}') = \frac{e^{ik|\mathbf{r}-\mathbf{r}'|}}{4\pi|\mathbf{r} - \mathbf{r}'|} = \frac{e^{ikr}}{4\pi r}. \tag{3.3}$$

In wireless communications, with the growth of antenna array size and the increase of carrier frequency, the Rayleigh distance will expand and the near-field effect has to be taken into account. Thus, the near-field components of the radiated field and spherical wavefronts are incorporated into the proposed channel model. To present a more precise manifestation of the near-field effect and bridge the radiation process with the communication process, spherical wave functions are employed to decompose source currents and dyadic Green's functions. This procedure is also named the singular value decomposition of the radiation operator. The spherical wave functions are defined as [23]

$$\mathbf{U}_{nml}(r, \theta, \varphi) = \begin{cases} \nabla \times \mathbf{r}h_n^{(1)}(kr)Y_{nm}(\theta, \varphi), & l = 1, \\ \frac{1}{k}\nabla \times \nabla \times \mathbf{r}h_n^{(1)}(kr)Y_{nm}(\theta, \varphi), & l = 2, \end{cases} \tag{3.4}$$

and

$$\mathbf{V}_{nml}(r, \theta, \varphi) = \begin{cases} \nabla \times \mathbf{r}j_n^{(1)}(kr)Y_{nm}(\theta, \varphi), & l = 1, \\ \frac{1}{k}\nabla \times \nabla \times \mathbf{r}j_n^{(1)}(kr)Y_{nm}(\theta, \varphi), & l = 2, \end{cases} \tag{3.5}$$

where $\mathbf{r} = r\hat{\mathbf{r}}$, $l = 1$, and $l = 2$ represent the transverse electric (TE) and transverse magnetic (TM) mode, respectively. The subscripts $n, m \in \mathbf{Z}^3$ are the orders of spherical harmonics, where $n = 1, 2, 3, \ldots, -n \leq m \leq n$. In (3.4), $h_n^{(1)}(x)$ is the nth order spherical Hankel function of the first kind defined as $h_n^{(1)}(x) = j_n(x) + iy_n(x)$, where $j_n(x)$ represents the nth order spherical Bessel function of the first kind, and $y_n(x)$ stands for the nth order spherical Bessel function of the second kind. The spherical harmonics are denoted as $Y_{nm}(\theta, \varphi)$.

Then, the dyadic Green's function can be broken down in terms of spherical wave functions $\mathbf{U}_{nml}$ and $\mathbf{V}_{nml}$ and is shown as [23,24]

$$\overline{\mathbf{G}}(\mathbf{r}, \mathbf{r}') = ik \sum_{n=1}^{\infty} \sum_{m=-n}^{n} \sum_{l=1}^{2} \frac{1}{n(n+1)} \mathbf{U}_{nml}(\mathbf{r}) \mathbf{V}_{nml}^{\dagger}(\mathbf{r}'), \tag{3.6}$$

where $|\mathbf{r}'| < |\mathbf{r}|$.

The 3D continuous-space electromagnetic channels are mainly constructed through Green's functions and basis functions, which can depict the connection between transmitted currents and received electric fields. First, for the modeling of electromagnetic channels, it is necessary to define the communication environments and pick a suitable Green's function to describe the electromagnetic features, like the near-field spherical wavefront. Second, basis functions are utilized to break down Green's functions, transmitted currents, and received electric fields. Different basis functions, such as the Fourier plane wave basis functions and spherical wave basis functions mentioned earlier, might possess diverse channel characteristics. Eventually, the communication process is associated with the radiation process.

The DoF is defined as the maximum number of independent communication subchannels that can be used to convey information. Here, we use the water-filling algorithm to determine the number of DoF, which is calculated by

$$DoF = \sum_{i=1}^{\min(N_t, N_r)} \mathcal{I}\left(\mu_l - \frac{N_0}{\sigma_i^2}\right), \tag{3.7}$$

where $\mathcal{I}(x)$ is an indicator function that is 1 if condition x is met, and 0 otherwise, N_t and N_r denote the number of transmitting and receiving antennas, respectively. Additionally, μ_l is the water level determined by total power P_t, σ_i^2 is the channel gain of the ith subchannel, and N_0 is the noise power spectral density.

The DoF of the continuous-space electromagnetic channel model was analyzed with different antenna apertures in Fig. 3.2. The results show that the DoF increases as the transmit antenna aperture expands, which can reflect the performance of 3D continuous-space channels.

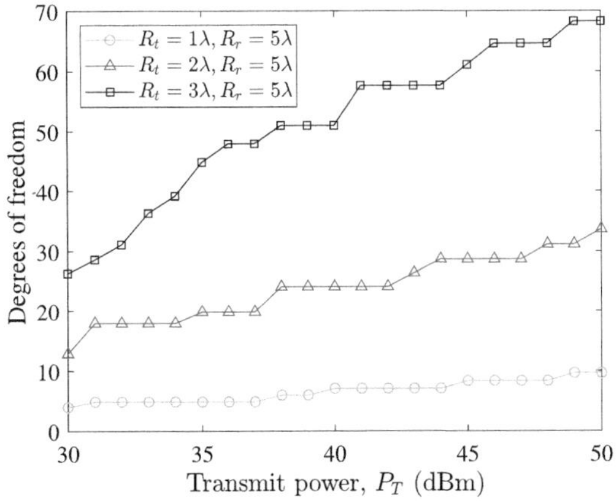

Figure 3.2:
Degrees of freedom of the continuous-space electromagnetic channel model with varying antenna apertures.

3.3.3 A circuit-based radio channel model

A circuit-based radio channel model jointly considering transceiver antennas and propagation channels is proposed. The input-output relationship of the multiport network is deduced. The correlation functions, power spectral densities, and channel capacities are computed and simulated [25].

For an antenna whose structure can be completely embedded in a sphere of radius a, L. J. Chu claimed that it is equivalent to a circuit model corresponding to the superposition of TM radiation modes [26]. Let n_c represent the mode index. For a vertically polarized omnidirectional lossless antenna, the impedance Z_{n_c} of the TM_{n_c} mode's equivalent circuit is a series connection of capacitances and a parallel connection of inductances, terminated with a unit resistance. In [26], it was also pointed out that Z_{n_c} is physically realizable and valid at all frequencies. A larger n_c means that there is more storage energy of the electromagnetic field. When $n_c > ka$ (k represents wavenumber), TM_{n_c} mode will quickly enter the cut-off region [27]. It is reasonable to suppose that a quite small antenna ($a < \lambda_c/10$, where λ_c stands for the corresponding wavelength of carrier frequency f_c) has only TM_1 mode, which is what we will focus on. Applications of such antenna are amplitude modulation broadcasting, mobile phones, etc. Its equivalent circuit consists of a resistance, an inductance, and a capacitance. Although this circuit is simple, the necessity of considering an antenna's inductive and capacitive impedances in channel modeling can be revealed.

In Fig. 3.3, a multiport circuit network is given to characterize the Tx, Rx, and wireless channel between them. The parameters in Fig. 3.3 are described as follows. The vector

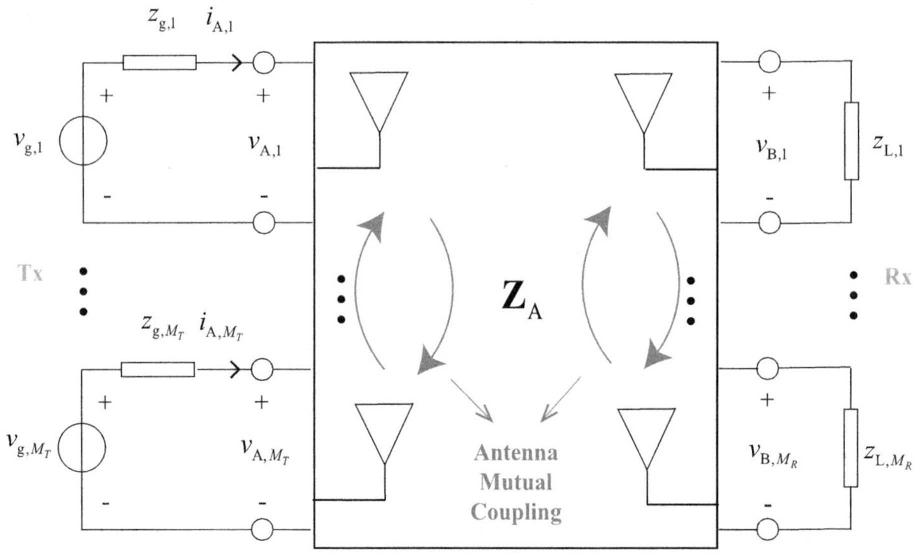

Figure 3.3:
Block diagram of the proposed circuit-based channel model.

$\boldsymbol{v}_g \in \mathbb{C}^{M_T \times 1}$ is source voltage, $\boldsymbol{z}_g \in \mathbb{C}^{M_T \times 1}$ is source impedance, $\mathbf{Z}_A \in \mathbb{C}^{(M_T+M_R)\times(M_T+M_R)}$ is impedance matrix, $\boldsymbol{v}_A \in \mathbb{C}^{M_T \times 1}$ is input voltage, $\boldsymbol{i}_A \in \mathbb{C}^{M_T \times 1}$ is input current, $\boldsymbol{v}_B \in \mathbb{C}^{M_R \times 1}$ is output voltage, $\boldsymbol{i}_B \in \mathbb{C}^{M_R \times 1}$ is output current, and $\boldsymbol{z}_L \in \mathbb{C}^{M_R \times 1}$ is load impedance. In the subsequent part, $\mathbb{C}^{d_1 \times d_2}$ represents the collection of $d_1 \times d_2$ complex matrices. The transpose operation is indicated by $(\cdot)^T$, and diag$\{\boldsymbol{x}\}$ signifies a diagonal matrix where each diagonal entry is the corresponding element within $\boldsymbol{x}$. What's more, $\mathbf{I}_N$ and $\mathbf{O}_{N \times M}$ denote the $N \times N$ identity matrix and $N \times M$ zero matrix, respectively. For a square matrix $\mathbf{Z}$, $\mathbf{Z}^{-1}$ stands for its inverse.

Before discussing details of the multiport network, the relationship between each pair of the transceiver antennas will be mentioned. The structure of the unterminated circuit network Port$_{qp}$ only relating A_p^T and A_q^R is shown in Fig. 3.4, in which the speed of light is denoted as c. In order to investigate the changes triggered by an equivalent circuit with capacitances and inductances, we take Rx antenna as an example and set its size as a such that its equivalent circuit encompasses resistance R_q, inductance aR_q/c, and capacitance $a/(cR_q)$. Regarding the Tx antenna, it is presumed that its size is sufficiently large to render the inductance an open circuit and the capacitance a short circuit, i.e., the equivalent circuit of it only has a resistance R_p so that the radiation efficiency could reach the maximum [28].

Based on a general GBSM and aforementioned assumptions, the wireless channel will be described in more details. Note that the omnidirectional, vertically polarized Tx and Rx

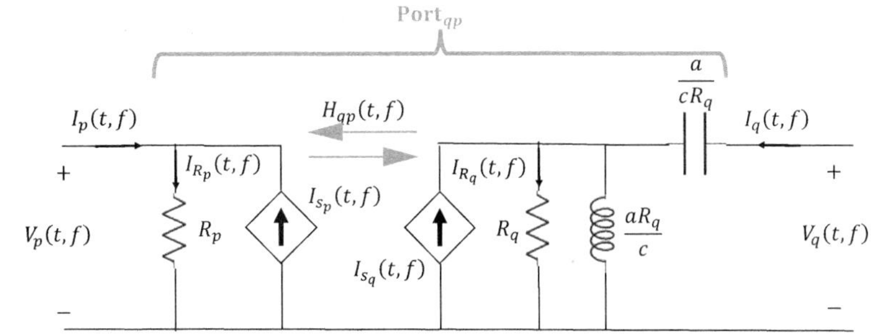

Figure 3.4:
Block diagram of circuit network Port$_{qp}$.

antennas possess gains G_t and G_r, respectively, with $G_t = G_r = 3/2$ [14,28]. Subsequently, the line-of-sight (LoS) and non-LoS (NLoS) components of the transfer function can be formulated as

$$H_{qp}^{L}(t, f) = \sqrt{G_r G_t} e^{j2\pi(f_c - f)\tau_{qp}^{L}(t)}, \tag{3.8}$$

$$H_{qp}^{N}(t, f) = \sum_{n=1}^{N_{qp}(t)} \sum_{m=1}^{M_n} \sqrt{G_r G_t} \sqrt{P_{qp,m_n}(t)} \cdot e^{j2\pi(f_c - f)\tau_{qp,mn}(t) + j\theta_{mn}}, \tag{3.9}$$

where $N_{qp}(t)$ is the total number of clusters in the link between A_p^T and A_q^R at time t, M_n is the total number of rays in cluster n ($n = 1, 2, \ldots, N_{qp}(t)$), θ_{m_n} is a random phase uniformly distributed over $(0, 2\pi]$, $P_{qp,m_n}(t)$ and $\tau_{qp,m_n}(t)$ stand for the powers and delays of the mth ray in the nth cluster between A_p^T and A_q^R at time t, respectively. Furthermore, $\tau_{qp}^{L}(t)$ represents the LoS path's propagation delay. In Fig. 3.4, the relationships between the currents flowing through the resistances at the antenna ports and the controlled current sources are described as

$$I_{s_p}(t, f) = \eta_t I_{R_q}(t, f)\sqrt{R_q/R_p} \cdot H_{qp}(t, f), \tag{3.10}$$

$$I_{s_q}(t, f) = \eta_r I_{R_p}(t, f)\sqrt{R_p/R_q} \cdot H_{qp}(t, f), \tag{3.11}$$

where both η_t and η_r stand for the nonideal amplitude loss and phase shift in the current conversion processes. For simplicity, let $\eta_t = \eta_r = 1$.

The impedance matrix $\mathbf{Z}_{qp}$ of Port$_{qp}$ establishes the connection between port voltages and port currents within this circuit network. Let $s = j2\pi f$ denote the Laplace factor, considering the unilateral approximation in far field wireless communication [29] and the relationship

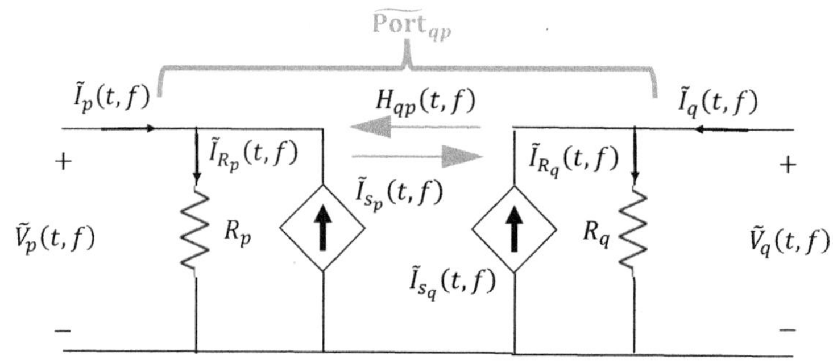

Figure 3.5:
Block diagram of resistive circuit network $\widetilde{\text{Port}}_{qp}$.

between some specified items [28], $\mathbf{Z}_{qp}$ can be determined as

$$\mathbf{Z}_{qp} = \begin{bmatrix} R_p & 0 \\ \frac{sa\sqrt{R_p R_q^3}H_{qp}(t,f)}{saR_q+R_q c} & \frac{(saR_q)^2+(cR_q)^2+sacR_q^2}{sa(saR_q+R_q c)} \end{bmatrix}. \tag{3.12}$$

As mentioned in [30], it is unwise to estimate an antenna's power through squaring the voltage or current since a realistic antenna's circuit cannot be seen as a strictly resistive one (indeed, there are inductive and capacitive impedances). For comparison, we introduce circuit network $\widetilde{\text{Port}}_{qp}$ whose antenna equivalent circuits at the Tx and Rx are strictly resistive, as shown in Fig. 3.5. Similarly, impedance matrix of $\widetilde{\text{Port}}_{qp}$ is

$$\tilde{\mathbf{Z}}_{qp} = \begin{bmatrix} R_p & 0 \\ \sqrt{R_p R_q} H_{qp}(t,f) & R_q \end{bmatrix}. \tag{3.13}$$

In order to further study the performance difference between Port_{qp} and $\widetilde{\text{Port}}_{qp}$, without considering matching networks, these circuit networks are terminated with voltage source v_g, source impedance z_g, and load impedance z_L. We use $\mathbf{X}_{qp}$ to represent $\mathbf{Z}_{qp}$ or $\tilde{\mathbf{Z}}_{qp}$, and introduce the concept of voltage transfer function (VTF) for better comparison. The VTF can be expressed as [31]

$$\text{VTF} = \frac{X_{qp}(2,1)z_L}{\left[X_{qp}(1,1)+z_g\right]\left[X_{qp}(2,2)+z_L\right]-X_{qp}(1,2)X_{qp}(2,1)}. \tag{3.14}$$

With the knowledge of the properties of Port_{qp} and $\widetilde{\text{Port}}_{qp}$, the multiport channel model in Fig. 3.3 will be discussed in detail. By segmenting $\mathbf{Z}_\text{A}$, the association between the output and the input of the multiport network is

$$\begin{bmatrix} \boldsymbol{v}_{\mathrm{A}} \\ \boldsymbol{v}_{\mathrm{B}} \end{bmatrix} = \mathbf{Z}_{\mathrm{A}} \begin{bmatrix} \boldsymbol{i}_{\mathrm{A}} \\ \boldsymbol{i}_{\mathrm{B}} \end{bmatrix} = \begin{bmatrix} \mathbf{Z}_{\mathrm{AT}} & \mathbf{O}_{M_T \times M_R} \\ \mathbf{Z}_{\mathrm{ART}} & \mathbf{Z}_{\mathrm{AR}} \end{bmatrix} \begin{bmatrix} \boldsymbol{i}_{\mathrm{A}} \\ \boldsymbol{i}_{\mathrm{B}} \end{bmatrix}, \tag{3.15}$$

where the upper right corner of matrix $\mathbf{Z}_{\mathrm{A}}$ is $\mathbf{O}_{M_T \times M_R}$ due to unilateral approximation. Each entry of the transimpedance matrix $\mathbf{Z}_{\mathrm{ART}}$ obeys $Z_{\mathrm{ART}}(q, p) = X_{qp}(2, 1)$. Since both R_p and R_q are unit resistances, $X_{qp}(1, 1)$ and $X_{qp}(2, 2)$ are defined as $X^{(1,1)}$ and $X^{(2,2)}$, respectively. Furthermore, $\mathbf{Z}_{\mathrm{AT}} = X^{(1,1)} \cdot \mathbf{C}_{M_T}^{\delta_T} \in \mathbb{C}^{M_T \times M_T}$ and $\mathbf{Z}_{\mathrm{AR}} = X^{(2,2)} \cdot \mathbf{C}_{M_R}^{\delta_R} \in \mathbb{C}^{M_R \times M_R}$ describe the antenna near-field mutual coupling effect at the Tx and Rx [29], respectively, in which

$$\mathbf{C}_M^d = \begin{bmatrix} 1 & j_0(kd) & j_0(2kd) & \cdots \\ j_0(kd) & 1 & j_0(kd) & \ddots \\ \vdots & \ddots & \ddots & \ddots \end{bmatrix} \in \mathbb{R}^{M \times M} \tag{3.16}$$

is a Toeplitz matrix and $j_0(x) = \sin x / x$. Additionally, $\mathbf{C}_{M_T}^{\delta_T}$ and $\mathbf{C}_{M_R}^{\delta_R}$ can be reduced to identity matrices when δ_T and δ_R are integer multiples of half the wavelength. If this is not the case, there will be antenna mutual coupling. To obtain the voltage transfer relationship of the proposed circuit-based channel model, source and load are considered and (3.15) is further expanded. Now we have

$$\begin{cases} \boldsymbol{v}_{\mathrm{A}} = \mathbf{Z}_{\mathrm{AT}} \boldsymbol{i}_{\mathrm{A}}, \\ \boldsymbol{v}_{\mathrm{B}} = \mathbf{Z}_{\mathrm{ART}} \boldsymbol{i}_{\mathrm{A}} + \mathbf{Z}_{\mathrm{AR}} \boldsymbol{i}_{\mathrm{B}}, \\ \boldsymbol{v}_{\mathrm{g}} = \mathrm{diag}\left\{ z_{\mathrm{g}} \right\} \boldsymbol{i}_{\mathrm{A}} + \boldsymbol{v}_{\mathrm{A}}, \\ \boldsymbol{v}_{\mathrm{B}} = -\,\mathrm{diag}\left\{ z_L \right\} \boldsymbol{i}_{\mathrm{B}}, \end{cases} \tag{3.17}$$

in which $\boldsymbol{v}_{\mathrm{B}}$ and $\boldsymbol{v}_{\mathrm{g}}$ can be related by

$$\boldsymbol{v}_{\mathrm{B}} = \left(\mathbf{I}_{M_R} + \mathbf{Z}_{\mathrm{AR}}^{-1} \mathrm{diag}\left\{ z_L \right\} \right) \mathbf{Z}_{\mathrm{ART}} \left(\mathrm{diag}\left\{ z_{\mathrm{g}} \right\}^{-1} + \mathbf{Z}_{\mathrm{AT}}^{-1} \right) \boldsymbol{v}_{\mathrm{g}} \triangleq \mathbf{H}^{\mathrm{CAS}}(t, f) \boldsymbol{v}_{\mathrm{g}}, \tag{3.18}$$

where $\mathbf{H}^{\mathrm{CAS}}(t, f)$ is the radio channel matrix which jointly considers the antenna equivalent circuit, the wireless channel, and the antenna mutual coupling.

Fig. 3.6 illustrates the probability density function (PDF) of VTF through 200,000 Monte-Carlo experiments. The standard deviation values for cluster delay spread, elevation spread, and azimuth spread are set to $\sigma_{DS} = 3$ m, $\sigma_{AS} = 5$ m, and $\sigma_{ES} = 4$ m, respectively. Fig. 3.6(a) represents the PDF of VTF's absolute value with different circuit types and antenna sizes. When a becomes larger, the VTF's absolute value will generally increase and Port_{qp}'s PDF will approach $\widetilde{\mathrm{Port}}_{qp}$'s PDF since the values of inductances and capacitances are different. Actually, Port_{qp} reduces to $\widetilde{\mathrm{Port}}_{qp}$ when a is sufficiently large. Fig. 3.6(b) reveals that the

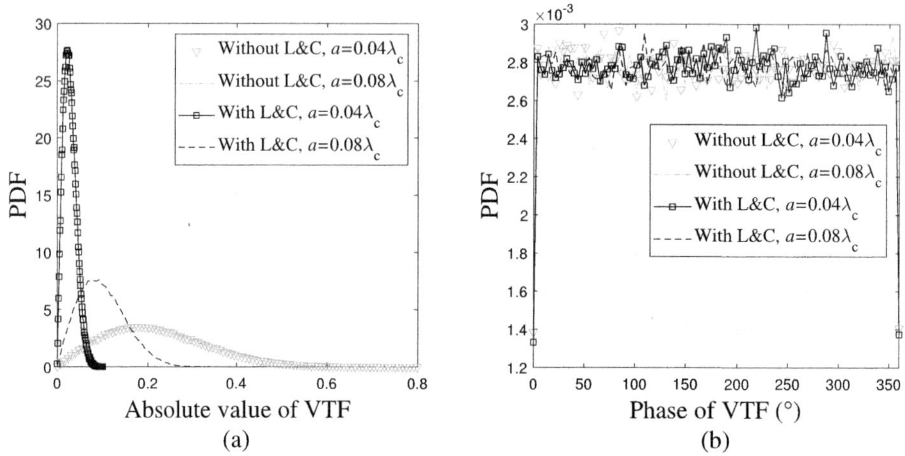

Figure 3.6:
(a) PDF of VTF's absolute value with different circuit types and antenna sizes, (b) PDF of VTF's phase with different circuit types and antenna sizes ($f_c = 3$ GHz, $p = 1$, $q = 1$, $n = 1$, $\sigma_{DS} = 3$ m, $\sigma_{AS} = 5$ m, $\sigma_{ES} = 4$ m).

VTF's phase obeys uniform distribution and does not change with different circuit types and antenna sizes. It results from the identical phase shift introduced by an antenna equivalent circuit at the same frequency.

3.3.4 A 3D continuous-space radio channel model

The idea of 3D continuous-space radio channel models is to combine the field-based channel model and circuit-based channel model. Here, the method for 3D continuous-space radio channel model is envisioned [32].

The wireless propagation environment including large- and small-scale fading between Tx and Rx is described using a GBSM [33]. The Tx and Rx have N_T and N_R antennas with antenna spacing δ_T and δ_R, respectively. Both Tx and Rx are uniform linear arrays. The antenna elements at Tx and Rx sides are symbolized as A_p^T $(p = 1, 2, \ldots, N_T)$ and A_q^R $(q = 1, 2, \ldots, N_R)$, respectively. The small-scale fading can be presented as $\mathbf{Hs} = [H_{qp}(t, f)]_{N_R \times N_T}$, where $H_{qp}(t, f)$ is the channel impulse response between A_p^T and A_q^R and is defined as [33]

$$H_{qp}(t, f) = \sqrt{\frac{K_R}{K_R + 1}} H_{qp}^L(t, f) + \sqrt{\frac{1}{K_R + 1}} H_{qp}^N(t, f), \tag{3.19}$$

where K_R is the Rician factor. The response function is the sum of the LoS components $H_{qp}^{L}(t, f)$ and NLoS components $H_{qp}^{N}(t, f)$.

In this part, $\mathbf{M}_{Rx}$ and $\mathbf{M}_{Tx}$ signify the impact of mutual coupling at the Rx and Tx, respectively. The continuous-space electromagnetic channel matrix $\mathbf{H}^{\mathrm{c}}$ can be expressed as

$$\mathbf{H}^{\mathrm{c}} = \mathbf{M}_{Rx} \cdot \mathbf{H} \cdot \mathbf{M}_{Tx}, \tag{3.20}$$

where $\mathbf{H} = [H_{qp} \cdot G_{qp}]_{N_R \times N_T}$ is considered as the propagation channel matrix and G_{qp} encompasses the influence of antenna response and large-scale fading [34].

For an arbitrary region S' with current distribution $\mathbf{J}\left(\mathbf{r}'\right)$, the radiated electric field $\mathbf{E}(\mathbf{r})$ can be expressed as [35]

$$\mathbf{E}(\mathbf{r}) = \frac{1}{j\omega\varepsilon}\int_{S'}\left(\bar{\mathbf{I}} + \frac{\nabla\nabla}{k^2}\right) g\left(\mathbf{r}, \mathbf{r}'\right)\mathbf{J}\left(\mathbf{r}'\right) d\mathbf{r}', \tag{3.21}$$

where $k = 2\pi/\lambda$ is free-space wavenumber, λ is the wavelength of electromagnetic wave, and $g\left(\mathbf{r}, \mathbf{r}'\right)$ is scalar Green's function in the free space.

To simplify electromagnetic analysis, a known electromagnetic calculation method, namely the method of moments (MoM), is utilized to build the continuous-space electromagnetic channel response between A_p^T and A_q^R. Here, the Tx antenna is supposed to be a thin linear array with a radius of d and composed of a perfect electric conductor (PEC). Thus, the array response (AR) incorporates certain details of the antenna configuration and can be presented as [35]

$$G_{qp} = \frac{1}{j\omega\varepsilon}\int_{s_p - \frac{\delta_T}{2}}^{s_p + \frac{\delta_T}{2}}\left[\frac{1 + jkR}{R^2}(2 - 3\left(\frac{d}{R}\right)^2 + (kd)^2)\right] g(\mathbf{r}, \mathbf{r}_q')d\mathbf{r}', \tag{3.22}$$

where $R = \left|\mathbf{r} - \mathbf{r}_q'\right|$. The expression of array response can be divided into real part and imaginary part. The real part contains the energy that can be propagated, while the imaginary part contains evanescent energy and it is named reactive field.

The excitation of an individual element is not directly proportional to the applied source in an antenna array. It results from antenna array elements' interaction with each other through the electromagnetic field, which induces currents on adjacent antennas. This interaction is named as mutual coupling [14].

Let us consider an antenna element A_p^T at the Tx side and suppose that this antenna element is driven by a current source with a complex amplitude α_p. If MC is ignored, the amplitude of $\mathbf{J}_p\left(\mathbf{r}'\right)$ is $\alpha_p = 1$ if it is excited identically. Due to the reflection within this antenna element, the actual amplitude γ_p is the sum of incident wave α_p and reflected wave β_p, that is, $\gamma_p = \beta_p + \alpha_p$. The actual amplitude in the antenna array is also affected by MC from other antenna elements. Therefore, as an antenna element extends to an antenna array, the aforementioned incident wave vector $\mathbf{A}$ and reflected wave vector $\mathbf{B} = [\beta_1, \ldots, \beta_{N_T}]^{\mathrm{T}}$ satisfies

$$\mathbf{B} = \mathbf{S}_{Tx}\mathbf{A}, \tag{3.23}$$

where $\mathbf{S}_{Tx}$ is reflection coefficient matrix at the Tx.

The total radiation field is expressed as

$$\mathbf{E}(\mathbf{r}) = \frac{1}{j\omega\varepsilon}\int_{S'}\left(\bar{\mathbf{I}} + \frac{\nabla\nabla}{k^2}\right) g\left(\mathbf{r}, \mathbf{r}'\right)\left(\mathbf{M}_{Tx}\mathbf{A}\right)^{\mathrm{T}} \cdot \mathbf{J}d\mathbf{r}', \tag{3.24}$$

where $\mathbf{I}_{N_R}$ is identity matrix with N_R columns and rows and $\mathbf{M}_{Tx} = \left(\mathbf{S}_{Tx} + \mathbf{I}_{N_T}\right)$ is a symmetric matrix.

In conclusion, a continuous-space radio channel model is proposed jointly considering array response, mutual coupling, and wireless propagation environment in (3.20).

Some important statistical properties can be derived based on the continuous-space electromagnetic channel matrix $\mathbf{H^c} = [H_{qp}^c(t, f)]_{N_R \times N_T}$.

The space-time-frequency correlation function between $H_{qp}^c(t, f)$ and $H_{\tilde{q}\tilde{p}}^c(t, f)$ in the continuous-space radio channel model can be obtained as

$$R_{qp,\tilde{q}\tilde{p}}^c(t, f; \Delta r, \Delta t, \Delta f) = \mathbb{E}\left\{H_{qp}^c(t, f)\left[H_{\tilde{q}\tilde{p}}^c(t - \Delta t, f - \Delta f)\right]^*\right\}, \tag{3.25}$$

where $\mathbb{E}(\cdot)$ denotes the expectation, $(\cdot)^*$ is the complex conjugate form of a matrix, and Δr, Δt, and Δf present space, time, and frequency separations, respectively.

The spatial cross-correlation functions (SCCFs) with four different situations are calculated using (3.25) and illustrated in Fig. 3.7. It shows that AR and MC reduce the spatial correlation between antenna elements because AR contains the reaction field and makes the distribution of the electromagnetic field more uneven, MC changes the impedance of antennas, causing the different impedance of antennas at different positions.

3.4 Capacity analysis for 3D continuous-space radio channel models

The performance of the system is crucial for antenna or communication system design. In this part, the DoF and channel capacity of 3D continuous-space electromagnetic channel models and radio channel models are explored.

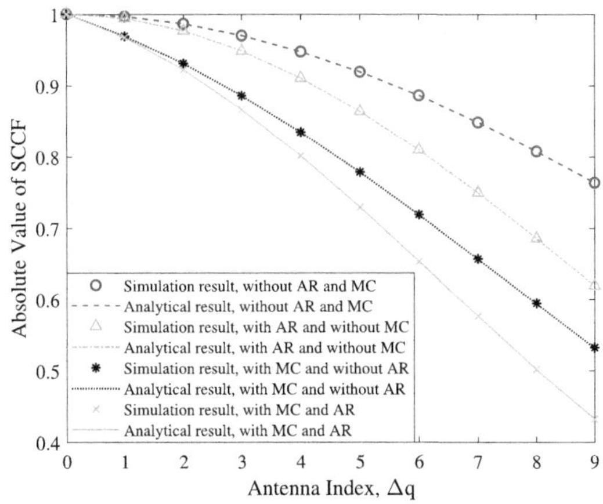

Figure 3.7:
Theoretical and simulation results of SCCFs under four situations ($\delta_T = \lambda/2$, $\delta_R = \lambda/4$, $p = 1$, $q = 1$).

3.4.1 Capacity analysis for 3D continuous-space electromagnetic channel models

DoFs are orthogonal subchannels in communications. The novel 3D continuous-space electromagnetic channel model considers dyadic Green's function, scattering environment, and spherical wave expansion. It is important to analyze its relationship between DoFs and channel capacity.

Define the self-adjoint operator $\mathbf{K}(\mathbf{r}, \mathbf{r}')$ as

$$\mathbf{K}(\mathbf{r}, \mathbf{r}') = \int \mathbf{G}(\mathbf{r}, \mathbf{s})\mathbf{G}^{\mathrm{H}}(\mathbf{r}', \mathbf{s})d\mathbf{s}, \tag{3.26}$$

where $(\cdot)^{\mathrm{H}}$ is Hermitian operator. Symmetric kernel $\mathbf{K}(\mathbf{r}, \mathbf{r}')$ is compact and can be expressed using countably infinite eigenfunctions $\phi_n(\mathbf{r}), n = 1, 2, \ldots, \infty$, with real eigenvalues $\xi_1 \geq \xi_2 \geq \cdots \geq \xi_\infty$ and $\xi_\infty = 0$. Since $\mathbf{K}(\mathbf{r}, \mathbf{r}')$ is compact, there is a level of accuracy ε and a specific integer N, satisfying $\xi_1 \geq \cdots \geq \xi_N \geq \varepsilon \geq \xi_{N+1} \geq \cdots \geq \xi_\infty$.

The eigenvalue ξ_n corresponding to the index n larger than N will exhibit a rapid decrease beyond N, which means these eigenvalues can be ignored. Therefore, the integer N is defined as the number of DoFs, and corresponding $\xi_n, n = 1, \ldots, N$ are gains of these DoFs. What's more, these nonnegligible eigenfunctions are defined as transmission modes. Thus, statistically independent information can be encoded on these transmission modes to achieve diversity [36].

Assume a scatterer at $\mathbf{r}_s = (R, \theta^R, \psi^R)$, where R is the distance between scatterers and the first element of Rx at $\mathbf{r}_1$, θ^R, and ψ^R stand for azimuth angle of arrival (AAoA) and elevation angle of arrival (EAoA) of the incident wave from this scatterer, respectively. The array response in the far field is

$$a_{FF} = \frac{1}{R}[1, \mathrm{e}^{-ik\delta_R \cos\theta}, \ldots, \mathrm{e}^{-ik(M_R-1)\delta_R \cos\theta}]^{\mathrm{T}}, \tag{3.27}$$

where k is the wavenumber of electromagnetic wave. The traditional array response in the near field is

$$a_{NF,tr} = \left[\frac{\mathrm{e}^{-ik|\mathbf{r_s}-\mathbf{r}_1|}}{|\mathbf{r_s}-\mathbf{r}_1|}, \ldots, \frac{\mathrm{e}^{-ik|\mathbf{r_s}-\mathbf{r}_{M_R}|}}{|\mathbf{r_s}-\mathbf{r}_{M_R}|}\right]^{\mathrm{T}}. \tag{3.28}$$

In the LoS scenario, eigenfunctions at the Rx is given as

$$\psi_i(\mathbf{r}) = F(\mathbf{r_s}, \mathbf{r})\alpha_i(\mathbf{r}), \tag{3.29}$$

where $F(\mathbf{r_s}, \mathbf{r})$ is focusing function, which reflects the spherical wave received at Rx, and $\alpha_i(\mathbf{r})$ is the ith prolate spheroidal harmonics function. The details of these two functions can be found in [37]. The proposed array response for near field is

$$a_{NF,pro} = [\psi_1(\mathbf{r}_1), \ldots, \psi_N(\mathbf{r}_1), \ldots, \psi_1(\mathbf{r}_{M_R}), \ldots, \psi_N(\mathbf{r}_{M_R})]^{\mathrm{T}}. \tag{3.30}$$

To fully extract DoFs in the near field, the dimension of this array response is NM_R. This array response can fully extract DoFs in the near field and the orthogonality of eigenfunctions guarantees low coherence of adjacent elements in (3.30), which was proved in [38]. The channel matrix $\mathbf{H}$ can be calculated using array responses, the details can be found in [39].

In the simulation, the path loss is considered, and only the location of Rx is changed. The N in (3.30) is set to 3. The near field and far field DoFs in the scattering environment with three different array responses in (3.27), (3.28), and (3.30) are illustrated in Fig. 3.8. The proposed near field array response can extract more DoF compared with other array responses as some scatterers are set in the near field in Fig. 3.8(a). As the Rx moves to the place where all scatterers are located in the far field of Rx, the distributions of DoF are similar in Fig. 3.8(b) because the electromagnetic wave received at Rx are plane wave and each wave has only one DoF so that the proposed array response cannot extract more DoFs. Channel capacities using three different array responses are drawn in Fig. 3.9. Since path loss is considered, channel capacities decrease with distance. The channel capacity using (3.30) is higher than that using (3.27) and (3.28) in the near field but they coincide at the far field because of the aforementioned distributions of DoFs.

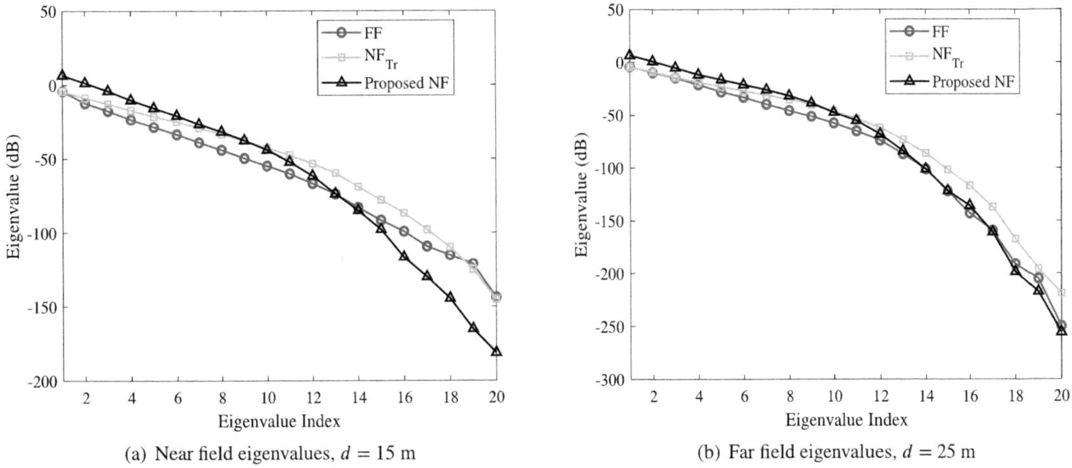

Figure 3.8:
Eigenvalues of $\mathbf{R}_r$: (a) near field and (b) far field.

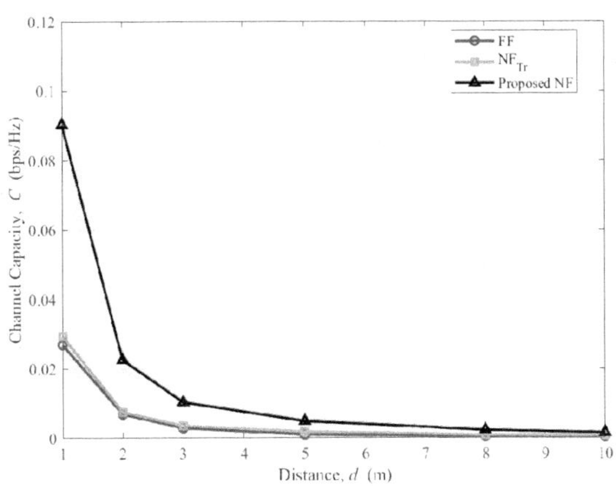

Figure 3.9:
Channel capacities with different array responses, $\rho = 10$ dB.

3.4.2 Capacity analysis for radio channel models

To study the channel capacity of radio channel models, the radio channel matrix $\mathbf{D} = \mathbf{C}_\mathrm{R}\mathbf{Z}_\mathrm{ART}\mathbf{C}_\mathrm{T}$ is expanded as

$$\begin{aligned} D_{qp} &= \sum_{i=1}^{N_\mathrm{R}}\sum_{j=1}^{N_\mathrm{T}} (\mathbf{Z}_\mathrm{ART})_{i,j}\cdot(\mathbf{C}_\mathrm{R})_{q,i}\cdot(\mathbf{C}_\mathrm{T})_{j,p} \\ &= \underbrace{(\mathbf{Z}_\mathrm{ART})_{q,p}\cdot(\mathbf{C}_\mathrm{R})_{q,q}\cdot(\mathbf{C}_\mathrm{T})_{p,p}}_{\text{self term}} + \underbrace{\sum_{i=1}^{N_\mathrm{R}}\sum_{j=1}^{N_\mathrm{T}} (\mathbf{Z}_\mathrm{ART})_{i,j}\cdot(\mathbf{C}_\mathrm{R})_{q,i}\cdot(\mathbf{C}_\mathrm{T})_{j,p}}_{(i,j)\neq(q,p),\ \text{mutual terms}}, \end{aligned} \tag{3.31}$$

and we notice that $(\mathbf{Z}_\mathrm{ART})_{qp}$ is proportional to H_{qp}. It can also be noticed that the computation of D_{qp} brings in mutual terms, as illustrated in (3.31), which are associated with the wireless propagation channels in other matrix elements.

The channel capacity represents the maximum error-free transmission rate of a communication system and acts as a significant criterion for assessing the system's performance. Taking into account the impact of noise with the noise variance σ_v^2, the input–output relation between the Tx and Rx is formulated as

$$\boldsymbol{v}_\mathrm{B} = \mathbf{D}\boldsymbol{v}_\mathrm{G} + \boldsymbol{n}, \tag{3.32}$$

where the noise vector $\boldsymbol{n}$ is supposed to follow a complex Gaussian distribution with a zero mean and a covariance matrix $\sigma_v^2\mathbf{I}_{N_\mathrm{R}}$. To precisely describe the transmit power, the power P_Tx is given as

$$\begin{aligned} P_\mathrm{Tx} &= \mathbb{E}\left\{\mathrm{Re}\left\{\boldsymbol{v}_\mathrm{A}^\mathrm{H}\boldsymbol{i}_\mathrm{A}\right\}\right\} \\ &= \mathbb{E}\left\{\mathrm{Re}\left\{\boldsymbol{i}_\mathrm{A}^\mathrm{H}\mathbf{Z}_\mathrm{AT}^\mathrm{H}\boldsymbol{i}_\mathrm{A}\right\}\right\} \\ &= \mathbb{E}\left\{\mathrm{Re}\left\{\boldsymbol{v}_\mathrm{G}^\mathrm{H}\left(\mathbf{Z}_\mathrm{AT} + z_\mathrm{G}\mathbf{I}_{M_\mathrm{T}}\right)^{-\mathrm{H}}\mathbf{Z}_\mathrm{AT}^\mathrm{H}\left(\mathbf{Z}_\mathrm{AT} + z_\mathrm{G}\mathbf{I}_{N_\mathrm{T}}\right)^{-1}\boldsymbol{v}_\mathrm{G}\right\}\right\} \\ &\triangleq \mathbb{E}\left\{\mathrm{Re}\left\{\boldsymbol{v}_\mathrm{G}^\mathrm{H}\mathbf{M}_\mathrm{Tx}\boldsymbol{v}_\mathrm{G}\right\}\right\}, \end{aligned} \tag{3.33}$$

where $\{\cdot\}^{-\mathrm{H}}$ represents the conjugate transpose of the inverse of a matrix. Therefore, the power coupling matrix can be expressed as

$$\mathbf{M}_\mathrm{Tx} = \left(\mathbf{Z}_\mathrm{AT} + z_\mathrm{G}\mathbf{I}_{M_\mathrm{T}}\right)^{-\mathrm{H}}\mathbf{Z}_\mathrm{AT}^\mathrm{H}\left(\mathbf{Z}_\mathrm{AT} + z_\mathrm{G}\mathbf{I}_{M_\mathrm{T}}\right)^{-1}. \tag{3.34}$$

Notably, the power $P_\mathrm{Tx} = \mathbb{E}\left\{\mathrm{Re}\left\{\boldsymbol{v}_\mathrm{G}^\mathrm{H}\mathbf{M}_\mathrm{Tx}\boldsymbol{v}_\mathrm{G}\right\}\right\} \neq \mathbb{E}\left\{\mathrm{Re}\left\{\boldsymbol{v}_\mathrm{G}^\mathrm{H}\boldsymbol{v}_\mathrm{G}\right\}\right\}$. Therefore, a linear transformation $\boldsymbol{x} = \mathbf{B}^{1/2}\boldsymbol{v}_\mathrm{G}$ is required, where the matrix $\mathbf{B}$ satisfies $\mathbf{M}_\mathrm{Tx} = \mathbf{B}^{\mathrm{H}/2}\mathbf{B}^{1/2}$,

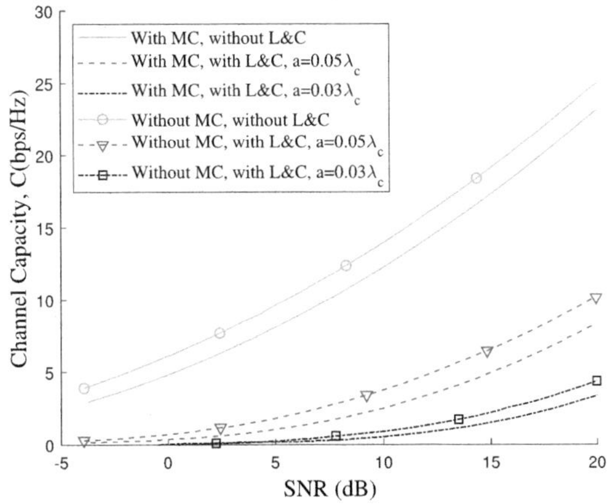

Figure 3.10:
The relation between channel capacity and SNR considering antenna mutual coupling, inductance and capacitance, and element size.

and consequently $P_{\mathrm{Tx}} = \mathbb{E}\left\{\mathrm{Re}\{\boldsymbol{x}^{\mathrm{H}}\boldsymbol{x}\}\right\}$, which is consistent with the representation of the transmission power in wireless propagation channel studies. We define $\mathbf{H}$ as the radio channel matrix in the sense of information theory. The information-theoretic input–output relationship between input $\boldsymbol{x}$ and output $\boldsymbol{y}$ can be expressed as

$$\boldsymbol{v}_{\mathrm{B}} = \mathbf{D}\mathbf{B}^{-1/2}\boldsymbol{x} + \boldsymbol{n} \triangleq \mathbf{H}\boldsymbol{x} + \boldsymbol{n}. \tag{3.35}$$

Assuming that the channel state information is unknown to the Tx, radio channel capacity can be expressed as

$$C = \mathbb{E}\left\{\log_2\left[\det\left(\mathbf{I}_{N_{\mathrm{T}}} + \frac{P_{\mathrm{Tx}}}{\sigma_v^2 N_{\mathrm{T}}}\mathbf{H}^{\mathrm{H}}\mathbf{H}\right)\right]\right\}, \tag{3.36}$$

where $\det(\cdot)$ denotes the determinant of a matrix.

Fig. 3.10 illustrates the impact of considering antenna mutual coupling, inductance and capacitance, and antenna element size on the radio channel capacity. Neglecting inductance and capacitance would mistakenly treat the antenna's equivalent circuit as a purely resistive circuit, thereby overestimating the radio channel capacity. This is because the impedance matrix element $\left(\mathbf{Z}_{qp}\right)_{2,1} = sa\sqrt{R_p R_q^3 \overline{H}_{qp}}/(saR_q + R_q c)$ of the two-port network considering inductance and capacitance in (3.12) is smaller than the impedance matrix element $\left(\tilde{\mathbf{z}}_{qp}\right)_{2,1} = \sqrt{R_p R_q H_{qp}}$ of the two-port network ignoring inductance and capacitance in (3.13), resulting in an increase in the values of the radio channel matrix when inductance and capacitance are ignored. Similarly, as the element size a increases, the elements $\left(\mathbf{Z}_{qp}\right)_{2,1}$ of the two-port circuit network impedance matrix also increase, resulting in a larger radio channel capacity. Moreover, neglecting antenna mutual coupling may overestimate the channel capacity to some extent. This is because the impedance matching at Rx is not considered in Fig. 3.3, and the presence of antenna mutual coupling at Rx leads to a loss in radio channel capacity. These phenomena highlight the importance of considering factors such as antenna size, equivalent circuit structure, and antenna mutual coupling when evaluating channel capacity.

3.5 General antenna design

Traditional antenna design only considers single-side antenna parameters, but it does not consider the characteristics of channels and communication systems. General antenna design jointly considers the double-side antenna parameters and the channel characteristics. In this section, a characteristic mode (CM) analysis based antenna pattern design method is proposed. The antenna parameters and channel statistical properties are analyzed.

3.5.1 CM analysis

CM analysis can provide more valuable information for feed structure design and is more suitable for small and medium sized electromagnetic structure [40]. In addition, the eigenmode theory can also be used in electrical multi-port antenna systems. Such a multiport antenna system can be regarded as an N-port network, and the corresponding eigenmode theory is called the eigenmode theory of N-port networks [41]. In this scenario, the CM is decided by the impedance matrix of the N-port network, and the quantity of CMs is identical to the number of ports. The characteristic radiation field is orthogonal on the radiation sphere. Thus for multiport antenna systems, antenna arrays, and multiport loaded scatterers, the far-field radiation performance and impedance properties are essentially determined by the radiation current on the antenna and the scatterer. Consequently, the port-based characteristic current basis is employed to form the current distribution on the antenna or scatterer. Characteristic theory can be used to realize the optimization objective of MIMO antenna ports to coupling, the use of Foster device loading port to achieve broadband impedance matching, realize all kinds of the far field radiation characteristics the channel capacity to optimize the main use of such methods, based on orthogonal properties far field control port current, achieve energy pattern synthesis. The calculation of CMs is similar to that of traditional CMs of ideally conducting objects. The conventional eigenmode is figured out by means of the impedance matrix of the moment method. However, the eigenmode of the N-port network is computed using its port impedance matrix, which can be acquired through a variety of different approaches, like full-wave electromagnetic simulation and vector network measurement as cited in [42]. A typical N-port network is illustrated in Fig. 3.11.

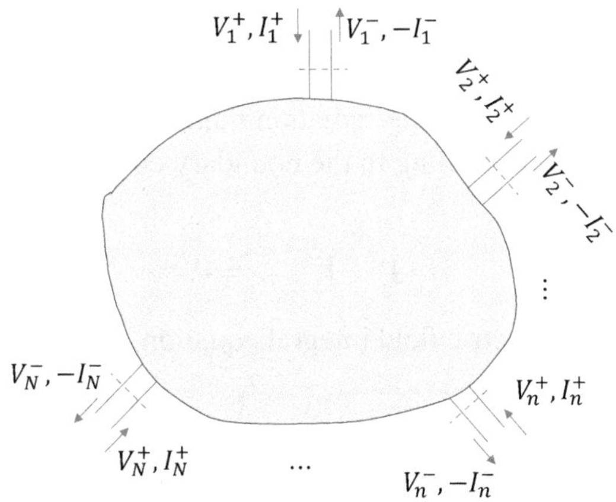

Figure 3.11:
N-port network.

The impedance matrix is determined by the current and voltage on a port. The port voltage and current comprise the incoming voltage and current V_{n+}, I_{n+}, as well as the reflection voltage and current V_{n-}, I_{n-}, and it can be represented as

$$V_n = V_n^+ + V_n^-, \tag{3.37}$$

$$I_n = I_n^+ - I_n^-. \tag{3.38}$$

Therefore, its impedance matrix is written as

$$\mathbf{V} = \mathbf{ZI}, \tag{3.39}$$

where $\mathbf{V} = [V_1, V_2, \ldots, V_N]^{\mathrm{T}}$, $\mathbf{I} = [I_1, I_2, \ldots, I_N]^{\mathrm{T}}$. Finally, the orthogonal decomposition of surface current is proved and the importance coefficient of parametric mode $\mathrm{MS} = 1/|1 + j\lambda_n|$ is introduced and this parameter maps the eigenvalues λ_n to the interval [0,1]. CM analysis is based on the MoM by solving the electrical field integration equation. It is proved that the characteristic magnetic fields of different modes are orthogonal to each other on the sphere at infinity, and the orthogonality of characteristic electric fields can be further proved. Thus, the total field can be written as a linear combination of pattern characteristic fields

$$\mathbf{E} = \sum_n \frac{V_n^i \mathbf{E}_n}{1 + j\lambda_n} = \sum_n \alpha_n \mathbf{E}_n, \tag{3.40}$$

$$\mathbf{H} = \sum_n \frac{V_n^i \mathbf{H}_n}{1 + j\lambda_n} = \sum_n \alpha_n \mathbf{H}_n, \tag{3.41}$$

where λ_n is the eigenvalue and V_n^i is the incoming voltage on the nth mode. For a perfect electric conductor with arbitrary shape, its surface is represented as S and a linear operator $L(\cdot)$ can be used to represent its scattering/radiation field, which means $L(\mathbf{J}) = \mathbf{E}^s$. If an incident electric field $\mathbf{E}^i$ exists, according to the boundary condition of PEC, it can be written as

$$[L(\mathbf{J}) + \mathbf{E}^i]_{tan} = 0. \tag{3.42}$$

Furthermore, according to the electric field integral equation, we can derive that

$$\mathbf{n} \times \mathbf{E}^i(\mathbf{r}) = -\mathbf{n} \times j\omega\mu \iint_{\mathbf{S}} \left[G(\mathbf{r}, \mathbf{r}')\mathbf{J}_s(\mathbf{r}') + \frac{\mathbf{1}}{k^2}(\nabla' \cdot \mathbf{J}_s(\mathbf{r}')) \cdot \nabla' G(\mathbf{r}, \mathbf{r}') \right] \mathrm{d}\mathbf{S}. \tag{3.43}$$

Select the appropriate basis functions to decompose the source current

$$\mathbf{J} = \sum_{i=1}^{M} \alpha_i \mathbf{J}_i = \begin{bmatrix} \mathbf{J}_1 & \mathbf{J}_2 & \dots & \mathbf{J}_M \end{bmatrix} \begin{bmatrix} \alpha_1 \\ \alpha_2 \\ \vdots \\ \alpha_M \end{bmatrix} = \widetilde{\mathbf{J}}\alpha. \tag{3.44}$$

Define a test function ω whose inner product with the basis functions can be expressed as

$$\left\langle \omega_j, \mathbf{J}_i \right\rangle = \iint_S \omega_j(\mathbf{r}) \cdot \mathbf{J}_i(\mathbf{r}') d\mathbf{S}. \tag{3.45}$$

Plug in the operator to get

$$\sum_{i=1}^{M} a_i \left\langle \omega_j, L\mathbf{J}_i \right\rangle = \left\langle \omega_j, \mathbf{E} \right\rangle. \tag{3.46}$$

Applying the electric field integral equation, the linear matrix equation is obtained as

$$[Z]\mathbf{J} = \mathbf{E}^{\mathrm{S}}, \tag{3.47}$$

$$[Z]_{i,j} = -j\omega\mu \iint_S (\mathbf{J}_i(\mathbf{r}) \cdot \mathbf{J}_j(\mathbf{r}') + \frac{1}{k^2}(\nabla \cdot \mathbf{J}_i(\mathbf{r}))(\nabla \cdot \mathbf{J}_j(\mathbf{r}')))G(\mathbf{r}, \mathbf{r}')\mathrm{d}\mathbf{S}\mathrm{d}\mathbf{S}'. \tag{3.48}$$

The voltage over each basis function is determined by the following formula:

$$[\mathbf{E}^{\mathbf{S}}]_j = \left\langle \mathbf{J}_j(\mathbf{r}), \mathbf{E}^{\mathbf{i}}(\mathbf{r}) \right\rangle = \iint_S \mathbf{J}_j(\mathbf{r}) \cdot \mathbf{E}^{\mathbf{i}}(\mathbf{r})\mathrm{d}\mathbf{S}. \tag{3.49}$$

The impedance matrix of any discretized PEC object can be found by using the MoM. By using different sets of integral equations, it is possible to find the impedance matrix of any object, not just the ideal conductor. In order to simplify the analysis, considering the characteristics of the ideal electric conductor PEC model defined by the previous derivation, the impedance matrix on the surface of the conductor is

$$Z(\mathbf{J}) = [L(\mathbf{J})]_{tan} = [\mathbf{E^i}]_{tan}. \tag{3.50}$$

Consider Rayleigh ratio

$$F(\mathbf{J}) = \frac{\langle \mathbf{J}, \mathbf{XJ} \rangle}{\langle \mathbf{J}, \mathbf{RJ} \rangle} = \frac{\text{Stored Energy} \times 2\omega}{\text{Radiated Power}}. \tag{3.51}$$

Solve the following generalized eigenvalue equation:

$$\mathbf{XJ}_i = \lambda_i \mathbf{RJ}_i. \tag{3.52}$$

Mathematically, it is not difficult to prove that the eigenvalue λ_i and characteristics of current J_i are real-valued. It is further found that the characteristic currents of different modes satisfy the orthogonal relationship with respect to the operators **R**, **X**, **Z**, which are expressed as

$$\left\langle \mathbf{J}_i, \mathbf{RJ}_j \right\rangle = \delta_{ij}, \tag{3.53}$$

$$\left\langle \mathbf{J}_i, \mathbf{XJ}_j \right\rangle = \lambda_j \delta_{ij}, \tag{3.54}$$

$$\left\langle \mathbf{J}_i, \mathbf{ZJ}_j \right\rangle = \left(1 + j\lambda_j\right) \delta_{ij}. \tag{3.55}$$

If two different modes have exactly the same eigenvalues, it is called a set of degenerate modes. In addition, it can be proved that different modes are orthogonal to each other. According to the power conservation, we have

$$\begin{aligned} P &= \left\langle \mathbf{J}^*, Z\mathbf{J} \right\rangle = \left\langle \mathbf{J}^*, R\mathbf{J} \right\rangle + j \left\langle \mathbf{J}*, X\mathbf{J} \right\rangle \\ &= \iint_{S'} (\mathbf{E} \times \mathbf{H}^*) \cdot \mathbf{n} ds + j\omega \iiint_{S'} (\mu \mathbf{H} \cdot \mathbf{H}^* - \mathbf{E} \cdot \mathbf{E}^*) d\tau'. \end{aligned} \tag{3.56}$$

Integration by substitution yields

$$\iint_{s'} (\mathbf{E} \times \mathbf{H}^*) \cdot \mathbf{n} ds + j\omega \iiint_{S'} (\mu \mathbf{H} \cdot \mathbf{H}^* - \mathbf{E} \cdot \mathbf{E}^*) d\tau' = \left(1 + j\lambda_j\right) \delta_{ij}. \tag{3.57}$$

Consider the sphere at infinity, S_∞, the far-field conditions, the characteristics of the electric field E_m, and characteristics of the magnetic field H_m satisfying the following relations:

$$\mathbf{E}_m = Z_0 \mathbf{H}_m \times \mathbf{n} = \frac{j\omega\mu}{4\pi r} e^{-j\mathbf{k}\cdot\mathbf{r}} F_m(\theta, \phi). \tag{3.58}$$

Conjugating the expressions and swapping m, n, we have

$$\iint_{S'} \left(\mathbf{E} \times \mathbf{H}^*\right) \cdot \mathbf{n} ds - j\omega \iiint_{S'} \left(\mu \mathbf{H} \cdot \mathbf{H}^* - \mathbf{E} \cdot \mathbf{E}^*\right) d\tau = (1 - j\lambda)\, \delta_{mn}. \tag{3.59}$$

Adding left and right terms, leads to

$$Z_0 \iint_{S'} \left[\mathbf{H}_m \cdot \mathbf{H}_n^* - (\mathbf{H}_m \cdot \mathbf{n}) \cdot \left(\mathbf{H}_n^* \cdot \mathbf{n}\right)\right] \mathrm{d}s = \delta_{mn}. \tag{3.60}$$

Considering that the radial component of the field at infinity is 0, it is not difficult to find that

$$Z_0 \iint_{S'} \left[(\mathbf{H}_m \cdot \mathbf{n}) \cdot \left(\mathbf{H}_n^* \cdot \mathbf{n}\right)\right] \mathrm{d}s = 0. \tag{3.61}$$

Thus we obtain

$$Z_0 \iint_{S'} \mathbf{H}_m \cdot \mathbf{H}_n^* ds = \delta_{mn}. \tag{3.62}$$

3.5.2 CM-based antenna response

Simplified from 6G pervasive channel model (6GPCM) [33], we define $A_{qp}^{\mathrm{LoS}}(t)$ and $A_{qp,m_n}^{\mathrm{NLoS}}(t)$ to characterize antenna and polarization response in LoS and NLoS conditions, respectively, as

$$\begin{aligned} A_{qp}^{\mathrm{LoS}}(t) = & \begin{bmatrix} F_{q,f_c,V}\left(\phi_{E,\mathrm{L}}^{R}(t), \phi_{A,\mathrm{L}}^{R}(t)\right) \\ F_{q,f_c,H}\left(\phi_{E,\mathrm{L}}^{R}(t), \phi_{A,\mathrm{L}}^{R}(t)\right) \end{bmatrix}^{\mathrm{T}} \begin{bmatrix} e^{j\theta_{\mathrm{L}}^{VV}} & 0 \\ 0 & e^{j\theta_{\mathrm{L}}^{HH}} \end{bmatrix} \\ & \times \begin{bmatrix} F_{p,f_c,V}\left(\phi_{E,L}^{T}(t), \phi_{A,L}^{T}(t)\right) \\ F_{p,f_c,H}\left(\phi_{E,L}^{T}(t), \phi_{A,L}^{T}(t)\right) \end{bmatrix}, \end{aligned} \tag{3.63}$$

$$\begin{aligned} A_{qp,m_n}^{\mathrm{NLoS}}(t) = & \begin{bmatrix} F_{q,f_c,V}\left(\phi_{E,m_n}^{R}(t), \phi_{A,m_n}^{R}(t)\right) \\ F_{q,f_c,H}\left(\phi_{E,m_n}^{R}(t), \phi_{A,m_n}^{R}(t)\right) \end{bmatrix}^{\mathrm{T}} \begin{bmatrix} e^{j\theta_{m_n}^{VV}} & \sqrt{\mu\kappa_{m_n}^{-1}(t)}e^{j\theta_{m_n}^{VH}} \\ \sqrt{\kappa_{m_n}^{-1}(t)}e^{j\theta_{m_n}^{HV}} & \sqrt{\mu}e^{j\theta_{m_n}^{HH}} \end{bmatrix} \\ & \times \begin{bmatrix} F_{p,f_c,V}\left(\phi_{E,m_n}^{T}(t), \phi_{A,m_n}^{T}(t)\right) \\ F_{p,f_c,H}\left(\phi_{E,m_n}^{T}(t), \phi_{A,m_n}^{T}(t)\right) \end{bmatrix}, \end{aligned} \tag{3.64}$$

where $F_{q(p),f_c,V}$ and $F_{q(p),f_c,H}$ are the antenna radiation patterns of the $q(p)$th Rx(Tx) antenna for vertical and horizontal polarizations at corresponding carrier frequency f_c, respectively. Also, κ_{m_n} is the cross-polarization power ratio and μ is copolar imbalance, while $\theta_{m_n}^{VV}, \theta_{m_n}^{VH}, \theta_{m_n}^{HV}, \theta_{m_n}^{HH}$ are initial phases modeled as random variables uniformly distributed over $(0, 2\pi]$. Assuming that the antenna type at the transmitting end is the same, and its current basis function is also the same, the difference is that each antenna has a different set of mode coefficients, and the mode coefficient $\alpha_{\mu} = [\alpha_{\mu,1}, \alpha_{\mu,2}, \ldots, \alpha_{\mu,M}]^{\mathrm{T}} \in C^{M\times 1}$. Tx surface current can be decomposed as $\mathbf{J}_{p,\mu}^{Tx} = \sum_{i=1}^{J_T} \alpha_{p,\mu,i}\mathbf{J}_i^{Tx} = \widetilde{\mathbf{J}}_{p,\mu}^{Tx}\cdot\alpha_{p,\mu}$, which represents the radiation mode of the pth Tx antenna is mode μ. Rx surface current can be decomposed as $\mathbf{J}_{q,\nu}^{Rx} = \sum_{i=1}^{J_R} \alpha_{q,\nu,i}\mathbf{J}_i^{Rx} = \widetilde{\mathbf{J}}_{q,\nu}^{Rx}\cdot\alpha_{q,\nu}$, which represents the radiation mode of the qth Rx antenna is mode ν. The radio channel model based on CMA can be obtained, and the response of the antenna part is

$$A_{qp}^{\mathrm{LoS}}(t,\nu,\mu) = \frac{(4\pi)\sqrt{\eta_p\eta_q}}{\sqrt{(\alpha_{p,\mu}^{\mathbf{H}}\cdot\alpha_{p,\mu})(\alpha_{q,\nu}^{\mathbf{H}}\cdot\alpha_{q,\nu})\iint_{4\pi}\sum_{i=1}^{J_R}|L(\mathbf{J}_i^{Rx})|^2\mathrm{d}\Omega\iint_{4\pi}\sum_{i=1}^{J_T}|L(\mathbf{J}_i^{Tx})|^2\mathrm{d}\Omega}}$$
$$\times\left(\begin{bmatrix} L(\hat{z}\widetilde{\mathbf{J}}_z^{Rx}) \\ L(\hat{x}\widetilde{\mathbf{J}}_x^{Rx}) + L(\hat{y}\widetilde{\mathbf{J}}_y^{Rx})\end{bmatrix}\alpha_{q,\nu}\right)^{\mathrm{T}}\begin{bmatrix} e^{j\theta_{\mathrm{L}}^{VV}} & 0 \\ 0 & e^{j\theta_{\mathrm{L}}^{HH}}\end{bmatrix}$$
$$\times\left(\begin{bmatrix} L(\hat{z}\widetilde{\mathbf{J}}_z^{Tx}) \\ L(\hat{x}\widetilde{\mathbf{J}}_x^{Tx}) + L(\hat{y}\widetilde{\mathbf{J}}_y^{Tx})\end{bmatrix}\alpha_{p,\mu}\right), \tag{3.65}$$

$$A_{qp,m_n}^{\mathrm{NLoS}}(t,\nu,\mu) = \frac{(4\pi)\sqrt{\eta_p\eta_q}}{\sqrt{(\alpha_{p,\mu}^{\mathbf{H}}\cdot\alpha_{p,\mu})(\alpha_{q,\nu}^{\mathbf{H}}\cdot\alpha_{q,\nu})\iint_{4\pi}\sum_{i=1}^{J_R}|L(\mathbf{J}_i^{Rx})|^2\mathrm{d}\Omega\iint_{4\pi}\sum_{i=1}^{J_T}|L(\mathbf{J}_i^{Tx})|^2\mathrm{d}\Omega}}$$
$$\times\left(\begin{bmatrix} L(\hat{z}\widetilde{\mathbf{J}}_z^{Rx}) \\ L(\hat{x}\widetilde{\mathbf{J}}_x^{Rx} + \hat{y}\widetilde{\mathbf{J}}_y^{Rx})\end{bmatrix}\alpha_{q,\nu}\right)^{\mathrm{T}}\begin{bmatrix} e^{j\theta_{m_n}^{VV}} & \sqrt{\mu\kappa_{m_n}^{-1}(t)}e^{j\theta_{m_n}^{VH}} \\ \sqrt{\kappa_{m_n}^{-1}(t)}e^{j\theta_{m_n}^{HV}} & \sqrt{\mu}e^{j\theta_{m_n}^{HH}}\end{bmatrix}$$
$$\times\left(\begin{bmatrix} L(\hat{z}\widetilde{\mathbf{J}}_z^{Tx}) \\ L(\hat{x}\widetilde{\mathbf{J}}_x^{Tx} + \hat{y}\widetilde{\mathbf{J}}_y^{Tx})\end{bmatrix}\alpha_{p,\mu}\right). \tag{3.66}$$

3.5.3 Correlation functions

Symbols $R_{A,qp,\tilde{q}\tilde{p}}^{\mathrm{LoS}}$ and $R_{A,m_n,qp,\tilde{q}\tilde{p}}^{\mathrm{NLoS}}$ are the correlation functions between antenna and polarization response under LoS and NLoS conditions, respectively, which can be written as

$$R_{A,qp,\tilde{q}\tilde{p}}^{\mathrm{LoS}}(t,s;\nu_k,\nu_l;\mu_i,\mu_j) = \frac{\mathbb{E}[A_{qp}^{\mathrm{LoS}}(t,\nu_k,\mu_i)A_{\tilde{q}\tilde{p}}^{\mathrm{LoS},*}(s,\nu_l,\mu_j)]}{\sqrt{\mathbb{E}[|A_{qp}^{\mathrm{LoS}}(t,\nu_k,\mu_i)|^2]\mathbb{E}[|A_{\tilde{q}\tilde{p}}^{\mathrm{LoS}}(s,\nu_l,\mu_j)|^2]}}, \tag{3.67}$$

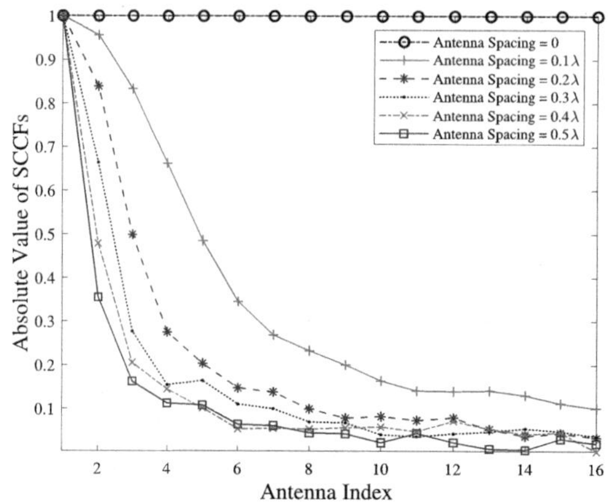

Figure 3.12:
SCCFs of fixed pattern antennas under different antenna spacing ($p = 1, q = \tilde{q} = 1$).

$$R^{\mathrm{NLoS}}_{A,m_n,qp,\tilde{q}\tilde{p}}(t,s;\nu_k,\nu_l;\mu_i,\mu_j) = \frac{\mathbb{E}[A^{\mathrm{NLoS}}_{m_n,qp}(t,\nu_k,\mu_i)A^{\mathrm{NLoS},*}_{m_n,\tilde{q}\tilde{p}}(s,\nu_l,\mu_j)]}{\sqrt{\mathbb{E}[|A^{\mathrm{NLoS}}_{m_n,qp}(t,\nu_k,\mu_i)|^2]\mathbb{E}[|A^{\mathrm{NLoS}}_{m_n,\tilde{q}\tilde{p}}(s,\nu_l,\mu_j)|^2]}}. \quad (3.68)$$

The impacts of antenna pattern are investigated separately, so that other antenna parameters are set to be consistent. The antenna total efficiency η is 100%. The antenna radiation patterns of different loaded impedance are obtained from the CST STUDIO SUITE. Unless otherwise specified, the following settings and parameters are used. Tx and Rx adopt reactively controlled antenna array (RCAA) and dipole uniform linear arrays, respectively. The basic parameters are given as $f_c = 5.6$ GHz, $N_T = 128$. Tx and Rx are both fixed.

As a comparison, spatial cross-correlation functions (SCCFs) of fixed pattern antenna are shown in Fig. 3.12. It can be seen that as the antenna index increases, SCCFs decrease sharply when the antenna spacing is larger than 0. When the antenna spacing increases, SCCFs drop more rapidly. The SCCFs stay to be 1 when the antenna spacing is 0. It is reasonable because SCCFs are mainly decided by the position difference between antennas. SCCFs of RCAA are shown in Fig. 3.13. Similar to those of fixed pattern antennas, SCCFs decrease sharply with the antenna index when the antenna spacing is larger than 0. But when the antenna spacing is 0, the SCCF of RCAA does not keep the same value. For further comparison, we put curves of fixed pattern antenna and RCAA together in Fig. 3.14. We can see that SCCFs of RCAA are always lower than those of fixed pattern antenna, because each RCAA is equipped with different radiation patterns, which brings about lower correlation between antennas. Although the performance of envelope correlation coefficient is not so good, it still brings lower SCCFs. In Fig. 3.14, when the antenna spacing increases, the differences between SCCFs of fixed pattern antenna and RCAA become smaller, which indicates that effects of deploying different

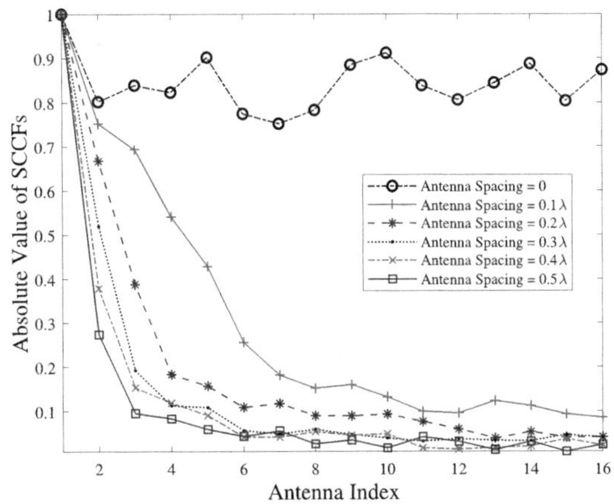

Figure 3.13:
SCCFs of RCAA with mode 1 under different antenna spacing ($p = 1$, $q = \tilde{q} = 1$).

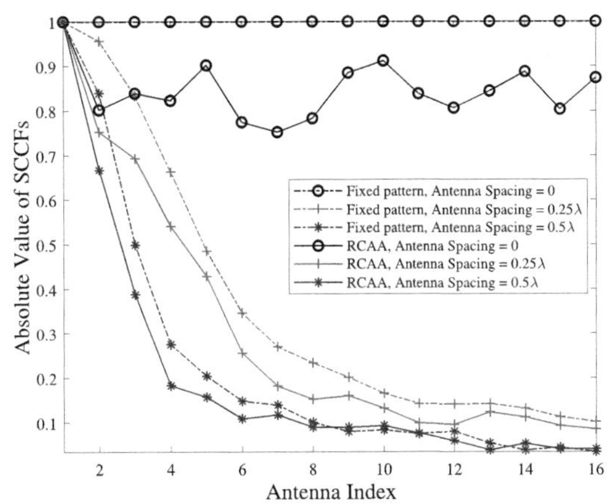

Figure 3.14:
SCCFs of fixed pattern antenna versus RCAA ($p = 1$, $q = \tilde{q} = 1$).

radiation patterns on different antennas may only work when the array is compact. For a conventional array whose antenna spacing is 0.5λ or larger, this method maybe unnecessary.

3.6 Channel map-based system design

With the emergence of 6G, relying solely on instantaneous channel measurements will swiftly deplete available spectrum resources, hindering the sustainable development of 6G MIMO wireless communications. An accurate and time-efficient channel model can provide real-time channel state information for pilot overhead reduction and transmission optimization, further enhancing system performance and ensuring effective spectrum management. As an important concept and tool, channel map has attracted extensive academic attention and research. Channel map offers a precise account of the wireless communication setting and channel characteristics in a particular geographical zone. It includes critical data such as received signal strength, parameters of multipath component (MPC), and interference distribution. Channel map can be constructed based on channel model, storing the channel information and channel characteristics. In this section, a 3D continuous-space dynamic channel map providing accurate and time-efficient channel information is proposed.

3.6.1 Channel map construction and challenges

Methods for constructing channel map can be categorized into two types: machine learning based on training data and channel modeling based on physical environments.

Among these, machine learning based on training data has seen numerous studies, which utilize machine learning algorithms and existing training data to learn predictive models for certain channel information regarding the positions of the Tx and Rx, such as path loss distribution maps and departure angle distribution maps.

Considering the 3D continuous-space, to obtain a predictive model that can predict the channel information of any Tx and Rx position within the space, the required amount of training data will be very large, making methods relying on machine learning unsuitable for scenarios where both the Tx and Rx move continuously.

Channel modeling according to physical surroundings can be further categorized into deterministic channel modeling and stochastic channel modeling. The approach of deterministic channel modeling is to reconstruct the surrounding environment of electromagnetic wave propagation and then physically calculate the attenuation of propagation. A widely applied deterministic channel model for channel-map construction is ray tracing (RT), which is an electromagnetic field strength prediction algorithm based on geometrical optics and unified diffraction theory. Based on geometrical optics, it simulates the geometric propagation paths of electromagnetic waves and then calculates the path loss or field strength attenuation of electromagnetic waves from the Tx to a certain receiving position along different propagation paths.

The modeling of material parameters is too uniform to reflect the differences of the same type materials used in different regions, different buildings, and different weather conditions in physical environment. Therefore, the RT method for channel map construction cannot support real-time and localized calibration of channel maps.

In stochastic channel modeling, the GBSM method has recently received significant attention. It has certain geometrical characteristics of multipath, which can describe the impact of environmental scatterers and reflectors on the link channel. GBSM performs certain modeling of the statistical distribution of obstacles, which, compared to models based on RT, can significantly reduce computational complexity at the cost of some accuracy.

GBSM only targets specific scenarios or limited link channel modeling. Although it considers the impact of scatterers in physical environment, the model's applicability and generalization in 3D continuous space for channel map are still not met. Meanwhile, some studies heavily rely on accurate channel measurement data, making it difficult to effectively predict without customized requirement-related measurement data.

3.6.2 A 3D continuous-space dynamic channel map construction method

We propose a novel hybrid channel modeling structure for the channel map construction. The static interaction objects (IOs) are modeled offline by RT. Meanwhile, MPCs of dynamic IOs are modeled and online updated by GBSM. We can obtain the channel transfer function (CTF) of the channel as

$$\mathbf{H}_{vu}(t, f, \ell) = \sqrt{\frac{K_D^{-1}}{K_S^{-1} + K_D^{-1} + 1}}\mathbf{H}_{vu}^D(t, f, \ell) + \sqrt{\frac{1}{K_S^{-1} + K_D^{-1} + 1}}\mathbf{H}_{vu}^L(f, \ell) + \sqrt{\frac{K_S^{-1}}{K_S^{-1} + K_D^{-1} + 1}}\mathbf{H}_{vu}^S(f, \ell), \tag{3.69}$$

where $\mathbf{H}_{vu}^D(t, f, \ell)$, $\mathbf{H}_{vu}^L(f, \ell)$, and $\mathbf{H}_{vu}^S(f, \ell)$ are the CTFs of dynamic scattering channel, LoS component, and NLoS component in static scattering channel, respectively; K_S and K_D denote the ratios of the power of LoS components to the power of static/dynamic scattering components. It is well known that the Rician K-factor K is the ratio of the power of LoS components to the power of non-line-of-sight (NLoS) components. Therefore, K can be calculated as

$$\frac{1}{K} = \frac{1}{K_S} + \frac{1}{K_D}. \tag{3.70}$$

Based on the parameters estimated from channel measurement and the parameters of rays in RT simulation, the MPCs in RT channel (RT-MPCs) and MPCs in measured channel (M-MPCs) can be illustrated in the joint AAoA and delay domain as shown in Fig. 3.15, where each marker denotes one MPC. Symbol colors indicate the power of the corresponding MPCs. RT-MPCs are considered to be matched with an M-MPC if the Euclidean distance

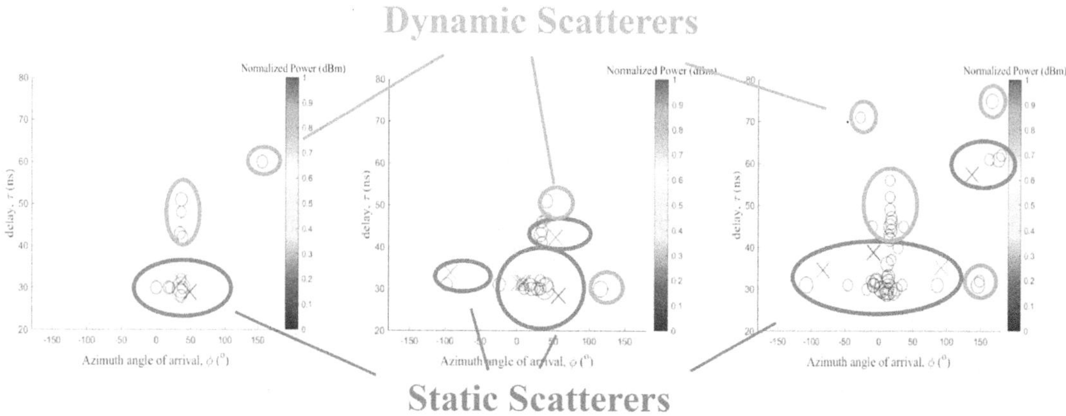

Figure 3.15:
Normalized AAoA-delay PSD of RT and measurement along the route: (a) location 3, (b) location 7, and (c) location 9.

between them is smaller than a defined threshold. The matching MPCs are labeled with red dotted circles and other MPCs are labeled with pink solid circles, standing for the dynamic clusters. It can be found from Fig. 3.15 that each RT-MPC can be matched with at least one M-MPC and the power of matching MPCs is almost the same. However, there are still some observable M-MPCs, which cannot be matched with RT-MPCs. Because the dynamic IOs in physical environment are not considered in RT simulation, these M-MPCs may be caused by diffraction and diffuse scattering from dynamic IOs, corresponding to the dynamic clusters in the proposed stochastic–deterministic hybrid channel model.

By matching the M-MPCs with different IOs in physical environment, we can obtain the values of K_S and K_D in simulation setting of the proposed hybrid model, which can be further used in the accuracy evaluation to ensure a fair comparison.

Because of the randomness of wireless channels, it is not suitable to assess the channel model by directly comparing the generated channel matrix with the measured data. As a feasible method, the statistical properties of wireless channels can be used instead to validate the effectiveness in a stochastic way. In this work, the accuracy of channel map is verified by validating the effectiveness of the proposed hybrid channel model. Therefore, Fig. 3.16 similarly compares the delay PSD across all locations. We can see that both the proposed hybrid model and GBSM align well with the measured results.

The cumulative distribution functions (CDFs) of root mean square (RMS) delay spread of the proposed hybrid channel model, channel measurement, RT, and 6GPCM are compared in Fig. 3.17. We can see from Fig. 3.17 that the RMS delay spread of the proposed hybrid

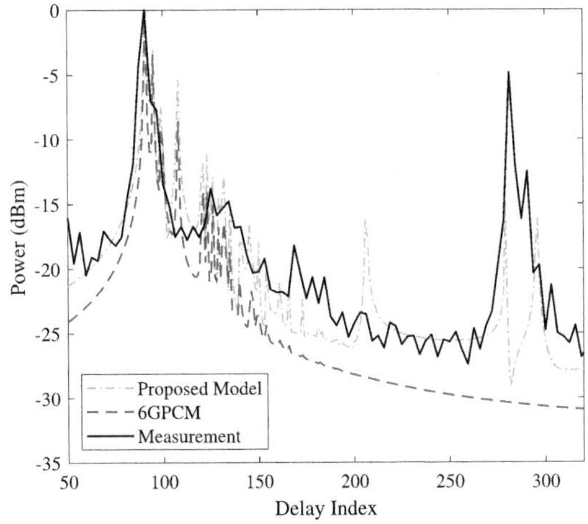

Figure 3.16:
Delay PSDs of the data generated by hybrid channel model, measurement results, and 6GPCM in all locations of the route.

channel model, RT method, and 6GPCM all accord well with the measurement results. This indicates that the proposed model achieves good model accuracy.

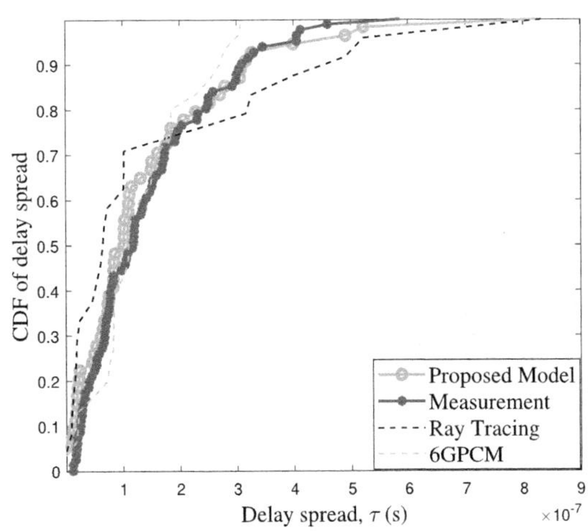

Figure 3.17:
CDFs of RMS delay spread of the proposed hybrid channel model, channel measurement, RT, and 6GPCM.

3.7 Future research directions and applications

In this section, we explore the future research directions and potential applications of EIT are envisioned. To address the challenges inherent in 3D continuous-space radio channels for 6G/B6G communication systems, it is essential to delve into various key issues. These challenges include accurately capturing the characteristics of the channel through precise measurement techniques, creating detailed and reliable channel models, increasing the speed and efficiency of data transmission, and enhancing the system's sensing capabilities to push beyond current performance limits. By conducting a comprehensive analysis, we can identify important future research directions, such as leveraging AI for predictive channel modeling. AI-based approaches can help predict channel changes and optimize communication performance, while other key technologies such as RIS, advanced antenna design, and innovative system architectures will play crucial roles. Finally, we introduce how channel maps, which provide a detailed understanding of channel characteristics, will facilitate these advancements. Combined with progress in antenna and system design, these technologies will offer strong support for the development of future communication systems.

3.7.1 Channel measurements and modeling for 3D continuous-space radio channel

As new technologies like terahertz communications, ultramassive MIMO, RIS, and AI continue to advance, the design of communication systems is facing a set of new challenges. One major challenge is that 3D continuous-space radio channels have a high density of multipath signals, making it difficult to accurately extract multipath information. Additionally, the complexity of modeling both antennas and wireless propagation channels together adds another layer of difficulty, as it complicates the process of isolating and measuring the response of each component. To address these issues, it is essential to develop a 3D continuous-space radio channel measurement platform. This platform would help researchers investigate high-resolution parameter extraction algorithms and adjust antenna parameters for thorough measurements and characterization of the radio channel. Early research has already explored measurement experiments for ISAC [43,44], RIS [45,46], and ultramassive MIMO [47,48]. In these experiments, massive MIMO and ultrawideband antenna arrays were used to conduct 3D continuous-space radio channel measurements. Virtual arrays were also employed to achieve continuous aperture measurements, which improves spatial resolution. Configuring different antenna element spacings and using adaptive antenna arrays are effective strategies to overcome the challenges in 3D continuous-space radio channel measurement and characterization.

As the number of antennas increases and the spacing between array elements decreases, the relationship between antennas and their propagation environments becomes increasingly coupled. Additionally, the electromagnetic characteristics of these environments, especially when they are cross-layer, multi-scale, and complexly nonuniform, remain poorly understood. This lack of clarity presents a significant challenge in developing a comprehensive 3D continuous-space radio channel model. To address this challenge, it is crucial to study both electromagnetic and radio channel modeling. We need to derive a unified expression for the channel impulse response in 3D continuous space and establish a unified framework for electromagnetic channel modeling. By integrating various conditions, the radio channel's behavior across different scenarios can be analyzed. For electromagnetic channel modeling, the MoM can be used to simulate the response of any antenna element and describe wireless propagation channels using GBSMs. This approach helps in calculating the electromagnetic responses for arbitrary antennas at both the Tx and Rx ends in 3D space. For radio channel modeling, we need to consider factors like source impedance, matching networks, antenna impedance matrices, and radiation patterns. By analyzing circuits and antennas, we can develop a comprehensive model for 3D continuous-space radio channels. Other challenges include ensuring that the channel model accounts for new antenna characteristics and determining whether traditional models like the GBSM remain suitable for these advanced systems. Addressing these challenges will create more accurate and reliable models for future communication systems.

3.7.2 Capacity theory for 3D continuous-space radio channel

For ISAC systems, it is crucial to thoroughly consider the collaboration between communication and sensing. This integration helps achieve the possible transmission capabilities and perception performance in 3D continuous-space channels. These channels combine antennas and propagation environments, expanding communication capabilities. However, the impact of antennas on capacity has often been overlooked. It is crucial to consider the electromagnetic characteristics of both antennas and propagation environments to analyze and calculate the theoretical limits of 3D continuous-space channel capacity. Analyzing channel capacity in complex, multiscale correlated environments is challenging. Using 3D continuous-space electromagnetic modeling, we need to study the DoF in these channels. This research will reveal the benefits of 6G/B6G technologies like aperture-limited MIMO, RIS, and ISAC. By leveraging electromagnetic models and measurement data, the capacity boundaries of multiscale channels can be estimated. For various application scenarios, including different environments and mobile states, accurate channel modeling requires a physical model that closely characterizes real-world conditions. Traditional theories often ignore the physical transmission process, leading to inaccuracies in complex electromagnetic environments. For instance, factors like interference and diffraction can either strengthen or weaken signals at different spatial locations, affecting channel capacity differently than free-space models predict. Therefore, a more precise characterization of the electromagnetic environment is essential for accurate 3D continuous-space radio channel capacity analysis.

The 6G/B6G network will introduce new perception services based on RF signals, such as imaging from ground base stations, reconnaissance by drones, and remote sensing by satellites. Unlike traditional communication, which focuses on data transmission rates, these perception services prioritize accuracy, leading to different performance limits. In wireless perception systems, performance is evaluated using concepts from traditional estimation theory and radar sensing. Key indicators include positioning accuracy, imaging accuracy, mutual information of perception, and the Cramér–Rao bound for imaging. Recent research has introduced metrics based on perception rate, similar to Shannon's capacity in communication, allowing both perception and communication to be optimized together using comparable units. However, the balance between these two functions in 3D continuous space remains unclear. To address this, practical wireless perception systems need to develop models inspired by electromagnetic effects to predict communication and sensing performance and establish the intrinsic relationship between sensing performance and channel state information, thereby effectively resolving the trade-offs between communication and perception.

3.7.3 General antenna design

With the development of continuous-space radio channel models, it has become crucial to explore antenna theories that can approach channel capacity limits by aligning antenna parameters with channel characteristics. The research in generalized antenna theory focuses on three main areas: understanding the interaction between antennas and propagation channels to optimize the matching between antenna parameters and channel features, mapping Tx and Rx antenna parameters to channel capacity to enhance overall performance, and building a multiscenario, multilink matching model for antenna parameters. This approach aims to advance a general antenna theory that integrates antenna parameters, channel characteristics, and channel capacity.

First, a model that aligns antenna parameters with channel characteristics needs to be created. This includes analyzing channel properties and identifying key antenna parameters that significantly influence performance. Once these critical parameters are identified, the next phase is to investigate methods for enhancing channel capacity. By using the set of key antenna parameters, researchers can derive channel capacity and evaluate it across multiple links, locations, and scenarios. This generalized antenna theory addresses the challenges of increasing channel capacity and provides a framework for optimizing communication systems in diverse environments.

Once the digital twin-channel is established, it becomes feasible to optimize multidimensional Tx and Rx antenna parameters comprehensively to enhance channel capacity. Environmental factors are considered in this holistic optimization process to design antennas that maximize channel capacity. In scenarios with limited aperture sizes, the optimization of multidimensional Tx–Rx antenna parameters is tailored to improve channel capacity within these constraints. Ultimately, by integrating both approaches and considering various deployment scenarios and mobility, the goal is to optimize multidimensional Tx–Rx antenna parameters to design antennas that boost channel capacity in 3D continuous space. Additionally, a generalized method for optimizing antenna array designs is proposed.

To achieve these objectives, it is essential to first develop a methodology for designing and optimizing ultramassive and aperture-limited generalized MIMO antenna arrays. This involves using techniques such as flexible electromagnetic control, advanced array optimization algorithms, and innovative array configurations to address challenges like external environmental coupling and mismatched RF DoF. Next, the design and optimization of aperture-limited generalized MIMO antenna arrays need to be addressed. This includes employing high-density spatial sampling, coupling suppression and exploitation, and extending into spatial dimensions to mitigate the significant challenge of extreme electromagnetic coupling within antennas under constrained apertures. The aim of this generalized antenna array design and optimization is to effectively match continuous spatial antennas with the channel characteristics.

3.7.4 System design enabled by 3D continuous-space channel map

Due to the high complexity of large-dimensional matrix computations in traditional transmission methods, there is a significant challenge in maintaining efficient communication, especially in 6G/B6G scenarios where high mobility leads to a considerable drop in transmission performance. To address these issues, there is an urgent need for a large-dimensional transmission theory enabled by digital twin channel maps. This approach first requires integrating geometric theory to explore low-complexity channel estimation and signal detection methods, which can simplify the computational load. Secondly, it involves leveraging customized posterior statistical models to accurately describe the non-ideal characteristics of channel information caused by mobility, ensuring more precise modeling of dynamic channel conditions. Ultimately, the goal is to develop theoretical methods for large-dimensional robust precoding, which will enhance the reliability and efficiency of transmissions in highly mobile and complex environments.

Currently, communication systems face significant challenges in areas such as beam selection, user scheduling, interference management, and coordination among multiple cells. Traditional resource management decision-making methods often struggle to meet the demands of real-time processing, leading to inefficiencies and suboptimal performance. The introduction of emerging technologies like RIS further complicates these issues by introducing stronger nonlinearity into problem structures, making solutions more complex and computationally intensive. To overcome these challenges, the establishment of digital twin channel models offers a promising approach. By leveraging preobtained channel characteristics, these models can significantly reduce the solution space for real-time decision-making problems. Digital twins provide a virtual replica of the physical communication environment, allowing for more accurate predictions and optimizations. This not only simplifies the computational burden but also enhances the precision of resource allocation decisions. Moreover, integrating intelligent processing techniques based on channel knowledge can further improve resource management efficiency. These techniques involve extracting key features from channel maps, which can then be used to formulate and solve optimization problems. For instance, machine learning algorithms can analyze historical data to identify patterns and trends in channel behavior, enabling more informed and dynamic decision-making. Additionally, deep learning-inspired heuristic algorithms can be employed to evaluate system performance and optimize multiagent information sharing. This ensures that each node in the network can efficiently coordinate with others, reducing interference and improving overall network throughput.

3.8 Conclusions

Compared with 5G, the emergence of crucial 6G technologies gives rise to novel applications and technical demands that go beyond the application range of each of the four individual theories. In 6G space–air–ground–sea networks, the underlying wireless channels have a continuous characteristic in 3D space, and the wireless propagation channels are closely intertwined with the antennas. Thus, the integration of electromagnetic theory, information theory, wireless propagation channel modeling theory, and antenna theory becomes an inescapable trend. As a result, we have described the associations and discrepancies among the four fundamental theories. Subsequently, channel measurements and modeling, channel capacity theory, general antenna design, and system design have been deliberated. Moreover, the latest advancements, potential solutions for theoretical unification, and four major research directions have been recapitulated. To sum up, EIT research will lay a firm theoretical foundation for 6G systems and promote remarkable breakthroughs.

References

[1] L.-W. Li, P.-S. Kooi, M.-S. Leong, T.-S. Yee, Electromagnetic dyadic Green's function in spherically multilayered media, IEEE Transactions on Microwave Theory and Techniques 42 (1994) 2302–2310.
[2] A. Pizzo, L. Sanguinetti, T.L. Marzetta, Spatial characterization of electromagnetic random channels, IEEE Open Journal of the Communications Society 3 (2022) 847–866.
[3] C. Huang, S. Hu, G.C. Alexandropoulos, A. Zappone, C. Yuen, R. Zhang, M. Di Renzo, M. Debbah, Holographic MIMO surfaces for 6G wireless networks: opportunities, challenges, and trends, IEEE Wireless Communications 27 (2020) 118–125.

[4] F. Guidi, D. Dardari, Radio positioning with EM processing of the spherical wavefront, IEEE Transactions on Wireless Communications 20 (2021) 3571–3586.
[5] J. Huang, C.-X. Wang, Y. Sun, R. Feng, J. Huang, B. Guo, Z. Zhong, T.J. Cui, Reconfigurable intelligent surfaces: channel characterization and modeling, Proceedings of the IEEE 110 (2022) 1290–1311.
[6] C.E. Shannon, Communication in the presence of noise, Proceedings of the IRE 37 (1949) 10–21.
[7] W. Jeon, S.-Y. Chung, Capacity of continuous-space electromagnetic channels with lossy transceivers, IEEE Transactions on Information Theory 64 (2017) 1977–1991.
[8] A. Pizzo, T.L. Marzetta, L. Sanguinetti, Degrees of freedom of holographic MIMO channels, in: 2020 IEEE 21st International Workshop on Signal Processing Advances in Wireless Communications (SPAWC), IEEE, 2020, pp. 1–5.
[9] Z. Wan, J. Zhu, Z. Zhang, L. Dai, C.-B. Chae, Mutual information for electromagnetic information theory based on random fields, IEEE Transactions on Communications 71 (2023) 1982–1996.
[10] M. Ptzold, Mobile Radio Channels, Wiley Publishing, 2012.
[11] C.-X. Wang, J. Huang, H. Wang, X. Gao, X. You, Y. Hao, 6G wireless channel measurements and models: trends and challenges, IEEE Vehicular Technology Magazine 15 (2020) 22–32.
[12] Z. Yun, M.F. Iskander, Ray tracing for radio propagation modeling: principles and applications, IEEE Access 3 (2015) 1089–1100.
[13] C.-X. Wang, M. Di Renzo, S. Stanczak, S. Wang, E.G. Larsson, Artificial intelligence enabled wireless networking for 5G and beyond: recent advances and future challenges, IEEE Wireless Communications 27 (2020) 16–23.
[14] C.A. Balanis, Antenna Theory: Analysis and Design, John Wiley & Sons, 2015.
[15] D.R. Smith, O. Yurduseven, L.P. Mancera, P. Bowen, N.B. Kundtz, Analysis of a waveguide-fed metasurface antenna, Physical Review Applied 8 (2017) 054048.
[16] P. Hannan, The element-gain paradox for a phased-array antenna, IEEE Transactions on Antennas and Propagation 12 (1964) 423–433.
[17] A.S. Poon, N. David, Degree-of-freedom gain from using polarimetric antenna elements, IEEE Transactions on Information Theory 57 (2011) 5695–5709.
[18] G. Wen, Multi-antenna information theory, Progress in Electromagnetics Research 75 (2007) 11–50.
[19] C.-X. Wang, Y. Yang, J. Huang, X. Gao, T.J. Cui, L. Hanzo, Electromagnetic information theory: fundamentals and applications for 6G wireless communication systems, IEEE Wireless Communications 31 (2024) 279–286.
[20] Y. Yang, C.-X. Wang, J. Huang, J. Thompson, H.V. Poor, A 3D continuous-space electromagnetic channel model for 6G tri-polarized multi-user communications, IEEE Transactions on Wireless Communications 23 (2024) 17354–17367.
[21] L.-W. Li, P.-S. Kooi, M.-S. Leong, T.-S. Yee, Electromagnetic dyadic Green's function in spherically multilayered media, IEEE Transactions on Microwave Theory and Techniques 42 (1994) 2302–2310.
[22] H.F. Arnoldus, Representation of the near-field, middle-field, and far-field electromagnetic Green's functions in reciprocal space, Journal of the Optical Society of America. B, Optical Physics 18 (2001) 547–555.
[23] W.C. Chew, Waves and Fields in Inhomogenous Media, vol. 16, John Wiley & Sons, 1999.
[24] S. Lin, Z. Peng, T.M. Antonsen, A stochastic Green's function for solution of wave propagation in wave-chaotic environments, IEEE Transactions on Antennas and Propagation 68 (2020) 3919–3933.
[25] J. Gao, C.-X. Wang, J. Huang, Y. Yang, A novel circuit-based MIMO channel model considering antenna size and mutual coupling, in: 2021 13th International Conference on Wireless Communications and Signal Processing (WCSP), 2021, pp. 1–5.
[26] L.J. Chu, Physical limitations of omnidirectional antennas, Journal of Applied Physics 19 (1948) 1163–1175.
[27] T.-H.E. Fields, Time Harmonic Electromagnetic Fields, McGraw-Hill, New York City, New York, 1961.
[28] V. Shyianov, M. Akrout, F. Bellili, A. Mezghani, R.W. Heath, Achievable rate with antenna size constraint: Shannon meets Chu and Bode, IEEE Transactions on Communications 70 (2021) 2010–2024.
[29] M.T. Ivrlač, J.A. Nossek, Toward a circuit theory of communication, IEEE Transactions on Circuits and Systems. I, Regular Papers 57 (2010) 1663–1683.

[30] T.K. Sarkar, S. Burintramart, N. Yilmazer, S. Hwang, Y. Zhang, A. De, M. Salazar-Palma, A discussion about some of the principles/practices of wireless communication under a Maxwellian framework, IEEE Transactions on Antennas and Propagation 54 (2006) 3727–3745.

[31] P. Jayathurathnage, M. Vilathgamuwa, C. Simovski, Revisiting two-port network analysis for wireless power transfer (WPT) systems, in: 2018 IEEE 4th Southern Power Electronics Conference (SPEC), IEEE, 2018, pp. 1–5.

[32] Q. Ma, C.-X. Wang, J. Huang, Y. Yang, A novel 6G continuous-space channel model based on electromagnetic information theory, in: 2024 IEEE Wireless Communications and Networking Conference (WCNC), IEEE, 2024, pp. 1–6.

[33] C.-X. Wang, Z. Lv, X. Gao, X. You, Y. Hao, H. Haas, Pervasive wireless channel modeling theory and applications to 6G GBSMs for all frequency bands and all scenarios, IEEE Transactions on Vehicular Technology 71 (2022) 9159–9173.

[34] A.S. Poon, R.W. Brodersen, D.N. Tse, Degrees of freedom in multiple-antenna channels: a signal space approach, IEEE Transactions on Information Theory 51 (2005) 523–536.

[35] A. Ishimaru, Electromagnetic Wave Propagation, Radiation, and Scattering: From Fundamentals to Applications, John Wiley & Sons, 2017.

[36] M.D. Migliore, On electromagnetics and information theory, IEEE Transactions on Antennas and Propagation 56 (2008) 3188–3200.

[37] D.A. Miller, Waves, modes, communications, and optics: a tutorial, Advances in Optics and Photonics 11 (2019) 679–825.

[38] M. Cui, L. Dai, Channel estimation for extremely large-scale MIMO: far-field or near-field?, IEEE Transactions on Communications 70 (2022) 2663–2677.

[39] S. Sun, H. Yan, Small-scale spatial-temporal correlation and degrees of freedom for reconfigurable intelligent surfaces, IEEE Wireless Communications Letters 10 (2021) 2698–2702.

[40] J. Mautz, R. Harrington, Modal analysis of loaded N-port scatterers, IEEE Transactions on Antennas and Propagation 21 (1973) 188–199.

[41] T. Lonsky, P. Hazdra, J. Kracek, Characteristic modes of dipole arrays, IEEE Antennas and Wireless Propagation Letters 17 (2018) 998–1001.

[42] M. Gustafsson, L. Jelinek, K. Schab, M. Capek, Unified theory of characteristic modes—part I: fundamentals, IEEE Transactions on Antennas and Propagation 70 (2022) 11801–11813.

[43] C. Ouyang, Y. Liu, H. Yang, Performance of downlink and uplink integrated sensing and communications (ISAC) systems, IEEE Wireless Communications Letters 11 (2022) 1850–1854.

[44] G. Jing, J. Hong, X. Yin, J. Rodríguez-Piñeiro, Z. Yu, Measurement-based 3-D channel modeling with cluster-of-scatterers estimated under spherical-wave assumption, IEEE Transactions on Wireless Communications 22 (2023) 5828–5843.

[45] A. Mudonhi, M. Lotti, A. Clemente, R. D'Errico, C. Oestges, RIS-enabled mmwave channel sounding based on electronically reconfigurable transmitarrays, in: 2021 15th European Conference on Antennas and Propagation (EuCAP), IEEE, 2021, pp. 1–5.

[46] R. Fara, D.-T. Phan-Huy, P. Ratajczak, A. Ourir, M. Di Renzo, J. De Rosny, Reconfigurable intelligent surface-assisted ambient backscatter communications–experimental assessment, in: 2021 IEEE International Conference on Communications Workshops (ICC Workshops), IEEE, 2021, pp. 1–7.

[47] Y. Lu, C. Tao, L. Liu, K. Liu, Spatial characteristics of the massive MIMO channel based on indoor measurement at 1.4725 GHz, IET Communications 12 (2018) 192–197.

[48] J. Li, B. Ai, R. He, M. Yang, Z. Zhong, Y. Hao, G. Shi, The 3D spatial non-stationarity and spherical wavefront in massive MIMO channel measurement, in: 2018 10th International Conference on Wireless Communications and Signal Processing (WCSP), IEEE, 2018, pp. 1–6.

CHAPTER 4

Electromagnetically consistent modeling, optimization, and scaling laws of RIS

Matteo Nerini[a], Hongyu Li[b], Shanpu Shen[c], and Bruno Clerckx[d]

[a]Imperial College London, London, United Kingdom [b]The Hong Kong University of Science and Technology (Guangzhou), Guangzhou, China [c]University of Macau, Taipa, Macau [d]University of Liverpool, Liverpool, United Kingdom

4.1 Introduction

Reconfigurable intelligent surface (RIS) is a revolutionary technology expected to transform wireless communication systems and thus allow us to control the propagation environment [1,2]. RISs are surfaces that comprise numerous passive elements whose scattering properties can be electronically manipulated to shape the reflection of electromagnetic (EM) waves. By strategically controlling these elements, RIS can steer signals toward the intended destination, boosting signal strength and coverage, suppressing interference, and ultimately achieving substantial gains in spectral and energy efficiency [1,2].

RISs have been conventionally implemented by reconfiguring the scattering properties of each RIS element individually through a tunable load impedance. This means that the EM wave impinging on one RIS element is purely reflected by that RIS element, mathematically resulting in the RIS being characterized with a diagonal scattering matrix, commonly known as phase shift matrix [1,2]. To improve the flexibility of this conventional RIS architecture, also referred to as diagonal RIS (D-RIS), beyond diagonal RIS (BD-RIS) represents a novel family of more general RIS architectures where the elements are interconnected, enabling signals incident on one element to be reflected from others [3]. These interconnections grant BD-RIS enhanced flexibility and capabilities compared to D-RIS [4–7].

To optimize D-RIS and BD-RIS, and accurately evaluate their performance, it is essential to accurately model the channel in RIS-aided wireless systems. A robust tool used to develop electromagnetically consistent channel models for multiple-input multiple-output (MIMO) wireless systems is multiport network theory [8–10]. Recently, this theory was also used to model RIS-aided systems [3,11,12], to capture the real-world interactions between RIS and EM waves, often neglected in channel models widely used in literature. Specifically, this

Electromagnetic Signal and Information Theory
https://doi.org/10.1016/B978-0-44-336354-2.00011-3
Copyright © 2027 Elsevier Inc. All rights are reserved, including those for text and data mining, AI training, and similar technologies.

approach treats the RIS-aided wireless channel as a multiport network, characterized by a scattering matrix relating the incident and reflected waves at each port [3], or, equivalently, by an impedance matrix [11] or admittance matrix [12] relating voltages and currents.[1]

EM-consistent models developed via multiport network theory encompass crucial physical phenomena often overlooked in simplified models, such as the specular reflection caused by the structural scattering of the RIS and the mutual coupling between the RIS elements. On the one hand, specular reflection refers to the reflection of signals from the RIS surface independent of the reconfiguration of individual elements, and was first highlighted for RIS-aided systems in [14]. This phenomenon is due to the so-called structural scattering of the RIS, which constitutes a component of the channel through RIS that is always present, even in the case of a fully obstructed direct link between transmitter and receiver [14–16]. On the other hand, mutual coupling describes the EM interaction between closely spaced RIS elements, whose currents are influenced by each other. Recent works focused on modeling and optimizing D-RIS with mutual coupling [17–22], remarking the necessity to account for these effects when optimizing the RIS. Moreover, considering the effects of mutual coupling, BD-RIS has been optimized iteratively in [23], and with globally optimal solutions in closed form in [24]. In light of works [14–24], incorporating real-world phenomena into the channel model is crucial for accurate performance analysis in realistic scenarios.

While most works focus on wireless systems aided by a single RIS, multi-RIS systems have been also studied, where multiple RISs are strategically deployed to collaboratively enhance communication performance and coverage. Double-RIS systems were first studied in [25,26] and the concept has been subsequently extended to multi-RIS systems in [27–29]. In such systems, the signal traverses a multihop path, propagating from one RIS to the next before reaching the receiver. EM-consistent modeling becomes even more critical in multi-RIS systems due to the increasing complexity and the interplay between signals reflected from different RISs. Only limited attention has been paid to the development of EM-consistent channel models for multi-RIS-aided systems. Nevertheless, [30] highlights that the impact of the specular reflection increases with the number of RISs in the system, suggesting that the need for EM-consistent models is even stronger in multi-RIS systems.

In this chapter, we provide a foundational analysis of EM-consistent channel models for both single- and multi-RIS-aided wireless systems. *First*, we introduce the widely used channel model and the scaling laws of its channel gain in the presence of D-RIS and BD-RIS. *Second*, we present more accurate models based on multiport network theory, carefully considering specular reflection and mutual coupling at the RIS. We derive scaling laws for these models, allowing for a comparative performance assessment with the widely used model. This analysis unveils the substantial impact of specular reflection and mutual coupling on channel

[1] For a comprehensive introduction to multiport network theory, the reader is referred to [13, Chapter 4].

gain, particularly with small interelement distances. *Third*, we extend our analysis to multi-RIS-aided systems, quantifying the impact of the specular reflection and mutual coupling on systems including multiple RISs.

Organization. In Sections 4.2, 4.3, and 4.4, we model the channel of RIS-aided systems, considering the widely used model, a model including the specular reflection of the RIS, and a model accounting for mutual coupling at the RIS, respectively. In Sections 4.5, 4.6, and 4.7, we consider multi-RIS-aided systems, presenting the widely used model, a model accounting for the specular reflection of the RIS, and mutual coupling, respectively. Finally, Section 4.8 concludes this chapter.

Notation. Vectors and matrices are denoted with bold lower and bold upper letters, respectively. Scalars are represented with letters not in bold font. Symbols $\Re\{a\}$, $|a|$, and $\arg(a)$ refer to the real part, modulus, and phase of a complex scalar a, respectively, while $[\mathbf{a}]_i$ and $\|\mathbf{a}\|$ refer to the ith element and l_2-norm of a vector $\mathbf{a}$, respectively. Further, $\Re\{\mathbf{A}\}$, $\mathbf{A}^T$, and $\mathbf{A}^H$ refer to the elementwise real part, transpose, and conjugate transpose of a matrix $\mathbf{A}$, respectively. Also, $\mathbb{C}$ denotes the complex number set and $j = \sqrt{-1}$ denotes imaginary unit; $\mathbf{0}$ and $\mathbf{I}$ denote an all-zero matrix and an identity matrix, respectively; $\mathcal{CN}(\mathbf{0}, \mathbf{I})$ denotes the distribution of a circularly symmetric complex Gaussian random vector with mean vector $\mathbf{0}$ and covariance matrix $\mathbf{I}$ and $\sim$ stands for "distributed as." Moreover, $\text{diag}(a_1, \ldots, a_N)$ refers to a diagonal matrix with diagonal elements being $a_1, \ldots, a_N$, while $\text{diag}(\mathbf{A}_1, \ldots, \mathbf{A}_N)$ refers to a block diagonal matrix with blocks being $\mathbf{A}_1, \ldots, \mathbf{A}_N$.

4.2 Widely used RIS-aided channel

In this section, we introduce the channel model of an RIS-aided MIMO system widely used in RIS literature [1,2]. We briefly review how D-RIS and BD-RIS can be globally optimized in a single-user single-input single-output (SISO) system and analyze the scaling laws of the corresponding channel gains.

4.2.1 Modeling

Consider an MIMO wireless system between an N_T-antenna transmitter and an N_R-antenna receiver, where the communication is aided by an N_I-element RIS, as represented in Fig. 4.1. In this MIMO system, the received signal $\mathbf{y} \in \mathbb{C}^{N_R \times 1}$ is given by $\mathbf{y} = \mathbf{H}\mathbf{x} + \mathbf{n}$, where $\mathbf{H} \in \mathbb{C}^{N_R \times N_T}$ is the wireless channel between transmitter and receiver, $\mathbf{x} \in \mathbb{C}^{N_T \times 1}$ is the transmitted signal, and $\mathbf{n} \in \mathbb{C}^{N_R \times 1}$ is the additive white Gaussian noise (AWGN). In the presence of an RIS with phase shift matrix $\mathbf{\Theta} \in \mathbb{C}^{N_I \times N_I}$, the wireless channel $\mathbf{H}$ reads as

$$\mathbf{H} = \mathbf{H}_{RT} + \mathbf{H}_{RI}\mathbf{\Theta}\mathbf{H}_{IT}, \tag{4.1}$$

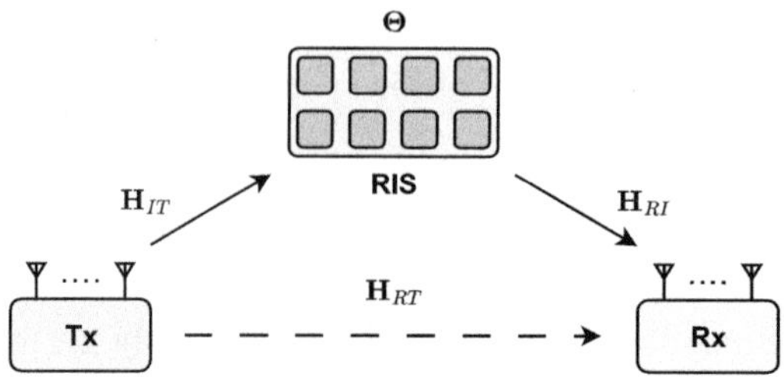

Figure 4.1:
RIS-aided MIMO system.

where $\mathbf{H}_{RT} \in \mathbb{C}^{N_R \times N_T}$ is the direct channel between transmitter and receiver, $\mathbf{H}_{RI} \in \mathbb{C}^{N_R \times N_I}$ is the channel between RIS and receiver, and $\mathbf{H}_{IT} \in \mathbb{C}^{N_I \times N_T}$ is the channel between transmitter and RIS [1,2].

Conventionally, an RIS has been characterized by a diagonal phase shift matrix

$$\boldsymbol{\Theta} = \text{diag}\left(e^{j\theta_1}, e^{j\theta_2}, \ldots, e^{j\theta_{N_I}}\right), \tag{4.2}$$

where $\theta_{n_I} \in [0, 2\pi)$ is the phase shift for the n_Ith RIS element, for $n_I = 1, \ldots, N_I$. Physically, the phase shift matrix is the so-called scattering matrix of the RIS, and represents how the incident signal is reflected by the RIS [3]. Thus, a diagonal scattering matrix means that the n_Ith RIS element individually reflects the incident signal with a reflection coefficient $e^{j\theta_{n_I}}$, where we implicitly assume the RIS to be lossless since the reflection coefficients have unit modulus. This conventional RIS architecture is also denoted as D-RIS since its matrix $\boldsymbol{\Theta}$ is diagonal.

To improve the capabilities of RIS, more general RIS architectures have been proposed under the name of BD-RIS. In these architectures, the RIS elements are connected to each other, and the signal incident on one element can also be reflected from other elements. Mathematically, a lossless BD-RIS is characterized by a unitary scattering matrix $\boldsymbol{\Theta}$, i.e., $\boldsymbol{\Theta}^H \boldsymbol{\Theta} = \mathbf{I}$ [3]. Furthermore, if the RIS is reciprocal, the scattering matrix is symmetric, i.e., $\boldsymbol{\Theta} = \boldsymbol{\Theta}^T$ [3]. Thus, the constraints on $\boldsymbol{\Theta}$ for a lossless and reciprocal BD-RIS are given by

$$\boldsymbol{\Theta}^H \boldsymbol{\Theta} = \mathbf{I}, \; \boldsymbol{\Theta} = \boldsymbol{\Theta}^T, \tag{4.3}$$

which are more general than (4.2), and, naturally, include (4.2) as a special case.

4.2.2 Optimization and scaling law

Consider an RIS-aided SISO system, i.e., $N_T = 1$ and $N_R = 1$, with a completely obstructed direct channel between transmitter and receiver, i.e., $h_{RT} = 0$. In this system, the expression of the channel in (4.1) boils down to

$$h = \mathbf{h}_{RI}\boldsymbol{\Theta}\mathbf{h}_{IT}, \tag{4.4}$$

and we are interested in optimizing a D-RIS or a BD-RIS to maximize the channel gain $|h|^2$. The channel gain maximization problem is given by

$$\max_{\boldsymbol{\Theta}} |\mathbf{h}_{RI}\boldsymbol{\Theta}\mathbf{h}_{IT}|^2 \quad \text{such that (4.2) or (4.3)}, \tag{4.5}$$

where the constraint in (4.2) is considered in the case of a D-RIS, while the constraint in (4.3) is considered for a BD-RIS.

For a D-RIS, it is well known that a globally optimal solution to (4.5) is $\theta_{n_I} = -\arg([\mathbf{h}_{RI}]_{n_I}[\mathbf{h}_{IT}]_{n_I})$, $\forall n_I$, which achieves a channel gain given by

$$|h_{\mathrm{D}}|^2 = \left(\sum_{n_I=1}^{N_I} \left|[\mathbf{h}_{RI}]_{n_I}[\mathbf{h}_{IT}]_{n_I}\right|\right)^2. \tag{4.6}$$

Besides, for a BD-RIS whose scattering matrix is constrained as in (4.3), a global optimal closed-form solution to (4.5) has been proposed in [6], which can achieve a channel gain

$$|h_{\mathrm{BD}}|^2 = \|\mathbf{h}_{RI}\|^2 \|\mathbf{h}_{IT}\|^2. \tag{4.7}$$

Remarkably, (4.7) is never lower than (4.6) because of the Cauchy–Schwarz inequality and this performance improvement is due to the higher flexibility of BD-RIS.

To compute the scaling laws of the average channel gains given in (4.6) and (4.7), we consider independent and identically distributed (i.i.d.) Rayleigh fading channels given by $\mathbf{h}_{RI} \sim \mathcal{CN}(\mathbf{0}, \mathbf{I})$ and $\mathbf{h}_{IT} \sim \mathcal{CN}(\mathbf{0}, \mathbf{I})$, where we assumed unit path gains with no loss of generality. Under Rayleigh distributed channels, it has been shown in [3] that we have

$$\mathrm{E}\left[|h_{\mathrm{D}}|^2\right] = N_I + \frac{\pi^2}{16} N_I (N_I - 1), \tag{4.8}$$

$$\mathrm{E}\left[|h_{\mathrm{BD}}|^2\right] = N_I^2, \tag{4.9}$$

for D-RIS and BD-RIS aided systems, respectively, giving the scaling laws of the channel gain in the case of the widely used channel model.

4.3 RIS-aided channel with specular reflection

The widely used channel model introduced in Section 4.2 provides a simple and intuitive model to describe the impact of an RIS on an MIMO channel. However, it is obtained by implicitly considering several assumptions and approximations. In this section, we relax one of these approximations and present a more EM-consistent channel model that accounts for the specular reflection of the RIS, effect of the structural scattering [14–16].

4.3.1 Modeling

4.3.1.1 Impedance parameter analysis

To obtain such an EM-consistent channel model, we exploit multiport network theory, which has been extensively adopted to characterize MIMO system [8–10] and RIS-aided systems [3,11,12,14–16]. An RIS-aided MIMO channel can be seen as an N-port network, where $N = N_T + N_I + N_R$, with impedance matrix $\mathbf{Z} \in \mathbb{C}^{N \times N}$, as represented in Fig. 4.2. According to multiport network theory [13, Chapter 4], the matrix $\mathbf{Z}$ links the voltages and currents at the N ports, denoted as $\mathbf{v} \in \mathbb{C}^{N \times 1}$ and $\mathbf{i} \in \mathbb{C}^{N \times 1}$, respectively, through

$$\mathbf{v} = \mathbf{Zi}. \tag{4.10}$$

To better grasp the meaning of (4.10), we can partition $\mathbf{v}$, $\mathbf{i}$, and $\mathbf{Z}$ as

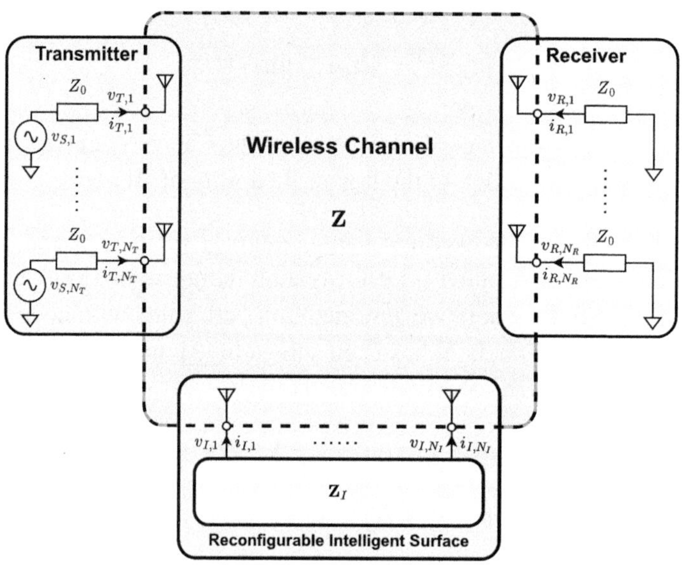

Figure 4.2:
RIS-aided MIMO system modeled with multiport network theory.

$$\mathbf{v}=\begin{bmatrix}\mathbf{v}_T\\ \mathbf{v}_I\\ \mathbf{v}_R\end{bmatrix},\quad \mathbf{i}=\begin{bmatrix}\mathbf{i}_T\\ \mathbf{i}_I\\ \mathbf{i}_R\end{bmatrix},\quad \mathbf{Z}=\begin{bmatrix}\mathbf{Z}_{TT} & \mathbf{Z}_{TI} & \mathbf{Z}_{TR}\\ \mathbf{Z}_{IT} & \mathbf{Z}_{II} & \mathbf{Z}_{IR}\\ \mathbf{Z}_{RT} & \mathbf{Z}_{RI} & \mathbf{Z}_{RR}\end{bmatrix}, \tag{4.11}$$

where $\mathbf{v}_X = [v_{X,1}, \ldots, v_{X,N_X}]^T \in \mathbb{C}^{N_X\times 1}$ and $\mathbf{i}_X = [v_{X,1}, \ldots, v_{X,N_X}]^T \in \mathbb{C}^{N_X\times 1}$ for $X \in \{T, I, R\}$ are the voltages and currents at the antennas of the transmitter, RIS, and receiver, respectively, $\mathbf{Z}_{XX} \in \mathbb{C}^{N_X\times N_X}$ for $X \in \{T, I, R\}$ are the impedance matrices of the antenna arrays at the transmitter, RIS, and receiver, respectively, and $\mathbf{Z}_{XY} \in \mathbb{C}^{N_X\times N_Y}$ for $X, Y \in \{T, I, R\}$ are the transmission impedance matrices [16]. Note that $\mathbf{v}_T$ and $\mathbf{v}_R$ are the transmitted and received signals, respectively, denoted as $\mathbf{x}$ and $\mathbf{y}$ in Section 4.2, which are denoted differently here to specify that they are voltage signals. Thus, our goal is to find the expression of the EM-consistent channel $\mathbf{H}'$ such that

$$\mathbf{v}_R = \mathbf{H}'\mathbf{v}_T, \tag{4.12}$$

i.e., $\mathbf{y} = \mathbf{H}'\mathbf{x}$, ignoring the presence of the noise $\mathbf{n}$. To this end, we solve a system of linear equations describing the model in Fig. 4.2, as illustrated in the following.

Assuming large transmission distances, the feedback channels from RIS to transmitter, receiver to transmitter, and receiver to RIS have negligible impact on the channel $\mathbf{H}'$, and we can set $\mathbf{Z}_{TI} = \mathbf{0}$, $\mathbf{Z}_{TR} = \mathbf{0}$, and $\mathbf{Z}_{IR} = \mathbf{0}$ [9,16]. Furthermore, assuming that the antennas at the transmitter, RIS, and receiver are perfectly matched to the reference impedance $Z_0 = 50\ \Omega$ and there is no mutual coupling, we have $\mathbf{Z}_{TT} = Z_0\mathbf{I}$, $\mathbf{Z}_{II} = Z_0\mathbf{I}$, and $\mathbf{Z}_{RR} = Z_0\mathbf{I}$ [9,16], yielding

$$\mathbf{Z}=\begin{bmatrix}Z_0\mathbf{I} & \mathbf{0} & \mathbf{0}\\ \mathbf{Z}_{IT} & Z_0\mathbf{I} & \mathbf{0}\\ \mathbf{Z}_{RT} & \mathbf{Z}_{RI} & Z_0\mathbf{I}\end{bmatrix}. \tag{4.13}$$

At the transmitter, the antennas are connected to voltage generators with their series impedance Z_0. Thus, the voltages $\mathbf{v}_T$ and currents $\mathbf{i}_T$ at the transmitting antennas are related to the voltages of the signal generators $\mathbf{v}_S = [v_{S,1}, \ldots, v_{S,N_T}]^T \in \mathbb{C}^{N_T\times 1}$ through

$$\mathbf{v}_T = \mathbf{v}_S - Z_0\mathbf{i}_T, \tag{4.14}$$

following Ohm's law. At the RIS, the voltages $\mathbf{v}_I$ and currents $\mathbf{i}_I$ are related by the impedance matrix of the RIS $\mathbf{Z}_I \in \mathbb{C}^{N_I\times N_I}$ via

$$\mathbf{v}_I = -\mathbf{Z}_I\mathbf{i}_I, \tag{4.15}$$

where the minus sign is needed to be consistent with the direction of the currents. Note that the impedance matrix $\mathbf{Z}_I$ and the scattering matrix $\mathbf{\Theta}$ are two equivalent representations for the effect of the RIS, and they are related to each other via the relationships

$$\mathbf{Z}_I = Z_0(\mathbf{I}+\mathbf{\Theta})(\mathbf{I}-\mathbf{\Theta})^{-1}, \tag{4.16}$$

$$\boldsymbol{\Theta} = (\mathbf{Z}_I + Z_0\mathbf{I})^{-1}(\mathbf{Z}_I - Z_0\mathbf{I}), \tag{4.17}$$

according to [13, Chapter 4]. At the receiver, the antennas are terminated with matched loads Z_0 and the voltages $\mathbf{v}_R$ and currents $\mathbf{i}_R$ are related by

$$\mathbf{v}_R = -Z_0\mathbf{i}_R, \tag{4.18}$$

according to Ohm's law.

Based on the above relationships, the channel $\mathbf{H}'$ can be derived by solving the system of the four linear Eqs. (4.10), (4.14), (4.15), and (4.18), i.e.,

$$\begin{cases} \begin{bmatrix} \mathbf{v}_T \\ \mathbf{v}_I \\ \mathbf{v}_R \end{bmatrix} = \begin{bmatrix} Z_0\mathbf{I} & \mathbf{0} & \mathbf{0} \\ \mathbf{Z}_{IT} & Z_0\mathbf{I} & \mathbf{0} \\ \mathbf{Z}_{RT} & \mathbf{Z}_{RI} & Z_0\mathbf{I} \end{bmatrix} \begin{bmatrix} \mathbf{i}_T \\ \mathbf{i}_I \\ \mathbf{i}_R \end{bmatrix}, \\ \mathbf{v}_T = \mathbf{v}_S - Z_0\mathbf{i}_T, \\ \mathbf{v}_I = -\mathbf{Z}_I\mathbf{i}_I, \\ \mathbf{v}_R = -Z_0\mathbf{i}_R. \end{cases} \tag{4.19}$$

System (4.19) can be readily solved, e.g., following [16], to find

$$\mathbf{H}' = \frac{1}{2Z_0}\left(\mathbf{Z}_{RT} - \mathbf{Z}_{RI}(\mathbf{Z}_I + Z_0\mathbf{I})^{-1}\mathbf{Z}_{IT}\right), \tag{4.20}$$

such that $\mathbf{v}_R = \mathbf{H}'\mathbf{v}_T$, which is the channel expression including the transmission impedance matrices $\mathbf{Z}_{RT}$, $\mathbf{Z}_{RI}$, and $\mathbf{Z}_{IT}$, and the RIS impedance matrix $\mathbf{Z}_I$, in agreement with [11, Corollary 1] and [14,16].

To rewrite (4.20) as a function of the RIS scattering matrix $\boldsymbol{\Theta}$, we notice that, by rewriting (4.17) as

$$\begin{aligned} \boldsymbol{\Theta} &= (\mathbf{Z}_I + Z_0\mathbf{I})^{-1}\mathbf{Z}_I - Z_0(\mathbf{Z}_I + Z_0\mathbf{I})^{-1} \\ &= (\mathbf{Z}_I + Z_0\mathbf{I})^{-1}\mathbf{Z}_I + Z_0(\mathbf{Z}_I + Z_0\mathbf{I})^{-1} - 2Z_0(\mathbf{Z}_I + Z_0\mathbf{I})^{-1} \\ &= \mathbf{I} - 2Z_0(\mathbf{Z}_I + Z_0\mathbf{I})^{-1}, \end{aligned} \tag{4.21}$$

we have

$$(\mathbf{Z}_I + Z_0\mathbf{I})^{-1} = -\frac{1}{2Z_0}(\boldsymbol{\Theta} - \mathbf{I}). \tag{4.22}$$

In addition, we introduce the wireless channels from transmitter to receiver, from RIS to receiver, and from transmitter to RIS as

$$\mathbf{H}_{RT} = \frac{\mathbf{Z}_{RT}}{2Z_0}, \quad \mathbf{H}_{RI} = \frac{\mathbf{Z}_{RI}}{2Z_0}, \quad \mathbf{H}_{IT} = \frac{\mathbf{Z}_{IT}}{2Z_0}, \tag{4.23}$$

following the discussion in [3]. Thus, by substituting (4.22) and (4.23) into (4.20), we obtain

$$\begin{aligned}\mathbf{H}' &= \mathbf{H}_{RT} + \mathbf{H}_{RI}\left(\boldsymbol{\Theta} - \mathbf{I}\right)\mathbf{H}_{IT} \\ &= \mathbf{H}_{RT} - \mathbf{H}_{RI}\mathbf{H}_{IT} + \mathbf{H}_{RI}\boldsymbol{\Theta}\mathbf{H}_{IT},\end{aligned} \tag{4.24}$$

which differs from the widely used model $\mathbf{H}$ in (4.1) because of the presence of the specular reflection term $-\mathbf{H}_{RI}\mathbf{H}_{IT}$. Remarkably, this term is always present even when the direct link is fully obstructed ($\mathbf{H}_{RT} = \mathbf{0}$) and regardless of the value of the RIS scattering matrix $\boldsymbol{\Theta}$.

4.3.1.2 Scattering parameter analysis

In alternative to the impedance parameter analysis, the RIS-aided MIMO channel can be equivalently represented by its scattering matrix $\mathbf{S} \in \mathbb{C}^{N \times N}$, which links the incident and reflected waves at the N ports, denoted as $\mathbf{a} \in \mathbb{C}^{N \times 1}$ and $\mathbf{b} \in \mathbb{C}^{N \times 1}$, respectively, via $\mathbf{b} = \mathbf{S}\mathbf{a}$ [13, Chapter 4]. The vectors $\mathbf{a}$ and $\mathbf{b}$ are related to $\mathbf{v}$ and $\mathbf{i}$ through $\mathbf{v} = \mathbf{a} + \mathbf{b}$ and $\mathbf{i} = (\mathbf{a} - \mathbf{b})/Z_0$, and can be similarly partitioned as $\mathbf{a} = [\mathbf{a}_T, \mathbf{a}_I, \mathbf{a}_R]^T$ and $\mathbf{b} = [\mathbf{b}_T, \mathbf{b}_I, \mathbf{b}_R]^T$ [3,15,16]. Besides, $\mathbf{S}$ is related to $\mathbf{Z}$ through $\mathbf{S} = (\mathbf{Z} + Z_0\mathbf{I})^{-1}(\mathbf{Z} - Z_0\mathbf{I})$, and, given a matrix $\mathbf{Z}$ written as in (4.13), the corresponding $\mathbf{S}$ reads as

$$\mathbf{S} = \begin{bmatrix} \mathbf{0} & \mathbf{0} & \mathbf{0} \\ \mathbf{S}_{IT} & \mathbf{0} & \mathbf{0} \\ \mathbf{S}_{RT} & \mathbf{S}_{RI} & \mathbf{0} \end{bmatrix}, \tag{4.25}$$

where

$$\mathbf{S}_{RI} = \frac{\mathbf{Z}_{RI}}{2Z_0}, \quad \mathbf{S}_{IT} = \frac{\mathbf{Z}_{IT}}{2Z_0}, \quad \mathbf{S}_{RT} = \frac{1}{2Z_0}\left(\mathbf{Z}_{RT} - \frac{\mathbf{Z}_{RI}\mathbf{Z}_{IT}}{2Z_0}\right). \tag{4.26}$$

At the transmitter, $\mathbf{b}_T$ is related to the voltages of the signal generators $\mathbf{v}_S$ through $\mathbf{b}_T = \mathbf{v}_S/2$, since their series impedance is matched to Z_0, as described in [3,16]. At the RIS, $\mathbf{a}_I$ and $\mathbf{b}_I$ are related by the scattering matrix of the RIS $\boldsymbol{\Theta}$ via $\mathbf{a}_I = \boldsymbol{\Theta}\mathbf{b}_I$, following [13, Chapter 4]. At the receiver, we have $\mathbf{a}_R = \mathbf{0}$ since the antennas are terminated with matched loads [3,16].

Based on the above relationships, the channel $\mathbf{H}'$ can be derived by solving the system of the four linear equations

$$\begin{cases} \begin{bmatrix} \mathbf{b}_T \\ \mathbf{b}_I \\ \mathbf{b}_R \end{bmatrix} = \begin{bmatrix} \mathbf{0} & \mathbf{0} & \mathbf{0} \\ \mathbf{S}_{IT} & \mathbf{0} & \mathbf{0} \\ \mathbf{S}_{RT} & \mathbf{S}_{RI} & \mathbf{0} \end{bmatrix} \begin{bmatrix} \mathbf{a}_T \\ \mathbf{a}_I \\ \mathbf{a}_R \end{bmatrix}, \\ \mathbf{a}_T = \mathbf{v}_S/2, \\ \mathbf{a}_I = \boldsymbol{\Theta}\mathbf{b}_I, \\ \mathbf{a}_R = \mathbf{0}. \end{cases} \tag{4.27}$$

System (4.27) can be readily solved, e.g., following [16], and, recalling that $\mathbf{v} = \mathbf{a} + \mathbf{b}$, we find

$$\mathbf{H}' = \mathbf{S}_{RT} + \mathbf{S}_{RI}\boldsymbol{\Theta}\mathbf{S}_{IT}, \tag{4.28}$$

such that $\mathbf{v}_R = \mathbf{H}'\mathbf{v}_T$, which is the channel expression including the transmission scattering matrices $\mathbf{S}_{RT}$, $\mathbf{S}_{RI}$, and $\mathbf{S}_{IT}$, and the RIS scattering matrix $\boldsymbol{\Theta}$, in agreement with [3,16]. Since impedance and scattering parameters are two equivalent representations, (4.28) is equivalent to (4.20), as it can be shown by substituting (4.26) into (4.28). In the following, we numerically quantify the difference between $\mathbf{H}'$ and $\mathbf{H}$ by analyzing the scaling laws of their channel gains.

4.3.2 Optimization and scaling law

In an SISO system with a completely obstructed direct link, i.e., $h_{RT} = 0$, the channel in (4.24) boils down to

$$h' = -\mathbf{h}_{RI}\mathbf{h}_{IT} + \mathbf{h}_{RI}\boldsymbol{\Theta}\mathbf{h}_{IT}, \tag{4.29}$$

and the channel gain $|h'|^2$ is maximized by solving the problem

$$\max_{\boldsymbol{\Theta}} |-\mathbf{h}_{RI}\mathbf{h}_{IT} + \mathbf{h}_{RI}\boldsymbol{\Theta}\mathbf{h}_{IT}|^2 \quad \text{such that (4.2) or (4.3)}, \tag{4.30}$$

where the constraint in (4.2) or (4.3) is considered for D-RIS or BD-RIS, respectively.

In the case of a D-RIS, problem (4.30) can be globally solved by setting $\theta_{n_I} = \arg(-\mathbf{h}_{RI}\mathbf{h}_{IT}) - \arg([\mathbf{h}_{RI}]_{n_I}[\mathbf{h}_{IT}]_{n_I})$, $\forall n_I$, which achieves a channel gain

$$|h'_{\mathrm{D}}|^2 = \left(|\mathbf{h}_{RI}\mathbf{h}_{IT}| + \sum_{n_I=1}^{N_I} \left|[\mathbf{h}_{RI}]_{n_I}[\mathbf{h}_{IT}]_{n_I}\right|\right)^2, \tag{4.31}$$

as observed in [16]. Besides, a BD-RIS constrained as in (4.3) can be globally optimized with the solution proposed in [6] to achieve a channel gain given by

$$|h'_{\mathrm{BD}}|^2 = (|\mathbf{h}_{RI}\mathbf{h}_{IT}| + \|\mathbf{h}_{RI}\|\,\|\mathbf{h}_{IT}\|)^2. \tag{4.32}$$

To derive the scaling laws of the average channel gains in (4.31) and (4.32), we again consider Rayleigh distributed channels, i.e., $\mathbf{h}_{RI} \sim \mathcal{CN}(\mathbf{0}, \mathbf{I})$ and $\mathbf{h}_{IT} \sim \mathcal{CN}(\mathbf{0}, \mathbf{I})$. Approximating the term $|\mathbf{h}_{RI}\mathbf{h}_{IT}|$ as uncorrelated with $|[\mathbf{h}_{RI}]_{n_I}|$ and $|[\mathbf{h}_{IT}]_{n_I}|$, the expected value of (4.31) writes as

$$\begin{aligned}\mathrm{E}\left[\left|h_{\mathrm{D}}^{\prime}\right|^{2}\right]=&\mathrm{E}\left[\left|\mathbf{h}_{RI}\mathbf{h}_{IT}\right|^{2}\right]+\mathrm{E}\left[\left(\sum_{n_I=1}^{N_I}\left|[\mathbf{h}_{RI}]_{n_I}[\mathbf{h}_{IT}]_{n_I}\right|\right)^{2}\right]\\&+2\mathrm{E}\left[\left|\mathbf{h}_{RI}\mathbf{h}_{IT}\right|\right]\sum_{n_I=1}^{N_I}\mathrm{E}\left[\left|[\mathbf{h}_{RI}]_{n_I}\right|\right]\mathrm{E}\left[\left|[\mathbf{h}_{IT}]_{n_I}\right|\right].\end{aligned}\tag{4.33}$$

Noticing that $\mathbf{h}_{RI}\mathbf{h}_{IT}\sim\mathcal{CN}(0,N_I)$ by the central limit theorem, we have $[|\mathbf{h}_{RI}\mathbf{h}_{IT}|]=\sqrt{\frac{\pi}{4}N_I}$ and $\mathrm{E}[|\mathbf{h}_{RI}\mathbf{h}_{IT}|^2]=N_I$. By also noticing that $\mathrm{E}[|[\mathbf{h}_{RI}]_{n_I}|]=\mathrm{E}[|[\mathbf{h}_{IT}]_{n_I}|]=\sqrt{\pi}/2$ and $\mathrm{E}[|[\mathbf{h}_{RI}]_{n_I}|^2]=\mathrm{E}[|[\mathbf{h}_{IT}]_{n_I}|^2]=1,\forall n_I$, following the moments of the Rayleigh distribution, we obtain

$$\mathrm{E}\left[\left|h_{\mathrm{D}}^{\prime}\right|^{2}\right]=2N_I+\frac{\pi^2}{16}N_I\left(N_I-1\right)+\frac{\pi}{4}\sqrt{\pi N_I}N_I.\tag{4.34}$$

Besides, in the case of a BD-RIS, the expected value of (4.32) is given by

$$\begin{aligned}\mathrm{E}\left[\left|h_{\mathrm{BD}}^{\prime}\right|^{2}\right]=&\mathrm{E}\left[\left|\mathbf{h}_{RI}\mathbf{h}_{IT}\right|^{2}\right]+\mathrm{E}\left[\left\|\mathbf{h}_{RI}\right\|^{2}\right]\mathrm{E}\left[\left\|\mathbf{h}_{IT}\right\|^{2}\right]\\&+2\mathrm{E}\left[\left|\mathbf{h}_{RI}\mathbf{h}_{IT}\right|\right]\mathrm{E}\left[\left\|\mathbf{h}_{RI}\right\|\right]\mathrm{E}\left[\left\|\mathbf{h}_{IT}\right\|\right],\end{aligned}\tag{4.35}$$

where we approximated $|\mathbf{h}_{RI}\mathbf{h}_{IT}|$ as uncorrelated with $\|\mathbf{h}_{RI}\|$ and $\|\mathbf{h}_{IT}\|$. Thus, by using $\mathrm{E}[|\mathbf{h}_{RI}\mathbf{h}_{IT}|]=\sqrt{\frac{\pi}{4}N_I}$ and $\mathrm{E}[|\mathbf{h}_{RI}\mathbf{h}_{IT}|^2]=N_I$, and $\mathrm{E}[\|\mathbf{h}_{RI}\|]=\mathrm{E}[\|\mathbf{h}_{IT}\|]=\Gamma(N_I+0.5)/\Gamma(N_I)$ and $\mathrm{E}[\|\mathbf{h}_{RI}\|^2]=\mathrm{E}[\|\mathbf{h}_{IT}\|^2]=N_I$, derived from the moments of the chi-squared distribution with $2N_I$ degrees of freedom, we obtain

$$\begin{aligned}\mathrm{E}\left[\left|h_{\mathrm{BD}}^{\prime}\right|^{2}\right]&=N_I+N_I^2+\sqrt{\pi N_I}\left(\frac{\Gamma\left(N_I+0.5\right)}{\Gamma\left(N_I\right)}\right)^2\\&\approx N_I+N_I^2+\sqrt{\pi N_I}N_I,\end{aligned}\tag{4.36}$$

where $\Gamma(\cdot)$ is the gamma function, and we used the approximation $(\Gamma(N_I+0.5)/\Gamma(N_I))^2\approx N_I$ following the Laurent series expansion.

To compare the scaling laws of the widely used model $\mathrm{E}[|h|^2]$ and the model with specular reflection $\mathrm{E}[|h'|^2]$, we introduce their relative difference as

$$\delta=\frac{\mathrm{E}\left[\left|h^{\prime}\right|^{2}\right]-\mathrm{E}\left[|h|^{2}\right]}{\mathrm{E}\left[|h|^{2}\right]}.\tag{4.37}$$

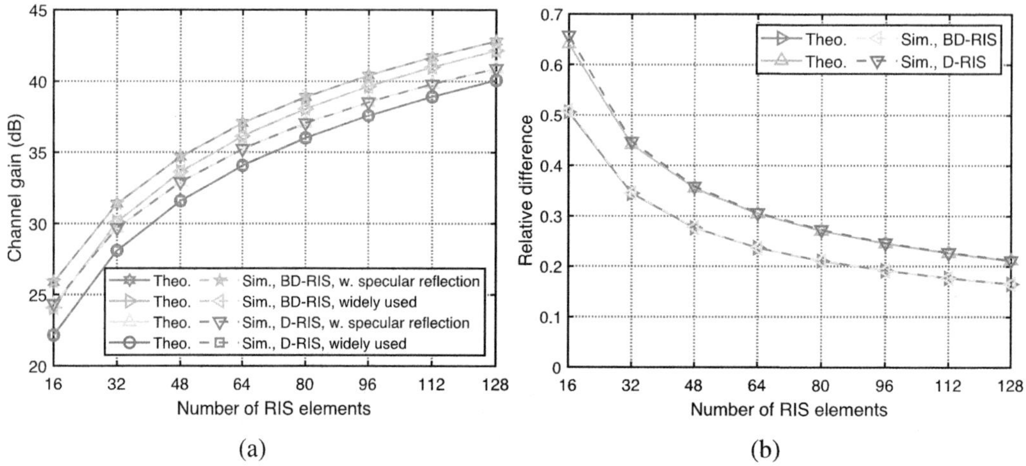

Figure 4.3:
(a) Channel gains of the widely used model and the model with specular reflection, and (b) their relative difference.

For D-RIS, this relative difference is obtained by substituting (4.8) and (4.34) into (4.37) and writes as

$$\delta_{\mathrm{D}} = \frac{\frac{\pi}{4}\sqrt{\pi N_I} N_I + N_I}{\frac{\pi^2}{16} N_I (N_I - 1) + N_I} \approx \frac{4}{\sqrt{\pi N_I}}, \tag{4.38}$$

converging to $4/\sqrt{\pi N_I}$ for $N_I \to \infty$. Besides, for BD-RIS, we substitute (4.9) and (4.36) into (4.37) to obtain the relative difference between the channel models as

$$\delta_{\mathrm{BD}} = \frac{\sqrt{\pi N_I} N_I + N_I}{N_I^2} \approx \frac{\sqrt{\pi}}{\sqrt{N_I}}, \tag{4.39}$$

which is $\sqrt{\pi}/\sqrt{N_I}$ in the limit $N_I \to \infty$.

In Fig. 4.3, we report the average channel gains of the widely used model and the model with specular reflection, and their relative difference δ. We make the following observations. *First*, when considering the channel model accounting for the specular reflection of the RIS, the channel gain is higher than that with the widely used model, since the additive term $|\mathbf{h}_{RI}\mathbf{h}_{IT}|$ is used to improve the gain, and this holds for both D-RIS and BD-RIS. *Second*, the relative difference between the two models scales with $1/\sqrt{N_I}$ as a function of the number of RIS elements. Thus, it vanishes as $N_I \to \infty$ but it is not negligible for practical values of the number of RIS elements. This shows the need for more EM-consistent channel models for communication theoretical analyses of RIS-aided systems.

4.4 RIS-aided channel with mutual coupling

We presented an RIS-aided channel model accounting for the specular reflection of the RIS, which is more EM-consistent than the widely used model. In this section, we derive an even more accurate RIS-aided channel model that accounts for the specular reflection and also the mutual coupling between the RIS elements.

4.4.1 Modeling

To derive the model accounting for mutual coupling, we resort to multiport network theory, as done in Section 4.3.1. We consider the same assumptions on the impedance matrix of the wireless channel $\mathbf{Z}$, while relaxing the assumption of perfect matching and no mutual coupling at the RIS, giving

$$\mathbf{Z} = \begin{bmatrix} Z_0\mathbf{I} & \mathbf{0} & \mathbf{0} \\ \mathbf{Z}_{IT} & \mathbf{Z}_{II} & \mathbf{0} \\ \mathbf{Z}_{RT} & \mathbf{Z}_{RI} & Z_0\mathbf{I} \end{bmatrix}. \tag{4.40}$$

Thus, the EM-consistent channel $\mathbf{H}''$ accounting for mutual coupling is still given by $\mathbf{v}_R = \mathbf{H}''\mathbf{v}_T$, and can be derived by solving the system of the four linear equations

$$\begin{cases} \begin{bmatrix} \mathbf{v}_T \\ \mathbf{v}_I \\ \mathbf{v}_R \end{bmatrix} = \begin{bmatrix} Z_0\mathbf{I} & \mathbf{0} & \mathbf{0} \\ \mathbf{Z}_{IT} & \mathbf{Z}_{II} & \mathbf{0} \\ \mathbf{Z}_{RT} & \mathbf{Z}_{RI} & Z_0\mathbf{I} \end{bmatrix} \begin{bmatrix} \mathbf{i}_T \\ \mathbf{i}_I \\ \mathbf{i}_R \end{bmatrix}, \\ \mathbf{v}_T = \mathbf{v}_S - Z_0\mathbf{i}_T, \\ \mathbf{v}_I = -\mathbf{Z}_I\mathbf{i}_I, \\ \mathbf{v}_R = -Z_0\mathbf{i}_R. \end{cases} \tag{4.41}$$

Note that the system in (4.41) is slightly more general than (4.19) since we relax the assumption that $\mathbf{Z}_{II} = Z_0\mathbf{I}$ in (4.19). System (4.41) can be readily solved, e.g., according to the steps in [16], to find

$$\mathbf{H}'' = \frac{1}{2Z_0}\left(\mathbf{Z}_{RT} - \mathbf{Z}_{RI}\left(\mathbf{Z}_I + \mathbf{Z}_{II}\right)^{-1}\mathbf{Z}_{IT}\right), \tag{4.42}$$

being the channel model based on the impedance parameters accounting for mutual coupling at the RIS, in agreement with [11, Corollary 1] and [16].

To rewrite (4.42) in a more convenient form, we introduce

$$\mathbf{H}_{II} = \frac{\mathbf{Z}_{II}}{Z_0}, \quad \mathbf{\Lambda} = \frac{\mathbf{Z}_I}{Z_0}, \tag{4.43}$$

as the mutual coupling matrix and the load matrix of the RIS, respectively. By substituting (4.23) and (4.43) into (4.42), we finally obtain the EM-consistent channel model accounting for mutual coupling as

$$\mathbf{H}'' = \mathbf{H}_{RT} - 2\mathbf{H}_{RI}\left(\boldsymbol{\Lambda} + \mathbf{H}_{II}\right)^{-1}\mathbf{H}_{IT}, \tag{4.44}$$

where the RIS is represented by its load matrix $\boldsymbol{\Lambda}$ instead of its scattering matrix.

Note that $\boldsymbol{\Lambda}$ and $\boldsymbol{\Theta}$ are equivalent representations. Following the definition of $\boldsymbol{\Lambda}$ in (4.43) as a function of $\mathbf{Z}_I$ and the relationships between $\mathbf{Z}_I$ and $\boldsymbol{\Theta}$ in (4.16) and (4.17), we have

$$\boldsymbol{\Lambda} = (\mathbf{I} + \boldsymbol{\Theta})(\mathbf{I} - \boldsymbol{\Theta})^{-1}, \tag{4.45}$$

$$\boldsymbol{\Theta} = (\boldsymbol{\Lambda} + \mathbf{I})^{-1}(\boldsymbol{\Lambda} - \mathbf{I}). \tag{4.46}$$

For a D-RIS, where $\boldsymbol{\Theta}$ is a phase shift matrix constrained as in (4.2), $\boldsymbol{\Lambda}$ is a diagonal purely imaginary matrix given by

$$\boldsymbol{\Lambda} = \mathrm{diag}\left(\lambda_1, \lambda_2, \ldots, \lambda_{N_I}\right), \ \lambda_{n_I} = j\cot\left(\frac{\theta_{n_I}}{2}\right), \ \forall n_I, \tag{4.47}$$

following (4.45). Besides, for a BD-RIS constrained as in (4.3), $\boldsymbol{\Lambda}$ is a symmetric purely imaginary matrix, i.e.,

$$\Re\{\boldsymbol{\Lambda}\} = \mathbf{0}, \ \boldsymbol{\Lambda} = \boldsymbol{\Lambda}^T. \tag{4.48}$$

The relationships between $\boldsymbol{\Theta}$ and $\boldsymbol{\Lambda}$ and their constraints are summarized in Table 4.1.

Table 4.1: Relationship between scattering matrix $\boldsymbol{\Theta}$ and load matrix $\boldsymbol{\Lambda}$ and their constraints for the different RIS architectures.

RIS architecture	$\boldsymbol{\Theta} = (\boldsymbol{\Lambda} + \mathbf{I})^{-1}(\boldsymbol{\Lambda} - \mathbf{I})$	$\boldsymbol{\Lambda} = (\mathbf{I} + \boldsymbol{\Theta})(\mathbf{I} - \boldsymbol{\Theta})^{-1}$
Lossless D-RIS	$\boldsymbol{\Theta} = \mathrm{diag}(e^{j\theta_1}, \ldots, e^{j\theta_{N_I}})$	$\boldsymbol{\Lambda} = \mathrm{diag}(\lambda_1, \ldots, \lambda_{N_I})$, $\Re\{\lambda_{n_I}\} = 0$
Lossless BD-RIS	$\boldsymbol{\Theta}^H\boldsymbol{\Theta} = \mathbf{I}$	$\Re\{\boldsymbol{\Lambda}\} = \mathbf{0}$
Reciprocal BD-RIS	$\boldsymbol{\Theta} = \boldsymbol{\Theta}^T$	$\boldsymbol{\Lambda} = \boldsymbol{\Lambda}^T$

4.4.2 Optimization and scaling law

Considering an SISO system with a completely obstructed direct link, i.e., $h_{RT} = 0$, the channel model with mutual coupling (4.44) boils down to

$$h'' = -2\mathbf{h}_{RI}\left(\boldsymbol{\Lambda} + \mathbf{H}_{II}\right)^{-1}\mathbf{h}_{IT}, \tag{4.49}$$

and the channel gain $|h''|^2$ maximization problem writes as

$$\max_{\boldsymbol{\Lambda}} 4\left|\mathbf{h}_{RI}\left(\boldsymbol{\Lambda} + \mathbf{H}_{II}\right)^{-1}\mathbf{h}_{IT}\right|^2 \quad \text{such that (4.47) or (4.48)}, \tag{4.50}$$

where we reconfigure the matrix $\mathbf{\Lambda}$ constrained as in (4.47) or (4.48), for a D-RIS or a BD-RIS, respectively. Interestingly, this problem has a known global optimal solution for BD-RIS, i.e., considering the constraint in (4.48), derived in [24]. By optimizing the BD-RIS through this solution, it has been proved in [24] that we can achieve the channel gain

$$|h''_{\text{BD}}|^2 = \left(\left|\mathbf{h}_{RI}\Re\{\mathbf{H}_{II}\}^{-1}\mathbf{h}_{IT}\right| + \left\|\mathbf{h}_{RI}\Re\{\mathbf{H}_{II}\}^{-1/2}\right\| \left\|\Re\{\mathbf{H}_{II}\}^{-1/2}\mathbf{h}_{IT}\right\|\right)^2. \tag{4.51}$$

Note that the channel gain $|h''_{\text{BD}}|^2$ in (4.51) boils down to $|h'_{\text{BD}}|^2$ given by (4.32) in the absence of mutual coupling, i.e., when $\mathbf{Z}_{II} = Z_0\mathbf{I}$, or $\mathbf{H}_{II} = \mathbf{I}$.

To obtain the scaling laws of the average channel gain in (4.51), we consider Rayleigh distributed channels, namely $\mathbf{h}_{RI} \sim \mathcal{CN}(\mathbf{0},\mathbf{I})$ and $\mathbf{h}_{IT} \sim \mathcal{CN}(\mathbf{0},\mathbf{I})$. By taking the expectation of (4.51), we obtain

$$\begin{aligned} \text{E}\left[|h''_{\text{BD}}|^2\right] &= \text{E}\left[\left|\mathbf{h}_{RI}\Re\{\mathbf{H}_{II}\}^{-1}\mathbf{h}_{IT}\right|^2\right] \\ &+ \text{E}\left[\left\|\mathbf{h}_{RI}\Re\{\mathbf{H}_{II}\}^{-1/2}\right\|^2\right]\text{E}\left[\left\|\Re\{\mathbf{H}_{II}\}^{-1/2}\mathbf{h}_{IT}\right\|^2\right] \\ &+ 2\text{E}\left[\left|\mathbf{h}_{RI}\Re\{\mathbf{H}_{II}\}^{-1}\mathbf{h}_{IT}\right|\right]\text{E}\left[\left\|\mathbf{h}_{RI}\Re\{\mathbf{H}_{II}\}^{-1/2}\right\|\right] \\ &\times \text{E}\left[\left\|\Re\{\mathbf{H}_{II}\}^{-1/2}\mathbf{h}_{IT}\right\|\right], \end{aligned} \tag{4.52}$$

where we approximated $|\mathbf{h}_{RI}\Re\{\mathbf{H}_{II}\}^{-1}\mathbf{h}_{IT}|$ as uncorrelated with

$$\|\mathbf{h}_{RI}\Re\{\mathbf{H}_{II}\}^{-1/2}\| \text{ and } \|\Re\{\mathbf{H}_{II}\}^{-1/2}\mathbf{h}_{IT}\|.$$

A closed-form expression for $\text{E}[|h''_{\text{BD}}|^2]$ can be obtained by individually studying all the expectation terms in (4.52). We have that, in the presence of Rayleigh distributed channels,

$$\text{E}\left[\left|\mathbf{h}_{RI}\Re\{\mathbf{H}_{II}\}^{-1}\mathbf{h}_{IT}\right|^2\right] = \text{Tr}\left(\Re\{\mathbf{H}_{II}\}^{-2}\right), \tag{4.53}$$

$$\text{E}\left[\left\|\mathbf{h}_{RI}\Re\{\mathbf{H}_{II}\}^{-1/2}\right\|^2\right]\text{E}\left[\left\|\Re\{\mathbf{H}_{II}\}^{-1/2}\mathbf{h}_{IT}\right\|^2\right] = \text{Tr}\left(\Re\{\mathbf{H}_{II}\}^{-1}\right)^2, \tag{4.54}$$

$$\text{E}\left[\left|\mathbf{h}_{RI}\Re\{\mathbf{H}_{II}\}^{-1}\mathbf{h}_{IT}\right|\right] = \sqrt{\frac{\pi}{4}\text{Tr}\left(\Re\{\mathbf{H}_{II}\}^{-2}\right)}, \tag{4.55}$$

$$\text{E}\left[\left\|\mathbf{h}_{RI}\Re\{\mathbf{H}_{II}\}^{-1/2}\right\|\right]\text{E}\left[\left\|\Re\{\mathbf{H}_{II}\}^{-1/2}\mathbf{h}_{IT}\right\|\right] = \text{Tr}\left(\Re\{\mathbf{H}_{II}\}^{-1}\right), \tag{4.56}$$

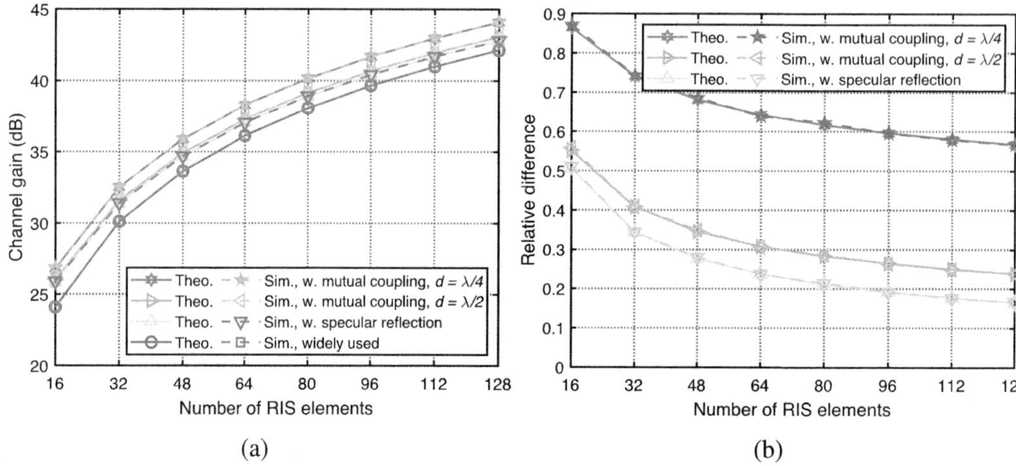

Figure 4.4:
(a) Channel gains of the widely used model, the model with specular reflection, and the model with mutual coupling, and (b) their relative difference with the widely used model.

as detailed in [24]. Thus, by substituting (4.53)–(4.56) into (4.52), we obtain the scaling law of the channel with mutual coupling as

$$\mathrm{E}\left[|h''_{\mathrm{BD}}|^2\right] = \mathrm{Tr}\left(\Re\{\mathbf{H}_{II}\}^{-2}\right) + \mathrm{Tr}\left(\Re\{\mathbf{H}_{II}\}^{-1}\right)^2 + \sqrt{\pi \mathrm{Tr}\left(\Re\{\mathbf{H}_{II}\}^{-2}\right)}\mathrm{Tr}\left(\Re\{\mathbf{H}_{II}\}^{-1}\right), \tag{4.57}$$

which is purely a function of the traces $\mathrm{Tr}(\Re\{\mathbf{H}_{II}\}^{-2})$ and $\mathrm{Tr}(\Re\{\mathbf{H}_{II}\}^{-1})$. Note that $\mathrm{E}[|h''_{\mathrm{BD}}|^2]$ in (4.57) boils down to $\mathrm{E}[|h'_{\mathrm{BD}}|^2]$ in (4.36) with no mutual coupling, i.e., when $\mathbf{Z}_{II} = Z_0\mathbf{I}$, or $\mathbf{H}_{II} = \mathbf{I}$.

In Fig. 4.4, we report the average channel gains of the widely used model described in Section 4.2, the model accounting for specular reflection developed in Section 4.3, and the model with mutual coupling, where the communication is aided by a BD-RIS. Furthermore, we also report their relative difference with the widely used model. For the model with mutual coupling, we consider interelement distances $d = \lambda/2$ and $d = \lambda/4$, where λ is the wavelength corresponding to frequency $f = 28$ GHz and the rest of the numerical setup to compute $\mathbf{H}_{II}$ is the same as in [24]. We make the following observations. *First*, when considering mutual coupling, the channel gain is higher, reaching its maximum for interelement distance $d = \lambda/4$ given the maximum strength of the mutual coupling in this scenario. *Second*, the relative difference with the widely used model is higher in the presence of mutual coupling. It decreases with the number of RIS elements, but it is significant for $N_I \leq 128$. This shows the

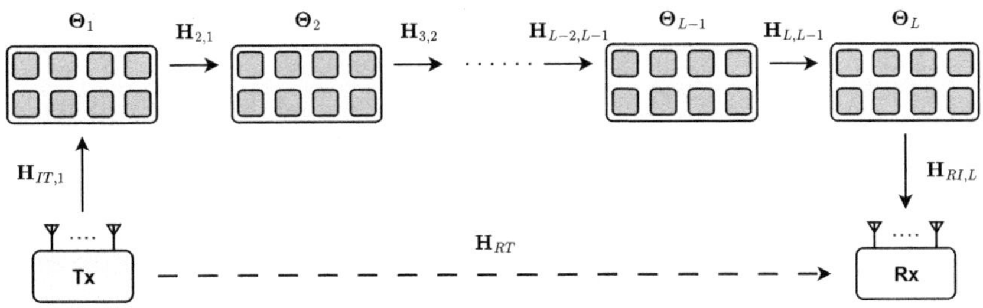

Figure 4.5:
Multi-RIS-aided MIMO system.

importance of accounting for mutual coupling in the channel model, especially for RIS with small interelement distances.

4.5 Widely used multi-RIS-aided channel

In Sections 4.2–4.4, we modeled and analyzed RIS-aided systems including only one RIS. In the rest of the chapter, we extend our models and analysis to systems having multiple RISs, also known as multi-RIS systems. This section introduces the multi-RIS-aided channel model widely used in RIS literature [25–29]. We review how RISs can be globally optimized to maximize the channel gain in an SISO system with line-of-sight (LoS) channels and analyze the channel gain scaling law.

4.5.1 Modeling

Consider an MIMO wireless system between an N_T-antenna transmitter and an N_R-antenna receiver, where the communication is aided by L RISs, each having N_I elements, as represented in Fig. 4.5. In this MIMO system, the received signal writes as $\mathbf{y} = \mathbf{H}\mathbf{x} + \mathbf{n}$, where $\mathbf{H} \in \mathbb{C}^{N_R \times N_T}$ is the wireless channel between transmitter and receiver, $\mathbf{x} \in \mathbb{C}^{N_T \times 1}$ is the transmitted signal, and $\mathbf{n} \in \mathbb{C}^{N_R \times 1}$ is the AWGN. As in [25–29], we assume that the signal reaches the receiver following a multihop path, from the first RIS to the Lth RIS. Thus, denoting as $\mathbf{\Theta}_\ell \in \mathbb{C}^{N_I \times N_I}$ the scattering matrix of the ℓth RIS, the wireless channel $\mathbf{H}$ is given by

$$\mathbf{H} = \mathbf{H}_{RT} + \mathbf{H}_{RI,L}\mathbf{\Theta}_L \prod_{\ell=L-1}^{1} \left(\mathbf{H}_{\ell+1,\ell}\mathbf{\Theta}_\ell\right) \mathbf{H}_{IT,1}, \tag{4.58}$$

where $\mathbf{H}_{RT} \in \mathbb{C}^{N_R \times N_T}$ is the direct channel between transmitter and receiver, $\mathbf{H}_{RI,L} \in \mathbb{C}^{N_R \times N_I}$ is the channel between the Lth RIS and receiver, $\mathbf{H}_{\ell+1,\ell} \in \mathbb{C}^{N_I \times N_I}$ is the channel between the ℓth and the $(\ell + 1)$th RIS, and $\mathbf{H}_{IT,1} \in \mathbb{C}^{N_I \times N_T}$ is the channel between transmitter and first RIS [25–29].

4.5.2 Optimization and scaling law

Consider an SISO system, i.e., $N_R = 1$ and $N_T = 1$, with a completely obstructed direct link, i.e., $h_{RT} = 0$, where (4.58) boils down to

$$h = \mathbf{h}_{RI,L}\boldsymbol{\Theta}_L \prod_{\ell=L-1}^{1} \left(\mathbf{H}_{\ell+1,\ell}\boldsymbol{\Theta}_\ell\right)\mathbf{h}_{IT,1}. \tag{4.59}$$

In this system, the channel gain maximization problem is written as

$$\max_{\boldsymbol{\Theta}_1,\ldots,\boldsymbol{\Theta}_L} \left| \mathbf{h}_{RI,L}\boldsymbol{\Theta}_L \prod_{\ell=L-1}^{1} \left(\mathbf{H}_{\ell+1,\ell}\boldsymbol{\Theta}_\ell\right)\mathbf{h}_{IT,1} \right|^2 \quad \text{such that (4.2) or (4.3)}, \tag{4.60}$$

where the constraint (4.2) is considered in the case of D-RISs, while (4.3) is considered for BD-RISs.

Under LoS channels with unit path gain, the channel between the Lth RIS and the receiver writes as $\mathbf{h}_{RI,L} = [e^{j\beta_{RI,L,1}}, \ldots, e^{j\beta_{RI,L,N_I}}]$ and $\mathbf{h}_{IT,1}$ is given by $\mathbf{h}_{IT,1} = [e^{j\alpha_{IT,1,1}}, \ldots, e^{j\alpha_{IT,1,N_I}}]^T$. Besides, $\mathbf{H}_{\ell+1,\ell} = \mathbf{a}_{\ell+1,\ell}\mathbf{b}_{\ell+1,\ell}^T$ consists of $\mathbf{a}_{\ell+1,\ell} = [e^{j\alpha_{\ell+1,\ell,1}}, \ldots, e^{j\alpha_{\ell+1,\ell,N_I}}]^T$ and $\mathbf{b}_{\ell+1,\ell} = [e^{j\beta_{\ell+1,\ell,1}}, \ldots, e^{j\beta_{\ell+1,\ell,N_I}}]^T$, for $\ell = 1, \ldots, L-1$. Considering the conventional D-RIS architecture (which is optimal under LoS channels [3]), we have $\boldsymbol{\Theta}_\ell = \mathrm{diag}(e^{j\theta_{\ell,1}}, \ldots, e^{j\theta_{\ell,N_I}})$, with θ_{ℓ,n_I} being the n_Ith phase shift of the ℓth RIS, for $n_I = 1, \ldots, N_I$ and $\ell = 1, \ldots, L$.

To solve the problem in (4.60) under LoS channels, there exists a global optimal solution for the scattering matrices $\boldsymbol{\Theta}_\ell$, as derived in [27]. Specifically, in the case of LoS channels, (4.59) can be expressed as

$$h = \mathbf{h}_{RI,L}\boldsymbol{\Theta}_L\mathbf{a}_{L,L-1} \prod_{\ell=L-1}^{2} \left(\mathbf{b}_{\ell+1,\ell}^T\boldsymbol{\Theta}_\ell\mathbf{a}_{\ell,\ell-1}\right)\mathbf{b}_{2,1}^T\boldsymbol{\Theta}_1\mathbf{h}_{IT,1}. \tag{4.61}$$

Then, (4.61) can be rewritten as a product of L scalars $K_\ell \in \mathbb{C}$ such that

$$h = \prod_{\ell=L}^{1} K_\ell, \tag{4.62}$$

where $K_L = \mathbf{h}_{RI,L}\boldsymbol{\Theta}_L\mathbf{a}_{L,L-1}$, $K_1 = \mathbf{b}_{2,1}^T\boldsymbol{\Theta}_1\mathbf{h}_{IT,1}$, and $K_\ell = \mathbf{b}_{\ell+1,\ell}^T\boldsymbol{\Theta}_\ell\mathbf{a}_{\ell,\ell-1}$, for $\ell = 2, \ldots, L-1$. As explained in [27], problem (4.60) can be globally solved by setting

$$\theta_{L,n_I} = -\arg([\mathbf{h}_{RI,L}]_{n_I}[\mathbf{a}_{L,L-1}]_{n_I}), \tag{4.63}$$

$$\theta_{1,n_I} = -\arg([\mathbf{b}_{2,1}]_{n_I}[\mathbf{h}_{IT,1}]_{n_I}), \tag{4.64}$$

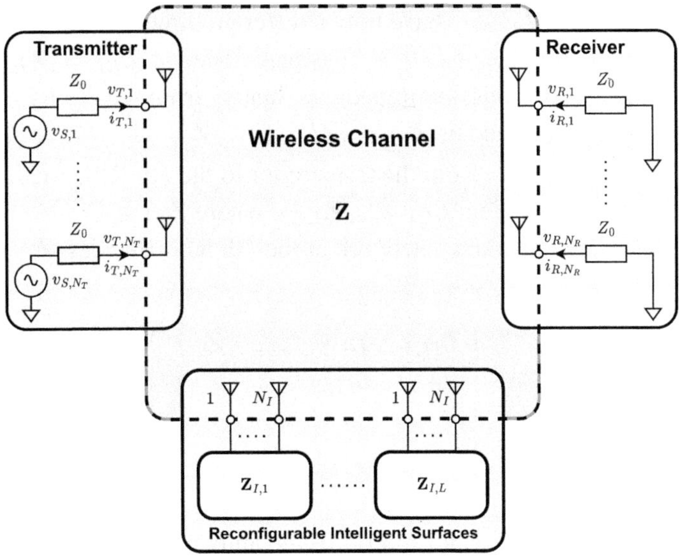

Figure 4.6:
Multi-RIS-aided MIMO system modeled with multiport network theory.

$$\theta_{\ell,n_I} = -\arg([\mathbf{b}_{\ell+1,\ell}]_{n_I}[\mathbf{a}_{\ell,\ell-1}]_{n_I}), \tag{4.65}$$

for $\ell = 2, \ldots, L-1$ and $n_I = 1, \ldots, N_I$. Accordingly, we have $|K_\ell|^2 = N_I^2$ for $\ell = 1, \ldots, L$, and the maximum channel gain and its expected value are given by

$$\mathrm{E}\left[|h|^2\right] = |h|^2 = N_I^{2L}, \tag{4.66}$$

in agreement with previous literature [27].

4.6 Multi-RIS-aided channel with specular reflection

We have introduced multi-RIS-aided systems and their channel model which is widely used in previous works. In this section, we present an EM-consistent channel model for multi-RIS systems accounting for the specular reflections at the RISs, as derived in [30].

4.6.1 Modeling

As in Sections 4.3 and 4.4, we model a multi-RIS MIMO system as an N-port network, where $N = N_T + LN_I + N_R$, with impedance matrix $\mathbf{Z} \in \mathbb{C}^{N \times N}$, as shown in Fig. 4.6. Considering the same assumptions as in Section 4.4, the channel of a multi-RIS-aided system is expressed as in (4.42), where $\mathbf{Z}_{RI}, \mathbf{Z}_{IT}, \mathbf{Z}_{I}, \mathbf{Z}_{II}$ have now a different dimensionality and meaning, as explained in [30]. Specifically, $\mathbf{Z}_{RI} \in \mathbb{C}^{N_R \times LN_I}$ is partitioned as $\mathbf{Z}_{RI} = [\mathbf{Z}_{RI,1}, \ldots, \mathbf{Z}_{RI,L}]$, and $\mathbf{Z}_{RI,\ell} \in \mathbb{C}^{N_R \times N_I}$ is the transmission impedance matrix from the ℓth RIS to the receiver, while $\mathbf{Z}_{IT} \in \mathbb{C}^{LN_I \times N_T}$ is partitioned as $\mathbf{Z}_{IT} = [\mathbf{Z}_{IT,1}^T, \ldots, \mathbf{Z}_{IT,L}^T]^T$, where $\mathbf{Z}_{IT,\ell} \in \mathbb{C}^{N_I \times N_T}$ is the transmission impedance matrix from the transmitter to the ℓth RIS. Also, $\mathbf{Z}_I \in \mathbb{C}^{LN_I \times LN_I}$ is a block diagonal matrix $\mathbf{Z}_I = \text{diag}(\mathbf{Z}_{I,1}, \ldots, \mathbf{Z}_{I,L})$, where $\mathbf{Z}_{I,\ell} \in \mathbb{C}^{N_I \times N_I}$ is the impedance matrix of the reconfigurable impedance network at the ℓth RIS. Besides, $\mathbf{Z}_{II} \in \mathbb{C}^{LN_I \times LN_I}$ can be partitioned as

$$\mathbf{Z}_{II} = \begin{bmatrix} \mathbf{Z}_{II,1} & \mathbf{Z}_{1,2} & \ldots & \mathbf{Z}_{1,L} \\ \mathbf{Z}_{2,1} & \mathbf{Z}_{II,2} & \ldots & \mathbf{Z}_{2,L} \\ \vdots & \vdots & \ddots & \vdots \\ \mathbf{Z}_{L,1} & \mathbf{Z}_{L,2} & \ldots & \mathbf{Z}_{II,L} \end{bmatrix}, \tag{4.67}$$

where $\mathbf{Z}_{II,\ell} \in \mathbb{C}^{N_I \times N_I}$ is the impedance matrix of the antenna array at the ℓth RIS, while $\mathbf{Z}_{i,j} \in \mathbb{C}^{N_I \times N_I}$ is the transmission impedance matrix from the jth RIS to the ith RIS.

To obtain an expression of the wireless channel of the multi-RIS-aided system in Fig. 4.5, we make the following simplifying assumptions:

1. We assume that the channel between the transmitter and RIS $2, \ldots, L$ is blocked, i.e., $\mathbf{Z}_{IT,\ell} = \mathbf{0}$, for $\ell = 2, \ldots, L$; and the channel between the receiver and RIS $1, \ldots, L-1$ is blocked, i.e., $\mathbf{Z}_{RI,\ell} = \mathbf{0}$, for $\ell = 1, \ldots, L-1$.
2. We assume large enough distances between the L RISs allowing us to set to zero the feedback links between the RISs, i.e., $\mathbf{Z}_{i,j} = \mathbf{0}$, $\forall i < j$ [9].
3. We assume that the wireless link between the ith and jth RIS is completely obstructed if $i - j \geq 2$, i.e., $\mathbf{Z}_{i,j} = \mathbf{0}$ if $i - j \geq 2$.
4. We assume that all the RIS antennas are all perfectly matched to Z_0 and no mutual coupling at the L RISs, i.e., $\mathbf{Z}_{II,\ell} = Z_0\mathbf{I}$, for $\ell = 1, \ldots, L$.

By applying assumptions 1–4 to the channel model in (4.42), we obtain the channel of a multi-RIS aided system accounting for the specular reflection at the RISs as

$$\begin{aligned} \mathbf{H}' = \frac{1}{2Z_0}\Bigg(&\mathbf{Z}_{RT} - (-1)^{L-1}\mathbf{Z}_{RI,L}\left(\mathbf{Z}_{I,L} + Z_0\mathbf{I}\right)^{-1} \\ &\times \prod_{\ell=L-1}^{1}\left(\mathbf{Z}_{\ell+1,\ell}\left(\mathbf{Z}_{I,\ell} + Z_0\mathbf{I}\right)^{-1}\right)\mathbf{Z}_{IT,1}\Bigg), \end{aligned} \tag{4.68}$$

following the steps in [30], as a function of the RIS impedance matrices $\mathbf{Z}_{I,\ell}$.

In alternative to the impedance matrices $\mathbf{Z}_{I,\ell}$, the L RISs can also be represented through their scattering matrices $\boldsymbol{\Theta}_\ell \in \mathbb{C}^{N_I \times N_I}$, as more popularly used in literature on multi-RIS systems [25–29]. To express the channel in (4.68) as a function of the RIS scattering matrices $\boldsymbol{\Theta}_\ell$, we exploit the relationship

$$\left(\mathbf{Z}_{I,\ell} + Z_0\mathbf{I}\right)^{-1} = -\frac{1}{2Z_0}\left(\boldsymbol{\Theta}_\ell - \mathbf{I}\right), \tag{4.69}$$

similar to (4.22), and introduce the wireless channels

$$\mathbf{H}_{RT} = \frac{\mathbf{Z}_{RT}}{2Z_0},\ \mathbf{H}_{RI,L} = \frac{\mathbf{Z}_{RI,L}}{2Z_0},\ \mathbf{H}_{IT,1} = \frac{\mathbf{Z}_{IT,1}}{2Z_0},\ \mathbf{H}_{\ell+1,\ell} = \frac{\mathbf{Z}_{\ell+1,\ell}}{2Z_0}, \tag{4.70}$$

similar to (4.23). Thus, by substituting (4.69) and (4.70) into (4.68), we obtain

$$\mathbf{H}' = \mathbf{H}_{RT} + \mathbf{H}_{RI,L}\left(\boldsymbol{\Theta}_L - \mathbf{I}\right) \prod_{\ell=L-1}^{1} \left(\mathbf{H}_{\ell+1,\ell}\left(\boldsymbol{\Theta}_\ell - \mathbf{I}\right)\right) \mathbf{H}_{IT,1}, \tag{4.71}$$

which differs from the widely used channel model $\mathbf{H}$ in (4.58) because the terms $\boldsymbol{\Theta}_\ell$ are now replaced by $(\boldsymbol{\Theta}_\ell - \mathbf{I})$ to account for the specular reflection at the RISs [30]. In the following, we numerically quantify the difference between $\mathbf{H}'$ and $\mathbf{H}$ for multi-RIS-aided SISO systems by comparing their channel gain scaling laws.

4.6.2 Optimization and scaling law

We consider a multi-RIS-aided SISO system, with a completely obstructed direct link, i.e., $h_{RT} = 0$, such that (4.71) boils down to

$$h' = \mathbf{h}_{RI,L}\left(\boldsymbol{\Theta}_L - \mathbf{I}\right) \prod_{\ell=L-1}^{1} \left(\mathbf{H}_{\ell+1,\ell}\left(\boldsymbol{\Theta}_\ell - \mathbf{I}\right)\right) \mathbf{h}_{IT,1}, \tag{4.72}$$

and the channel gain maximization problem is given by

$$\max_{\boldsymbol{\Theta}_1,\ldots,\boldsymbol{\Theta}_L} \left| \mathbf{h}_{RI,L}\left(\boldsymbol{\Theta}_L - \mathbf{I}\right) \prod_{\ell=L-1}^{1} \left(\mathbf{H}_{\ell+1,\ell}\left(\boldsymbol{\Theta}_\ell - \mathbf{I}\right)\right) \mathbf{h}_{IT,1} \right|^2 \text{ such that (4.2) or (4.3)}, \tag{4.73}$$

where (4.2) and (4.3) are considered for D-RISs and BD-RISs, respectively.

Under LoS channels, there exists a globally optimal solution to problem (4.73), as proposed in [30]. More precisely, when the channels $\mathbf{H}_{\ell+1,\ell}$ are LoS channels, i.e., $\mathbf{H}_{\ell+1,\ell} = \mathbf{a}_{\ell+1,\ell}\mathbf{b}_{\ell+1,\ell}^T$ as discussed in Section 4.5, (4.72) can be rewritten as

$$h' = \mathbf{h}_{RI,L}\left(\boldsymbol{\Theta}_L - \mathbf{I}\right)\mathbf{a}_{L,L-1} \prod_{\ell=L-1}^{2} \left(\mathbf{b}_{\ell+1,\ell}^T\left(\boldsymbol{\Theta}_\ell - \mathbf{I}\right)\mathbf{a}_{\ell,\ell-1}\right) \mathbf{b}_{2,1}^T\left(\boldsymbol{\Theta}_1 - \mathbf{I}\right)\mathbf{h}_{IT,1}, \tag{4.74}$$

which can be seen as a product of L scalars $K'_\ell \in \mathbb{C}$ such that

$$h' = \prod_{\ell=L}^{1} K'_\ell, \tag{4.75}$$

where $K'_L = \mathbf{h}_{RI,L}(\boldsymbol{\Theta}_L - \mathbf{I})\mathbf{a}_{L,L-1}$, $K'_1 = \mathbf{b}_{2,1}^T(\boldsymbol{\Theta}_1 - \mathbf{I})\mathbf{h}_{IT,1}$, and $K'_\ell = \mathbf{b}_{\ell+1,\ell}^T(\boldsymbol{\Theta}_\ell - \mathbf{I})\mathbf{a}_{\ell,\ell-1}$, for $\ell = 2, \dots, L-1$. Thus, each $\boldsymbol{\Theta}_\ell$ can be individually and globally optimized to maximize $|K'_\ell|^2$, respectively, by setting

$$\theta_{L,n_I} = -\arg(\mathbf{h}_{RI,L}\mathbf{a}_{L,L-1}) - \arg([\mathbf{h}_{RI,L}]_{n_I}[\mathbf{a}_{L,L-1}]_{n_I}), \tag{4.76}$$

$$\theta_{1,n_I} = -\arg(\mathbf{b}_{2,1}^T\mathbf{h}_{IT,1}) - \arg([\mathbf{b}_{2,1}]_{n_I}[\mathbf{h}_{IT,1}]_{n_I}), \tag{4.77}$$

$$\theta_{\ell,n_I} = -\arg(\mathbf{b}_{\ell+1,\ell}^T\mathbf{a}_{\ell,\ell-1}) - \arg([\mathbf{b}_{\ell+1,\ell}]_{n_I}[\mathbf{a}_{\ell,\ell-1}]_{n_I}), \tag{4.78}$$

for $\ell = 2, \dots, L-1$ and $n_I = 1, \dots, N_I$, as explained in [30]. With this globally optimal solution, the achievable channel gain $|h'|^2$ is given by

$$\begin{aligned} |h'|^2 &= \left(\left|\mathbf{h}_{RI,L}\mathbf{a}_{L,L-1}\right| + N_I\right)^2 \\ &\times \prod_{\ell=L-1}^{2} \left(\left|\mathbf{b}_{\ell+1,\ell}^T\mathbf{a}_{\ell,\ell-1}\right| + N_I\right)^2 \left(\left|\mathbf{b}_{2,1}^T\mathbf{h}_{IT,1}\right| + N_I\right)^2, \end{aligned} \tag{4.79}$$

depending on the channels $\mathbf{h}_{RI,L}$, $\mathbf{H}_{\ell+1,\ell}$, for $\ell = 1, \dots, L-1$, and $\mathbf{h}_{IT,1}$.

To obtain the expectation $\mathrm{E}[|h'|^2]$ in closed form, we regard the dot products $\mathbf{h}_{RI,L}\mathbf{a}_{L,L-1}$, $\mathbf{b}_{\ell+1,\ell}^T\mathbf{a}_{\ell,\ell-1}$, and $\mathbf{b}_{2,1}^T\mathbf{h}_{IT,1}$ as independent sums of N_I i.i.d. random variables with mean 0 and variance 1, which can be approximated as $\mathcal{CN}(0, N_I)$ by the central limit theorem, for large values of N_I. Thus, we obtain

$$\mathrm{E}\left[|h'|^2\right] = \mathrm{E}\left[(|c| + N_I)^2\right]^L, \tag{4.80}$$

where $c \sim \mathcal{CN}(0, N_I)$. By exploiting $\mathrm{E}[|c|] = \sqrt{\frac{\pi}{4}N_I}$ and $\mathrm{E}[|c|^2] = N_I$, which follow from the moments of the Rayleigh distribution, we can write

$$\mathrm{E}\left[|h'|^2\right] = \left(N_I^2 + \sqrt{\pi N_I}N_I + N_I\right)^L, \tag{4.81}$$

giving the scaling law of the channel gain of the EM-consistent model accounting for the specular reflection at the RISs.

To compare the scaling laws of the widely used model $\mathrm{E}[|h|^2]$ in (4.66) and the model with specular reflection $\mathrm{E}[|h'|^2]$ in (4.81), we consider their relative difference in (4.37). By

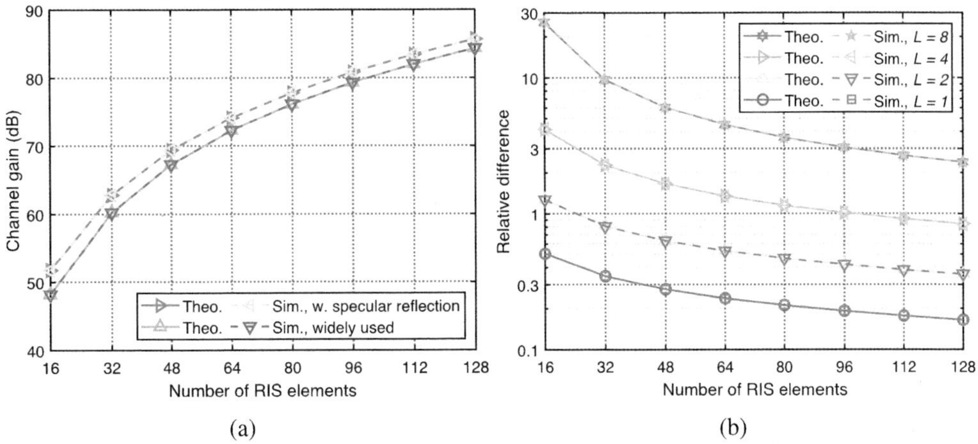

Figure 4.7:
(a) Channel gains of the widely used channel model and the channel model with specular reflection for $L = 2$ RISs, and (b) their relative difference.

substituting (4.66) and (4.81) into (4.37), we obtain

$$\begin{aligned} \delta &= \frac{\left(N_I + \sqrt{\pi N_I} + 1\right)^L - N_I^L}{N_I^L} = \left(1 + \frac{\sqrt{\pi N_I} + 1}{N_I}\right)^L - 1 \\ &\approx L\frac{\sqrt{\pi N_I} + 1}{N_I} \approx L\frac{\sqrt{\pi}}{\sqrt{N_I}}, \end{aligned} \tag{4.82}$$

as $N_I \to \infty$, where we used the first-order Taylor approximation of the binomial series $(1 + x)^L \approx (1 + Lx)$ as $x \to 0$, giving the relative difference in closed form as a function of the number of RISs L and their elements N_I. Interestingly, this shows analytically that the impact of the specular reflection increases linearly with the number of RIS in the system.

In Fig. 4.7, we consider multi-RIS SISO systems and report the scaling law of the channel gain for the widely used model (4.59) and the model accounting for specular reflection (4.72). We observe from Fig. 4.7(a) that accounting for the specular reflection results in higher channel gain, since the specular reflection term can be exploited constructively. In addition, Fig. 4.7(b) shows that the relative difference between the widely used model and the model with specular reflection increases with the number of RISs L, reaching a significant level of mismatch for $L = 8$ RISs, as the EM-consistent channel gain is 30 times higher than the widely used channel gain when the number of RIS elements is $N_I = 16$.

4.7 Multi-RIS-aided channel with mutual coupling

For completeness, we consider in this section a multi-RIS-aided system in the presence of mutual coupling at the RISs. The lengthy derivations are omitted since they directly follow from the discussions carried out for RIS-aided systems with mutual coupling (see Section 4.4 and [24]) and multi-RIS-aided systems (see Section 4.6 and [30]). It can be shown that the EM-consistent channel model for a multi-RIS-aided system with mutual coupling at the RISs is given by

$$\mathbf{H}'' = \mathbf{H}_{RT} + (-2)^L \mathbf{H}_{RI,L} (\mathbf{\Lambda}_L + \mathbf{H}_{II})^{-1} \prod_{\ell=L-1}^{1} \left(\mathbf{H}_{\ell+1,\ell} (\mathbf{\Lambda}_\ell + \mathbf{H}_{II})^{-1} \right) \mathbf{H}_{IT,1}, \quad (4.83)$$

where $\mathbf{\Lambda}_\ell \in \mathbb{C}^{N_I \times N_I}$ is the load matrix of the ℓth RIS, defined as $\mathbf{\Lambda}_\ell = \mathbf{Z}_{I,\ell}/Z_0$ as a function of the ℓth RIS impedance matrix $\mathbf{Z}_{I,\ell}$, and we assumed that all the RISs have the same mutual coupling matrix $\mathbf{H}_{II}$.

In a SISO system with obstructed direct link, i.e., $h_{RT} = 0$, the channel (4.83) boils down to

$$h'' = (-2)^L \mathbf{h}_{RI,L} (\mathbf{\Lambda}_L + \mathbf{H}_{II})^{-1} \prod_{\ell=L-1}^{1} \left(\mathbf{H}_{\ell+1,\ell} (\mathbf{\Lambda}_\ell + \mathbf{H}_{II})^{-1} \right) \mathbf{h}_{IT,1}, \quad (4.84)$$

and we aim at optimizing the matrices $\mathbf{\Lambda}_\ell$, for $\ell = 1, \ldots, L$, to maximize the channel gain $|h''|^2$. Under LoS channels, h'' in (4.84) is given by a product of scalars as h in (4.59) and h' in (4.72), whose moduli can be individually maximized. Thus, there exists a global optimal solution for optimizing L BD-RISs to maximize $|h''|^2$, consisting of individually optimizing each BD-RIS as done in Section 4.4. Speculating that the scaling law of the channel gain with mutual coupling (4.57) is valid also for LoS channels,[2] the scaling law of the multi-RIS-aided channel model accounting for mutual coupling writes as

$$\mathrm{E}\left[|h''|^2\right] = \left(\mathrm{Tr}\left(\Re\{\mathbf{H}_{II}\}^{-2}\right) + \mathrm{Tr}\left(\Re\{\mathbf{H}_{II}\}^{-1}\right)^2 + \sqrt{\pi \mathrm{Tr}\left(\Re\{\mathbf{H}_{II}\}^{-2}\right)} \mathrm{Tr}\left(\Re\{\mathbf{H}_{II}\}^{-1}\right) \right)^L, \quad (4.85)$$

which is given by (4.57) raised to the Lth power.

To summarize the content of this chapter, we report in Table 4.2 the derived channel models for RIS-aided and multi-RIS-aided systems. Furthermore, we report in Table 4.3 the scaling laws of the channel gains of the derived models in the case of globally optimized BD-RISs.

2 Despite (4.57) being theoretically derived for Rayleigh distributed channels in [24], it can be shown numerically that it is also accurate for LoS channels.

Table 4.2: RIS-aided MIMO channel models, with $\mathbf{H}_{RT} = \mathbf{0}$.

Channel model	$\mathbf{H}$
RIS-aided channel widely used	$\mathbf{H}_{RI}\boldsymbol{\Theta}\mathbf{H}_{IT}$
RIS-aided channel with specular reflection	$\mathbf{H}_{RI}(\boldsymbol{\Theta} - \mathbf{I})\mathbf{H}_{IT}$
RIS-aided channel with mutual coupling	$-2\mathbf{H}_{RI}(\boldsymbol{\Lambda} + \mathbf{H}_{II})^{-1}\mathbf{H}_{IT}$
Multi-RIS-aided channel widely used	$\mathbf{H}_{RI,L}\boldsymbol{\Theta}_L \prod_{\ell=L-1}^{1}(\mathbf{H}_{\ell+1,\ell}\boldsymbol{\Theta}_\ell)\mathbf{H}_{IT,1}$
Multi-RIS-aided channel with specular reflection	$\mathbf{H}_{RI,L}(\boldsymbol{\Theta}_L - \mathbf{I})\prod_{\ell=L-1}^{1}(\mathbf{H}_{\ell+1,\ell}(\boldsymbol{\Theta}_\ell - \mathbf{I}))\mathbf{H}_{IT,1}$
Multi-RIS-aided channel with mutual coupling	$(-2)^L\mathbf{H}_{RI,L}(\boldsymbol{\Lambda}_L + \mathbf{H}_{II})^{-1}\prod_{\ell=L-1}^{1}\left(\mathbf{H}_{\ell+1,\ell}(\boldsymbol{\Lambda}_\ell + \mathbf{H}_{II})^{-1}\right)\mathbf{H}_{IT,1}$

Table 4.3: Scaling laws of RIS-aided SISO channel gains, with $h_{RT} = 0$.

Channel model	$\mathrm{E}[\lvert h\rvert^2]$
RIS-aided channel widely used	N_I^2
RIS-aided channel with specular reflection	$N_I^2 + \sqrt{\pi N_I}N_I + N_I$
RIS-aided channel with mutual coupling	$\mathrm{Tr}(\Re\{\mathbf{H}_{II}\}^{-1})^2 + \sqrt{\pi \mathrm{Tr}(\Re\{\mathbf{H}_{II}\}^{-2})}\mathrm{Tr}(\Re\{\mathbf{H}_{II}\}^{-1}) + \mathrm{Tr}(\Re\{\mathbf{H}_{II}\}^{-2})$
Multi-RIS-aided channel widely used	N_I^{2L}
Multi-RIS-aided channel with specular reflection	$\left(N_I^2 + \sqrt{\pi N_I}N_I + N_I\right)^L$
Multi-RIS-aided channel with mutual coupling	$\left(\mathrm{Tr}(\Re\{\mathbf{H}_{II}\}^{-1})^2 + \sqrt{\pi \mathrm{Tr}(\Re\{\mathbf{H}_{II}\}^{-2})}\mathrm{Tr}(\Re\{\mathbf{H}_{II}\}^{-1}) + \mathrm{Tr}(\Re\{\mathbf{H}_{II}\}^{-2})\right)^L$

4.8 Conclusion

This chapter has examined electromagnetically consistent models for RIS-aided and multi-RIS-aided wireless communication systems. We introduced the widely used RIS-aided channel model and its scaling laws for D-RIS and BD-RIS. We then explored two more EM-consistent models that account for the specular reflection at the RIS (an effect that is always present regardless of the RIS reconfiguration) and mutual coupling between the RIS elements. We analyzed the scaling laws of these models, highlighting the importance of considering these factors for accurate system performance evaluation. We also extended the analysis to multi-RIS-aided systems, again comparing the widely used model with EM-consistent models. Analytical results and numerical simulations show that the effect of the specular reflection is amplified in multi-RIS-aided systems, emphasizing the need for more sophisticated channel models that incorporate real-world physical phenomena to accurately model RIS-aided systems.

References

[1] E. Basar, M. Di Renzo, J. De Rosny, M. Debbah, M.S. Alouini, R. Zhang, Wireless communications through reconfigurable intelligent surfaces, IEEE Access 7 (2019) 116753–116773.

[2] Q. Wu, S. Zhang, B. Zheng, C. You, R. Zhang, Intelligent reflecting surface-aided wireless communications: a tutorial, IEEE Transactions on Communications 69 (5) (2021) 3313–3351.

[3] S. Shen, B. Clerckx, R. Murch, Modeling and architecture design of reconfigurable intelligent surfaces using scattering parameter network analysis, IEEE Transactions on Wireless Communications 21 (2) (2022) 1229–1243.

[4] H. Li, S. Shen, M. Nerini, B. Clerckx, Reconfigurable intelligent surfaces 2.0: beyond diagonal phase shift matrices, IEEE Communications Magazine 62 (3) (2024) 102–108.

[5] H. Li, S. Shen, B. Clerckx, Beyond diagonal reconfigurable intelligent surfaces: from transmitting and reflecting modes to single-, group-, and fully-connected architectures, IEEE Transactions on Wireless Communications 22 (4) (2023) 2311–2324.

[6] M. Nerini, S. Shen, B. Clerckx, Closed-form global optimization of beyond diagonal reconfigurable intelligent surfaces, IEEE Transactions on Wireless Communications 23 (2) (2024) 1037–1051.

[7] M. Nerini, S. Shen, H. Li, B. Clerckx, Beyond diagonal reconfigurable intelligent surfaces utilizing graph theory: modeling, architecture design, and optimization, IEEE Transactions on Wireless Communications 23 (8) (2024) 9972–9985.

[8] J.W. Wallace, M.A. Jensen, Mutual coupling in MIMO wireless systems: a rigorous network theory analysis, IEEE Transactions on Wireless Communications 3 (4) (2004) 1317–1325.

[9] M.T. Ivrlač, J.A. Nossek, Toward a circuit theory of communication, IEEE Transactions on Circuits and Systems. I, Regular Papers 57 (7) (2010) 1663–1683.

[10] M.T. Ivrlač, J.A. Nossek, The multiport communication theory, IEEE Circuits and Systems Magazine 14 (3) (2014) 27–44.

[11] G. Gradoni, M. Di Renzo, End-to-End Mutual coupling aware communication model for reconfigurable intelligent surfaces: an electromagnetic-compliant approach based on mutual impedances, IEEE Wireless Communications Letters 10 (5) (2021) 938–942.

[12] R.J. Williams, P. Ramírez-Espinosa, J. Yuan, E. de Carvalho, Electromagnetic based communication model for dynamic metasurface antennas, IEEE Transactions on Wireless Communications 21 (10) (2022) 8616–8630.

[13] D.M. Pozar, Microwave Engineering, John Wiley & Sons, 2011.

[14] J.A. Nossek, D. Semmler, M. Joham, W. Utschick, Physically consistent modelling of wireless links with reconfigurable intelligent surfaces using multiport network analysis, IEEE Wireless Communications Letters 13 (8) (2024) 2240–2244.

[15] A. Abrardo, A. Toccafondi, M. Di Renzo, Design of reconfigurable intelligent surfaces by using s-parameter multiport network theory – optimization and full-wave validation, IEEE Transactions on Wireless Communications 23 (11) (2024) 17084–17102.

[16] M. Nerini, S. Shen, H. Li, M. Di Renzo, B. Clerckx, A universal framework for multiport network analysis of reconfigurable intelligent surfaces, IEEE Transactions on Wireless Communications 23 (10) (2024) 14575–14590.

[17] X. Qian, M. Di Renzo, Mutual coupling and unit cell aware optimization for reconfigurable intelligent surfaces, IEEE Wireless Communications Letters 10 (6) (2021) 1183–1187.

[18] A. Abrardo, D. Dardari, M. Di Renzo, X. Qian, MIMO interference channels assisted by reconfigurable intelligent surfaces: mutual coupling aware sum-rate optimization based on a mutual impedance channel model, IEEE Wireless Communications Letters 10 (12) (2021) 2624–2628.

[19] M. Akrout, F. Bellili, A. Mezghani, J.A. Nossek, Physically consistent models for intelligent reflective surface-assisted communications under mutual coupling and element size constraint, in: 57th Asilomar Conference on Signals, Systems, and Computers, 2023, pp. 1589–1594.
[20] D. Semmler, J.A. Nossek, M. Joham, W. Utschick, Performance analysis of systems with coupled and decoupled RISs, in: 2024 19th International Symposium on Wireless Communication Systems (ISWCS), 2024, pp. 1–6.
[21] P. Mursia, S. Phang, V. Sciancalepore, G. Gradoni, M. Di Renzo, SARIS: Scattering aware reconfigurable intelligent surface model and optimization for complex propagation channels, IEEE Wireless Communications Letters 12 (11) (2023) 1921–1925.
[22] P. Zheng, R. Wang, A. Shamim, T.Y. Al-Naffouri, Mutual coupling in RIS-aided communication: model training and experimental validation, IEEE Transactions on Wireless Communications 23 (11) (2024) 17174–17188.
[23] H. Li, S. Shen, M. Nerini, M. Di Renzo, B. Clerckx, Beyond diagonal reconfigurable intelligent surfaces with mutual coupling: modeling and optimization, IEEE Communications Letters 28 (4) (2024) 937–941.
[24] M. Nerini, H. Li, B. Clerckx, Global optimal closed-form solutions for intelligent surfaces with mutual coupling: is mutual coupling detrimental or beneficial?, arXiv preprint, arXiv:2411.04949, 2024.
[25] Y. Han, S. Zhang, L. Duan, R. Zhang, Cooperative double-IRS aided communication: beamforming design and power scaling, IEEE Wireless Communications Letters 9 (8) (2020) 1206–1210.
[26] B. Zheng, C. You, R. Zhang, Double-IRS assisted multi-user MIMO: cooperative passive beamforming design, IEEE Transactions on Wireless Communications 20 (7) (2021) 4513–4526.
[27] W. Mei, R. Zhang, Cooperative beam routing for multi-IRS aided communication, IEEE Wireless Communications Letters 10 (2) (2021) 426–430.
[28] C. Huang, et al., Multi-hop RIS-empowered terahertz communications: a DRL-based hybrid beamforming design, IEEE Journal on Selected Areas in Communications 39 (6) (2021) 1663–1677.
[29] W. Mei, B. Zheng, C. You, R. Zhang, Intelligent reflecting surface-aided wireless networks: from single-reflection to multireflection design and optimization, Proceedings of the IEEE 110 (9) (2022) 1380–1400.
[30] M. Nerini, G. Gradoni, B. Clerckx, Physics-compliant modeling and scaling laws of multi-RIS aided systems, arXiv preprint, arXiv:2410.13089, 2024.

CHAPTER 5

The design and performance study of antenna array elements with more than two independent ports

Rui Ni[a], Wen Tong[a], Ross Murch[b], Frankie Chiu[b], and Qian Zhu[c]

[a]Advanced Wireless Technology Lab, Huawei Technologies Canada Co., Ltd. Kanata, Ontario, Canada
[b]Department of Electronic and Computer Engineering, The Hong Kong University of Science and Technology, Hong Kong, Hong Kong [c]Advanced Wireless Technology Lab, Huawei Technologies Co., Ltd. Shenzhen, Guangdong, China

5.1 Challenges in the design of MIMO antennas

5.1.1 Review of multiport antenna design principles

Interest in multiport antennas exploded with the development of multiple-input multiple-output (MIMO) systems in the late 1990s [1–4]. These systems could directly increase capacity by a factor equal to the minimum number of antennas at the transmitter or receiver. It was a significant development as it allowed very large increases in capacity that far exceeded previous approaches based on diversity only [5,6]. One bottleneck, however, was that it was difficult to include many antennas in handsets and therefore multiuser MIMO (MU-MIMO) was proposed in 2004 [7,8]. MU-MIMO could obtain the same capacity increases as with MIMO, but with only one or two antennas per handset. This allowed practical exploitation of MIMO in wireless communication systems. Further development of MU-MIMO increased the number of base station antennas massively and led to massive MIMO [9,10] in around 2010. With the development of MU-MIMO in 2004 and massive MIMO in 2010, MIMO system design became a mainstay of all wireless communication design and is now included in Wi-Fi, 4G, 5G, and will play an important role in 6G.

With the development of MIMO systems, a significant need was created for the design of multiport antennas that can meet the requirements of the proposed MIMO systems. One of the first integrated designs for a multiport antenna where both ports were designed together was provided in [11]. Since then, many multiport designs have been proposed, and some seminal designs include [12] and [13].

Electromagnetic Signal and Information Theory
https://doi.org/10.1016/B978-0-44-336354-2.00012-5

Copyright © 2027 Elsevier Inc. All rights are reserved, including those for text and data mining, AI training, and similar technologies.

One important consideration in the design of multiport antennas is whether it is a 2-port antenna or >2-port antenna. In general, 2-port antennas are more straightforward to design. When there are only two antennas, reducing mutual coupling from only one adjacent antenna is required, and a variety of unique techniques have been developed. However, in two-dimensions, when one antenna can be surrounded by 8 antennas, mutual coupling from all 8 antennas must be considered. For example, if mutual coupling from one adjacent antenna is −10 dB, then, when there are 8 adjacent antennas, the effective coupling also increases by a factor of 8. This significantly reduces efficiency, and effectively makes the coupling equivalent to −1 dB. In addition, many of the structures utilized in 2-port antenna design cannot be used in >2-port antenna design because they have been uniquely developed for providing isolation to only one other antenna and not to up to 8 antennas surrounding it.

In this book chapter we focus on multiport antennas with >2 ports. This is the most challenging scenario and also the most useful for the development of compact base station antenna arrays for MIMO.

In general, there are two approaches for designing multiport antennas. The first is to rely on fundamental antenna characteristics to accommodate multiple ports in one single radiating element. For example, using orthogonal polarizations or characteristic modes in producing one integrated element with multiple ports [11,12]. This approach can be extended to more ports by packing elements together and relying on their separations to keep coupling within required limits. This is an intrinsic and natural design concept. The second approach is to make use of certain artificial decoupling techniques to suppress mutual coupling [13]. The advantage is that a single design technique can be applied to different antenna types, but it requires additional structures or components to implement. In the following, examples of the two approaches will be provided. In addition, it should also be remembered that the two approaches can also be utilized together to obtain further designs.

5.1.2 Fundamental MIMO antenna designs

The tree diagram in Fig. 5.1 illustrates various approaches to MIMO antenna designs. On the left of the diagram, the techniques utilize the first approach described above, using fundamental properties of radiating elements such as polarization. It starts with 2-port antennas having low mutual coupling which can be achieved in three different ways utilizing different diversity:

- **Polarization diversity.** When a current is flowing on a patch antenna in a vertical and horizontal pair or rotating in a clockwise and anticlockwise pair, the orthogonality can ensure low mutual coupling between two ports. Another example is when two identical half-wavelength slot antennas are interlaced together, forming a "cross" structure, and

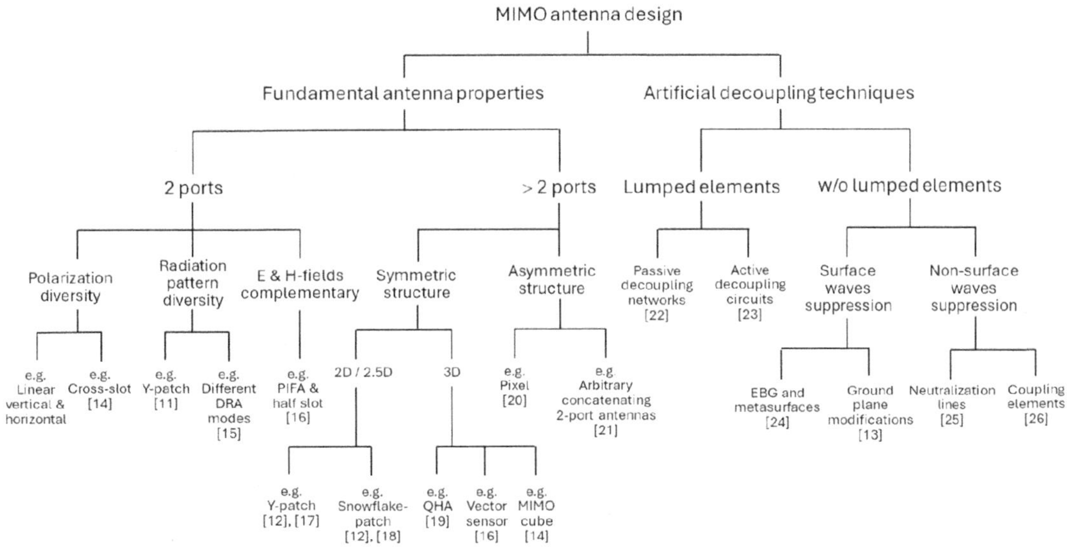

Figure 5.1:
Hierarchy chart of MIMO antenna design approaches.

share an entire common slot along their centerline, the mutual coupling can be maintained low, due to orthogonal feeds, and highly polarized [14].

- **Radiation pattern diversity.** A 2-port Y-patch antenna is reported to have low mutual coupling, though the magnitudes of the radiation patterns from Ports 1 and 2 are similar. However, due to the antenna symmetry, the phases are sufficiently different to produce a low cross-correlation result [11]. Another example is that of a dielectric resonator antenna (DRA) consisting of a block of substrate that can be excited by different ports for generating different modes of radiation. If two modes of radiation are omni-directional in the horizontal plane and broadside direction, respectively, the two antenna ports will exhibit low mutual coupling [15].
- **Complementary E- & H-fields.** A narrow half-wavelength strip electric antenna, stacked directly above and in parallel with a half-wavelength slot magnetic dipole can form a complementary pair. The resultant magnitude patterns are essentially similar, but have orthogonal field diversity [16].

When an antenna embedded with three ports or more is considered, the antenna is more difficult realize in a 2-dimensional (2D) structure since, by intuition, the spatial freedom appears to have been already been exploited. However, antennas that are compact with 3, 4, or more ports have been reported by utilizing the theory of characteristic modes (TCM). On the other hand, it is shown that a 3-dimensional (3D) structure can provide more design freedom to accommodate multiple ports with low mutual coupling. Furthermore, by introducing a structural symmetry in 2D, 2.5D, or 3D design, multiple multiport antennas are more convenient to be concatenated (scalable), forming a larger array including:

- **Symmetric 2D structure.** A Y-shaped patch was originally designed for supporting two-port radiation. Recently, it was found that the Y-shaped patch can also support three characteristic modes, but three ring-slot-coupled probe feeds, resembling capacitive feedings, are required to excite the three modes properly [12,17].
- **Symmetric 2.5D structure.** A snowflake-shaped patch can be thought of as two Y-shaped patches overlapped, but with 180° rotation. A patch with 6 radiating branches helps generate more broadside radiation patterns by the three antenna ports. To reduce the lateral aperture size, the branches can be folded downward, towards the ground plane [12,18].
- **Symmetric 3D structure.** Quadrifilar helix antenna (QHA) is a type of antenna that consists of four helical arms wound around a central axis. Four independent ports can be connected to the arms, providing four separated closely spaced antenna elements. Additional capacitive strips (rings) can be loaded to the antennas to reduce the overall physical height [19]. In addition, a vector sensor antenna consists of three orthogonal half-length slots, and a planar inverted-F antenna (PIFA) pair is capable of receiving electric and magnetic field vectors in three orthogonal directions [16]. It is also natural to utilize vector sensors to form MIMO cubes by rotating and aligning antenna elements in a cubic structure [14].
- **Asymmetric structure.** Pixel antennas are regarded as highly reconfigurable elements which consist of many configurations to select from. The N-port characteristic mode analysis (CMA) has been successfully utilized to characterize the maximum number of independent or orthogonal resonant modes that can be formed in the limited space constrained by the size of the pixel antenna [20]. Although symmetry constraint can also be introduced in objective function, it is not a compulsory optimization process. Another example is that a 20-port antenna can be formed by concatenating ten canonical 2-port elements [21]. It is shown that the antenna elements are repeated symmetrically with respect to the vertical centerline of the structure; however, arbitrary elements arrangement is also possible.

5.1.3 Artificial decoupling techniques

The second approach for designing multiport antennas is to introduce artificial decoupling techniques on multiple conventional antennas. These are represented by the right box in the diagram of Fig. 5.1. The technique is to directly mitigate mutual coupling, therefore, more ordinary antennas can be placed in a compact area. These techniques can be classified as utilizing passive and active decoupling networks, neutralization lines, electromagnetic bandgap (EBG) structures and metasurfaces, ground plane modifications, and coupling elements.

- **Passive decoupling networks.** These involve the use of resistors, capacitors, and inductors to create a network that helps isolate ports. For example, a resistive network can absorb some of the energy from one port, preventing it from affecting another [22].
- **Active decoupling circuits.** These circuits utilize amplifiers or other active components to manage and isolate signals. Active components can dynamically adjust to changes in the environment or signal conditions, providing adaptive isolation [23].
- **EBG structures and metasurfaces.** These are artificially engineered materials, usually having a mushroom-like structure, to create specific frequency bands where electromagnetic waves (EM waves) cannot propagate or to provide precise phase control of the waves, enabling the design of antennas with specific radiation patterns. By suppressing surface waves or controlling radiation patterns, the antenna coupling effects can be reduced [24].
- **Ground plane modifications.** Incorporating ground planes can enhance isolation by reflecting and absorbing stray EM waves or stopping current flowing on the ground plane from one antenna to another. A well-designed ground plane can also help direct the fields away from neighboring ports [13].
- **Neutralization lines.** The neutralization line is typically implemented as a transmission line or a specific geometric structure that connects the feed points of the antennas. Its length, width, and other parameters are specially designed to counteract the coupling effects between closely spaced antenna elements [25].
- **Coupling elements.** The idea is to use field cancellation to enhance isolation by placing a coupling element that artificially creates an additional coupling path between antenna elements. If the two EM waves are identical in magnitude and out of phase, the fields can be cancelled effectively [26].

5.1.4 Other considerations in designing MIMO antenna array

When the number of ports in an MIMO antenna array is large, more antenna design challenges need to be addressed. First, the system can employ a substantially larger number of antennas compared to traditional MIMO systems, which may be up to a thousand ports. From the antenna design point of view, it is neither practical nor efficient to simulate and optimize the entire antenna array in a single operation. Instead, it is desirable to focus on some multiport antenna structures which are scalable. For example, a multiport antenna with regular hexagonal structure, such as snowflake-shaped antenna, can be thought of as a unit cell that can be easily concatenated to form a large surface.

Managing the physical dimensions and weight of a large antenna array can also be very difficult, particularly for deployment on towers or rooftops. An efficient size reduction method is to utilize a high port number antenna as the unit cell. The more ports accommodated in each unit cell, the smaller the antenna array form factor can be achieved. However, the trade-off is that the severity of intraelement coupling is usually proportional to the number of ports in each unit cell. Special design techniques, such as the utilization of TCM searching for more orthogonal characteristic modes in a given antenna area, are usually required to balance both aspects.

When many ports are considered, the antenna feed network could be intricate and costly. Therefore, it is suggested to avoid the use of any antenna matching or decoupling networks in designing large MIMO systems. A simple 50-ohm probe-fed with radio frequency (RF) connector as the interface between the antenna and the RF chain is recommended. However, if the operating frequency is increasing, e.g., centimeter-wave (cmWave) range, the size of a conventional SMA (subminiature version A) will be comparable to a quarter-wavelength of the interested frequency. The area under the MIMO antenna ground plane would not be sufficient to accommodate multiple SMA connectors. Then, smaller RF connectors, e.g., SMP (subminiature push-on) or modifying probe-fed to side fed with transmission line extensions will be required.

Due to the large size of antenna arrays, the fabrication, testing and maintenance can be inconvenient and unmanageable. An attractive and practical approach is that if the unit cell can be modified, repaired or replaced individually when necessary. It can also be viewed as a sustainable solution as less waste would be produced starting from development to maintenance stages.

The consideration of the radiation pattern could be critical in some applications. If unidirectional beams are required to be generated from all ports, e.g., pointing to broadside direction, then radiation diversity is difficult to achieve, resulting in less antenna ports that can be supported by each unit cell. In other words, it is not fair to compare the number of ports that can be accommodated in two-unit antenna cells without paying attention to their radiation patterns.

5.2 Quadrifilar helix antenna

5.2.1 Design principles

The quadrifilar helix antenna (QHA) originates from the helix antenna and consists of four helices as four arms. Compared with the single helix antenna, QHA exhibits circular symmetry with a higher gain and better on-axis axial ratio. Conventionally, QHA is used to

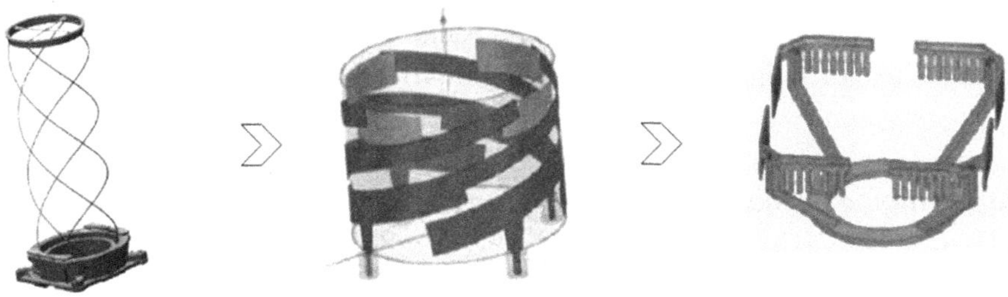

Figure 5.2:
The evolution of QHA design for base station applications: (left) original QHA with combined helices for circular polarization; (middle) modified QHA with four separately excited helices to generate broadside beams; and (right) improved QHA with four miniaturized dipoles in bowl-shape. Reprint from Communications of Huawei Research [27] and IEEE [28].

generate stable circular polarization with high purity and wide bandwidth, which has been widely used in satellite communications. To improve the MIMO system performance, the four arms of QHA are excited individually for more design freedom. The evolution of QHA design is shown in Fig. 5.2. The left picture is the original QHA with combined helices for circular polarization. The middle shoes the modified QHA with four separately excited helices to generate broadside beams. Moreover, the right picture is the improved QHA with four miniaturized dipoles in bowl-shape.

In the following section, the QHA is introduced in detail, including antenna design principles and MIMO system performance. The system simulations and field tests show that QHA can potentially improve channel capacity and provide a new design methodology for MIMO antennas.

5.2.2 4-Port antenna element with circular polarization

To generate better circular polarization, the four arms of QHA are fed by current with equal amplitude and 90° phase difference for adjacent ports. There are two ways to generate 90° phase difference. On the one hand, the length of a pair of helices is slightly longer than the resonate length, and the length of another pair of helices is slightly shorter than the resonate length to generate the 90° phase difference. Although the structure is straightforward to construct using this self-phase shift method, phase control is quite difficult to realize and the bandwidth is also limited. On the other hand, a more resilient way is to use a feed network with a 90° phase shifter where the four output ports of the feed network are connected to the four input ports of QHA. Both left- and right-hand circular polarizations can be realized, depending on the sign of the phase difference. To reduce the size of the feed network and provide more balanced excitations, a compact structure is designed by stacking two similar

Wilkinson power dividers with a common ground back-to-back and using parallel strip line in the feeding port to excite antiphase input signals for two power dividers [29].

5.2.3 4-Port antenna element with ±45° polarization

Compared to circular polarization, dual linear polarizations are more widely used for base stations. As a result, the four ports of QHA are excited separately to generate broadside beams, as shown in Fig. 5.2. When combining any two opposite ports, linearly polarized waves with high purity can be radiated. Different linear polarization is generated by different combinations of port pairs. When combining all four ports, more polarizations can be obtained, providing more design freedom.

When looking at each port, however, due to the off-axis excitation, the polarization and beam direction vary for different arms. The isolation between ports is less than 10 dB [28], especially for opposite ports as they share similar polarization. An improved deign for the four-port antenna is proposed to solve the problems above. Four minimized dipoles with spoof structures surround a central axis and form a bowl shape. Each dipole is centered at the corner of the bowl. When excited independently, each port can generate 45° or −45° polarization. Compared with conventional dual-polarized antenna, the four dipoles are separated, and same for the corresponding phase center. By introducing the difference in phase center, the correlation of radiation patterns between ports is reduced, greatly improving the MIMO performance.

5.2.4 System considerations and field trial results

The improved QHA is used to make an array and used to evaluate its system performance. The details of the antenna array specifications and field tests in the sub-1-GHz frequency band mainly. This band was selected for two reasons. On the one hand, the 690–960 MHz is the golden band for wireless communication system because of its excellent long-range coverage and wall-penetration capability. On the other hand, the aperture for sub-1-GHz is more limited than in other bands for mobile communications, because the electromagnetic wave at this band has a longer wavelength. However, the width in the horizontal direction of the base station antenna is strictly restricted by the wind load on the tower. Therefore, only one or two columns of antenna elements can be placed within the regulated width, placing an upper limit on utilizing the horizontal degree of freedom of a communication channel. Through the proper design of the improved QHA array in the sub-1-GHz, more antenna ports are enabled, and thus more data streams can be supported with the same time and frequency resources.

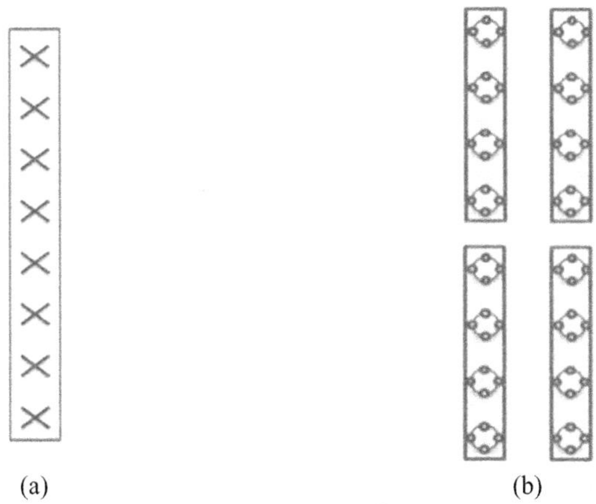

Figure 5.3:
The array configurations in the sub-1-GHz band: (a) conventional dual-polarized antenna and (b) improved QHA. Reprint from Communications of Huawei Research [27].

Fig. 5.3 shows the array configurations in the sub-1-GHz band. There are 16 improved QHA elements with four ports for each element. Four adjacent elements with the same polarizations in a column are combined by a feed network and then connected to a transceiver module. Therefore, 16 transceiver modules are needed in total, using 16T to stand for this configuration. In comparison, 8 conventional dual-polarized antenna elements are arranged in a column and each polarization is connected to one transceiver module via a feed network, so there are 2T in total. In the following session, the field tests are presented based on these array configurations.

Field tests were performed to verify the performance of QHA. The cell range is over 1 km. There are 8 user terminal (UE) positions in total. Two UEs are nearby the base station, labeled as UEs 1 and 2. UEs 3, 5, 7, and 8 are located 300, 800, 980, and 1770 m away from the base station, respectively. There are buildings or trees around for wave scattering and forming non-line-of-sight (NLOS) condition. UE 4 is placed indoors at a distance of 800 m. UE 6 is situated in the open field for LOS scenario, and the distance is 1400 m. All UEs are evenly distributed within the 120° azimuthal angle around the base station. Field tests in different conditions are performed, such as driving test for basic access capability, fixed point test for coverage gain, and single-user (SU)/multiuser (MU)-MIMO throughput gain.

For the driving test, the synchronization signal and physical broadcast channel block (SSB) signal-to-noise ratio (SNR) gain is around 4.2 dB on average for 16T QHA compared with 2T dual-polarized one. The coverage outage test results show that the signal is lost at a distance

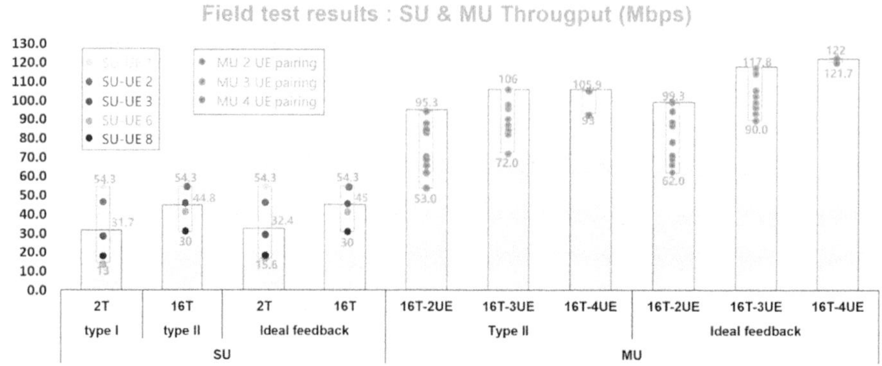

Figure 5.4:
Throughput field tests for SU and MU scenarios. Reprint from Communications of Huawei Research [27].

of 3.26 km with SSB SNR 4 dB for 16T QHA, while the signal is lost at a distance of 2.84 km with SSB SNR –3 dB for 2T comparison case.

For the fixed point test, New Radio (NR) release 15 (R15) type II codebook is used for 16T QHA, whereas NR R15 type I codebook is used for 2T dual-polarized antenna. The SSB SNR for 8 UE locations are tested and the average gain for 16T is 5.2 dB.

For the capacity test, both SU and MU-MIMO cases are considered for ideal and codebook feedback. A summary of the thoughput test results is shown in Fig. 5.4. The baseline is the SU results for UE locations 1, 2, 3, 6, and 8. In the SU scenario, the average throughput of 16T QHA is around 40% higher than that of 2T baseline. In the MU scenario, 2 to 4 UEs are grouped together. The average throughput of 4 UEs for 16T QHA is about 200% and 280% higher than that of 2T baseline for the codebook and ideal feedback, respectively.

All field tests above show that QHA has great potential for improving system performance of cellular communication systems, especially when the number of transceiver channels is limited, such as for sub-1-GHz band applications.

5.3 TCM inspired designs

The theory of characteristic modes (TCM) is one of the analytical frameworks used in the design and analysis of antennas, particularly in the context of multiport antenna systems. This theory provides insights into how electromagnetic fields resonate within an antenna structure, enabling designers or engineers understand and manipulate the modes of operation to achieve desired performance characteristics.

5.3.1 TCM principles

TCM has a long history which can be dated back to 1965 [30]. Characteristic modes are defined as the natural resonant modes of an antenna, which can be excited by incident electromagnetic waves. Each mode is characterized by its own current distribution and radiation pattern, and the excitation of these modes can be controlled through the antenna's geometry, feeding mechanisms, and surrounding environment. By analyzing these modes, engineers can predict how an antenna will behave in terms of radiation efficiency, bandwidth, impedance, and radiation pattern.

In multiport antenna designs, TCM becomes particularly significant. Multiport antennas consist of multiple feeding points, allowing for simultaneous transmission and reception of signals across different ports. Understanding the coupling and interaction between the characteristic modes of these ports is essential for optimizing performance, minimizing interference, and enhancing overall system capacity. The characteristic modes found from TCM are orthogonal to each other, which therefore naturally fits the design of multiport antennas. It allows for the systematic exploration of design parameters, leading to improved isolation between ports and enhanced diversity in signal reception. Additionally, it also facilitates the integration of advanced techniques such as beamforming and polarization diversity, which are critical in modern communication systems.

When the surface of a conducting body S is impressed by an electric field (E-field), a surface current $\boldsymbol{J}$ is obtained. In TCM, an impedance operator $Z(\mathbf{J})$ is introduced as [31]

$$Z(\mathbf{J}) = [L(\mathbf{J})]_{\tan} = R(\mathbf{J}) + jX(\mathbf{J}), \tag{5.1}$$

where $[\cdot]_{\tan}$ denotes the tangential components on S, and L is a linear operator.

Because the $L(\mathbf{J})$ operator has the same unit of impedance, R and X are the real and imaginary parts of Z, which correspond to resistance and reactance, respectively. To obtain orthogonal modal far-field patterns, the weighted eigenvalue

$$X(\mathbf{J}_n) = \lambda_n R(\mathbf{J}_n) \tag{5.2}$$

is considered, where $\mathbf{J}_n$ denotes the nth eigenfunction, also called the nth characteristic current in the TCM, which is associated with the nth eigenvalue λ_n. The smaller the magnitude of λ_n, the smaller the difference in stored magnetic and electric energies and the nth mode is more resonant. In addition, if $\lambda_n > 0$, the mode is inductive, and if $\lambda_n < 0$, the mode is capacitive. In summary, the three key aspects of the TCM used for developing multiport antennas are (i) characteristic current orthogonality; (ii) characteristic far-field orthogonality; and (iii) modal solutions supported by the given radiating structure [32].

5.3.2 Pixel-based designs

Pixel antennas represent a novel and innovative approach to antenna design, characterized by their modular and scalable architecture. Inspired by the concept of pixels in digital imaging, these antennas are composed of an array of small and discrete radiating elements that can be individually controlled and configured. This design enables a high degree of flexibility and adaptability, allowing for the optimization of performance across various frequency bands and applications [33–36].

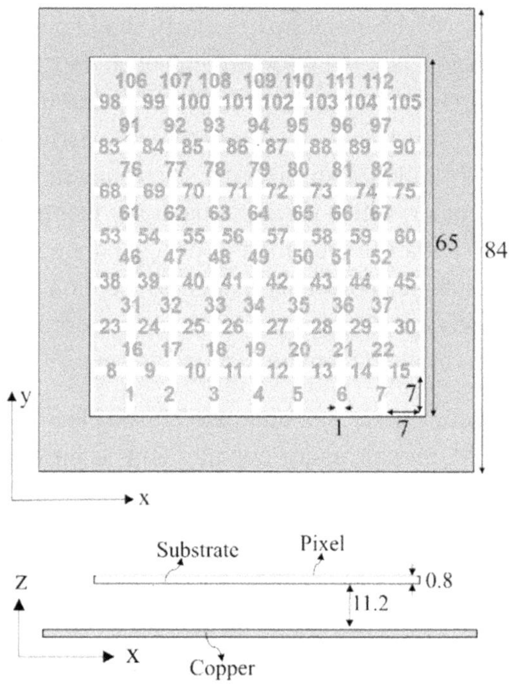

Figure 5.5:
Geometry of pixel antenna targeting at 2.45 GHz. Ports between pixels are numbered bold using red numbers and lengths are marked in black numbers. Reprint from IEEE TAP [33].

Considering a planar 8 × 8-pixel surface printed on a substrate, a total of 112 separation gaps between any two adjacent pixels can be defined as depicted in Fig. 5.5. Imagined that any of the two adjacent pixels can be either in connected (shorted) or disjointed (open), this allows 2^{112} different antenna configurations to be formed. To avoid any resonance within each pixel, the pixel size is always assumed to be much less than one wavelength or not larger than $\lambda/10$ as a reference dimension. As the characteristic modes found in CMA are orthogonal to each other, this naturally can be applied to design MIMO antennas. One advantage of using CMA is that it can separate antenna shape optimization from feeding port selection which helps reduce the design search space significantly.

In the CMA, we can evaluate the example pixel surface performance of many combinations in the total of 2^{112} candidates to find the optimal geometry configuration that meets the design requirements for the necessary multiple modes required. The pixel structure is now thought of as an N-port network and the relation between the port currents and voltages can be expressed as

$$\mathbf{v} = \mathbf{Z}\mathbf{i} \tag{5.3}$$

where $\mathbf{v} = [v_1, v_2, \cdots, v_{112}]^{\mathrm{T}}$ and $\mathbf{i} = [i_1, i_2, \cdots, i_{112}]^{\mathrm{T}}$, and

$$\mathbf{Z} = \begin{bmatrix} Z_{1,1} & \cdots & Z_{1,112} \\ \vdots & \ddots & \vdots \\ Z_{112,1} & \cdots & Z_{112,112} \end{bmatrix}.$$

For the N-port pixel surface structure, the formulation of N-port CMA can also be expressed as an eigenvalue problem as

$$\mathbf{X}\mathbf{i}_m = \lambda_m \mathbf{R}\mathbf{i}_m \tag{5.4}$$

where $\mathbf{Z} = \mathbf{R} + j\mathbf{X}$, where $\mathbf{R}$ and $\mathbf{X}$ are the real part and imaginary parts of $\mathbf{Z}$, respectively, and m $(m = 1, \cdots, N)$ is the index of the characteristic mode.

Then, we can find a set of pixel connections that provide K characteristic modes with small eigenvalues across the interested bandwidth. Here K is the number of modes that the optimal geometry and feed point positions can support. Since the characteristic modes are orthogonal and produce orthogonal radiation patterns, they will produce good MIMO antenna characteristics if excited. The optimization can be performed using genetic algorithm (GA) or successive exhaustive Boolean optimization (SEBO), for instance.

After the optimal geometry of pixel structure is obtained, the K feed ports need to be added vertically between the pixel layer and the ground. The example optimized geometry, and the 64 possible feed port locations are illustrated in Fig. 5.6, where K equals to 3. It should be emphasized that certain separation gaps are now linked up after optimization. One approach of optimizing feed port locations is to perform sequential search by adding feed port one by one to the antenna during the search process. We first search Port 1 location by evaluating all $|S_{ii}|$where $i = 1, \ldots, 64$. The ports with $|S_{ii}|$ below a threshold are considered as potential candidates and then intersected with all candidate ports $|S_{ij}|$ where $i \neq j$, to meet requirements of both input impedance matching and isolation. Once Port 1 location is fixed, the search process can be repeated until all K port locations are determined.

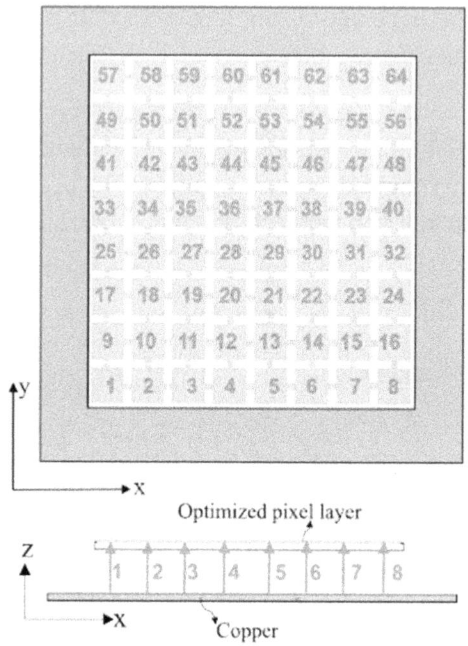

Figure 5.6:
Geometry of an optimized 3-port pixel antenna (top view and side view) showing 64 ports placed between pixel layer and ground for feeding port search. Reprint from IEEE TAP [33].

5.3.3 *Snowflake-based designs*

As mentioned in earlier section, the more ports accommodated in each unit cell, the smaller the antenna array form factor can be achieved. Considering that an antenna element consists of three ports and its size is approximately the same as a conventional dual-polarized patch antenna allows 50% more antenna ports to be constructed for the same array aperture. The advantage of this approach could be more remarkable when the number of ports is going large, e.g., 300 vs 200 ports. In addition, if the antenna element structure is in rotational symmetry as well as in regular (or periodic) tessellation, then it is easy to build up an extremely large multiple-input multiple-output (XL-MIMO) antenna array with balanced performance among all ports. It also allows a straightforward evaluation of the entire array properties from certain few unit cells.

Recently, snowflake-shaped radiators have been proposed for realizing trimodal broadside antennas [37–41]. The snowflake-shaped antenna represents an innovative approach to antenna design, characterized by its unique geometric structure that resembles the intricate patterns of snowflakes. This distinctive configuration not only offers aesthetic appeal but

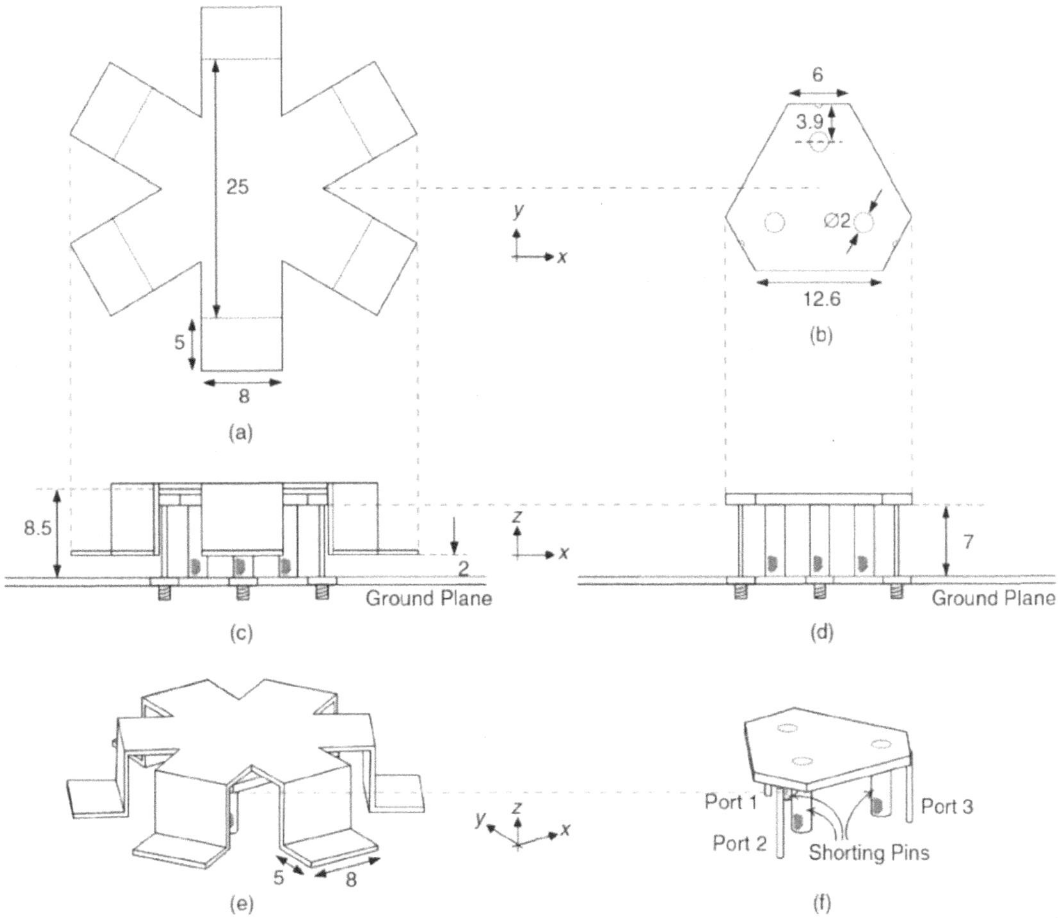

Figure 5.7:
The proposed compact three-port patch antenna in the (a) plan view (units: mm), (b) plan view without the folded snowflake-shaped radiator, (c) side view, (d) side view without the folded snowflake-shaped radiator, (e) perspective view, and (f) perspective view without the folded snowflake-shaped radiator. Reprint from IEEE TAP [38].

also provides significant advantages in terms of performance, particularly when excited by three antenna ports simultaneously. The inherent symmetry and fractal-like properties of the snowflake shape contribute to the ease of concatenation, making it suitable for multiple element integration.

Fig. 5.7 shows the basic configuration of a compact 3-port patch antenna. The antenna consists of a folded snowflake-shaped radiator located at the top, which is capacitively excited by a hexagonal patch underneath. The patch is connected to three shorting pins and the overall

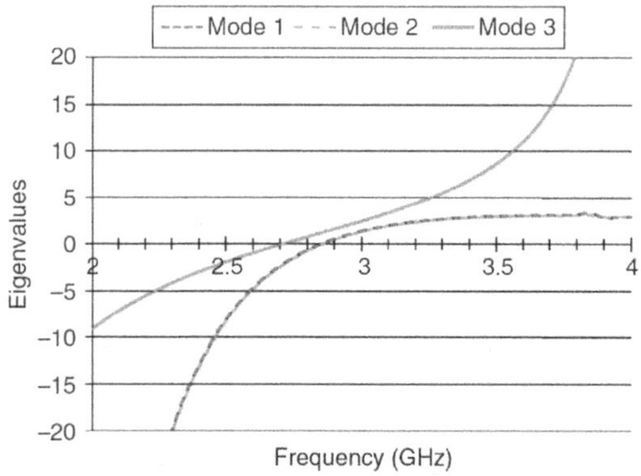

Figure 5.8:
Eigenvalues of the first three CMs. Reprint from IEEE TAP [38].

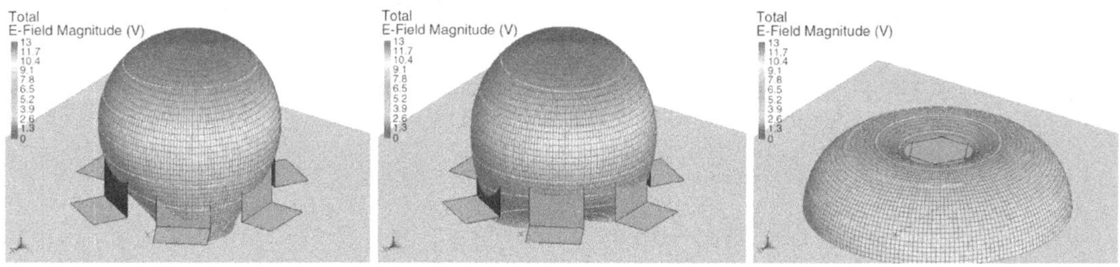

(a) (b) (c)

Figure 5.9:
The characteristic radiation patterns of the folded snowflake-shaped plate at 2.85 GHz. Modes 1–3 correspond to (a)–(c). Reprint from IEEE TAP [38].

antenna structure is rotationally symmetric by 120°. When CMA is performed, it is observed that three characteristic modes (CMs) have relatively small eigenvalues around 2.8 GHz as depicted in Fig. 5.8. This means that the antenna geometry can support three resonant modes that are orthogonal to each other. In theory, the structure can be excited by three individual ports to generate three orthogonal far-field radiation patterns. The corresponding three characteristic radiation patterns are given in Fig. 5.9, showing that the first two modes of radiation exhibit broadside patterns, whereas the third mode of radiation is in monopole-like pattern.

To construct a three-port antenna, three individual feeding probes are placed next to the shorting pins. Simulation results show that the 10-dB impedance bandwidth is 5.2% and the coupling coefficient between any two ports is lower than −16 dB within the band. When the complex correlation between embedded radiation and characteristic far-field patterns of the

Table 5.1: The complex correlation between embedded radiation patterns and characteristic far-field patterns of the snowflake-shaped patch. Reprint from IEEE TAP [38].

	Mode 1	Mode 2	Mode 3
Port 1	$0.541\angle 44.6°$	$0.587\angle -137.4°$	$0.574\angle -143.6°$
Port 2	$0.784\angle -137.2°$	$0.231\angle -140.9°$	$0.586\angle -144.5°$
Port 3	$0.257\angle 40.6°$	$0.780\angle 41.1°$	$0.589\angle -144°$

(a) (b)

Figure 5.10:
Testing scenario for a three-port antenna and a dual-polarized antenna element in (a) anechoic chamber and (b) reverberation chamber. Reprint from IEEE [42].

snowflake-shaped patch is calculated, it is found that the three CMs have contributed to the three port radiations, and the corresponding proportions are provided in Table 5.1. The third characteristic pattern has the same contribution to each port, roughly 0.33 of the total power, subject to small numerical errors.

5.3.4 System considerations

As an important metric for MIMO system, the effective degree of freedom (EDoF) for snowflake based three-port antenna is evaluated experimentally. In Fig. 5.10, two different scenarios, namely with anechoic and reverberation chambers, are used in the tests to simulate pure line-of-sight (LOS) and NLOS scenarios, respectively. One three-port antenna element functions as the transmitter and another as the receiver, and similarly for the dual-polarized antenna elements which are used for comparison. A vector network analyzer is used to obtain the channel response between transmitting and receiving ports. Thereafter, the EDoF can be calculated based on the obtained channel matrices [42].

As shown in Fig. 5.11, the peak EDoF for the three-port antenna is 50% higher than that of the dual-polarized antenna for both scenarios. However, very high SNR levels are required, especially for pure LOS scenarios. Furthermore, when the SNR is 20 dB, the EDoF of the three-port antenna is 36% higher than the dual-polarized antenna in reverberation chamber,

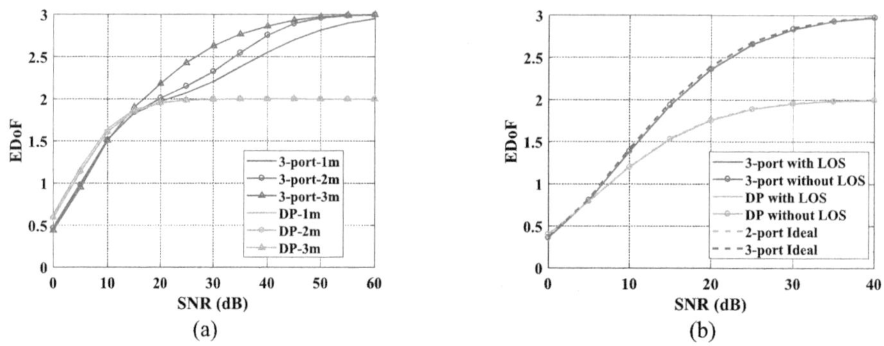

Figure 5.11:
EDoF of channel matrix for both three-port and dual-polarized antennas in (a) anechoic chamber and (b) reverberation chamber. Reprint from IEEE [42].

whereas only slightly improved in anechoic chamber. The real channel performance of multiport antennas should fall between these two extreme cases. In addition, the mutual coupling should also be reduced using the techniques mentioned above to further improve the performance of multiport antennas, particularly in an array configuration.

5.4 Future directions and challenges

5.4.1 Design bounds and EIT

One important question in the design of MIMO antennas is the density of the ports we can achieve in a certain area or volume. It is a key parameter as it determines the physical size of the MIMO arrays and the number of antennas that can be placed into a handset.

Port density can be expressed straightforwardly using the formula $D =$ Number of Ports / Area or $D =$ Number of Ports / Volume, where the area or volume is expressed in wavelengths. One of the highest claimed is 22 antennas per square wavelength, with isolations between all ports around 10 dB [43]. Another example is that 21 antenna ports have been achieved in a circle with radius of 0.6 wavelength which is also equivalent to the interelement center-to-center separation [38]. Natural questions to ask then are what the maximum we can achieve is and what the fundamental limits are.

From an intuitive perspective, there is a fundamental limit related to mutual coupling becoming too large. That is, incident energy in one port is coupled to the adjacent ports and reflected. From another perspective, the limit is related to not having enough degrees of freedom in the channel to support more individual subchannels [44].

To obtain bounds on antenna density, a framework has been developed and become known as electromagnetic information theory (EIT) [45,46]. EIT is an attempt to integrate the concepts of channel capacity developed by Shannon with electromagnetics. In particular, antenna models are integrated with the channel model to simulate the complete wireless communication system. Optimizing the channel capacity then leads to a bound on the maximum number of ports possible. One of the key parameters in EIT is determining the degrees-of-freedom (DoF) to estimate the number of orthogonal basis functions required for describing electromagnetic fields around antenna systems [44–46]. The number of DoF can then be used as a direct measure of the number of ports in a certain area or volume.

One possible approach to finding an antenna design that can obtain the necessary DoF is to use continuous current excitations restricted to a given volume or area. This current distribution is then used to estimate the channel capacity, which is then optimized to obtain the optimal current distribution [47]. However, the geometry for the antenna to obtain the required current distribution has not been provided. Another approach to estimating DoF is to use Nyquist sampling [48]. This reveals that when we assume incident radiation or that scattering can occur over the full 4π sphere, then half-wavelength separation is the maximum density that can be achieved for each polarization. That is, if we consider a 1×1 wavelength square area, 9 independent positions can be found for each polarization, giving a total of 18 (2D-polarized) or 27 (3D-polarized) ports in total. For a $1 \times 1 \times 1$ wavelength cube, this could then provide 26 independent positions or 52 ports (2D-polarized) and 78 (3D-polarized) ports. This is far greater than most practical antenna designs have achieved. A number of other EIT developments have also been found, and a selection is given in references [49–60].

One of the challenges in the development of the antenna density bound is finding methods that can link the EIT approach with the design and prototyping of an antenna that can reach that bound. The work [61] is one attempt to do that, and a design is provided with simulation results showing it can reach around 18 ports in a square wavelength area. In this approach a loaded N-port structure is utilized as the framework and analyzed using a technique closely related to TCM and known as the pattern correlation decomposition method (PCDM) [62]. The loaded N-port approach naturally allows feed ports into the formulation and thus can allow a link to the antenna geometry to be developed. Additionally, TCM and PCDM are useful in EIT because they can provide a natural orthogonal basis for the description of the radiating fields. A beam space channel model can then be exploited to obtain expressions for overall channel capacity. DoF bounds can then be obtained for specific structures under various practical constraints. This DoF estimate is then used to find feed positions that can reach the optimized capacity and provide antenna designs. Further work in this area is required to include bandwidth or Q-factor, and good research in this area can be found in [58].

In summary, a useful rule of thumb for determining the maximum number of antennas in a certain area is that each port needs to be separated by 0.5 wavelength for each polarization. Therefore, in an area of 1×1 square wavelengths, 18 or 27 ports are possible, and for 2×2

square wavelengths, 50 or 75 ports are possible. An antenna design method that can reach that limit is provided in [61] which is based on a technique related to PCDM [62].

5.4.2 Future approaches

In the future, 6G is anticipated to integrate deeply with AI-driven applications [63], giving rise to diverse communication demands, such as ultrahigh data rates, extensive coverage, reliability, and deterministic performance. To address these needs effectively, AI-managed heterogeneous networking across sub-6-GHz for frequency range (FR1) and millimeter-wave bands for FR2 emerges as a practical solution, leveraging the strengths of each frequency range.

For existing FR1 band and potential FR3 for cmWave band, the physical aperture of base station (BS) antennas is strictly constrained by engineering considerations such as wind resistance, weight, and construction limitations. Multiport antennas stand out as a promising technology in scenarios with stringent aperture constraints. By enabling more independent MIMO subchannels within the same physical aperture, these antennas can significantly enhance system throughput. This improvement can be realized either through fundamental antenna properties or advanced decoupling techniques.

The increase in port density brings challenges to the design of the related RF circuits, such as size, insertion loss, reliability, and so on. Apart from using smaller RF connectors to accommodating denser connections, integration between antennas and feed network, or even RF front end circuit, is in demand. One way is to design both antenna and RF circuit on the same board to reduce the loss from feeder and interconnection, known as active integrated antennas. An even higher-level of integration codesigns both RF circuit and antenna by embedding the impedance matching network into antennas. Therefore, lossy interconnections and complicated impedance matching network can be eliminated [64]. This technology can also be applied to multiport antenna for better system performance. More investigations need to be done in this field, for example, the integration of multiport antenna with feed network and active RF circuit for beam steering.

Different from traditional dual-polarized antennas, multiport antennas show more complex radiation patterns and performance characteristics. These differences introduce distinct end-to-end channel response features, which necessitate modifications to existing precoding and channel feedback strategies defined in 3GPP standards, e.g., precoding matrix indicator (PMI) in TS-38.214 [65]. In some cases, compensations must be implemented to counteract varying levels of interelement coupling. In a broad sense, multiport antenna schemes also belong to a sophisticated class of programmable surfaces based on electromagnetic wave manipulation [66]. These schemes aim to control electromagnetic waveforms by generating multiple orthogonal beams through a special current distribution. To achieve a greater system capacity gain through multiport antennas, it is necessary to integrate electromagnetic signal and information theory (ESIT) considerations [67].

Moreover, the combination of multiport antennas with hybrid beamforming (HBF) architectures offers a dual benefit: increasing the number of logical ports and subchannels while reducing the number of RF chains and associated power consumption. However, this also introduces new algorithmic challenges and standardization requirements for HBF beam management, particularly considering the differentiated radiation patterns of multiport antennas. For the multiuser beam management problem, a low-complexity two-stage codebook search algorithm based on angle and distance was proposed in [68]. This solution does not increase the training overhead as the number of users increases, making it an optional codebook solution for multiport antennas.

5.4.3 Conclusion

Integrating multiport antennas with 3GPP MIMO technologies offers a promising avenue for applying EIT to practical systems. This synergy highlights the cross-disciplinary convergence of antenna design innovation with architecture advancements in communication and signal processing, fostering continuous evolution in both domains. Furthermore, from the above discussion, it is noted that the fundamental limits or the antenna port density strongly links to the number of dimensions in a structure we can utilize. Therefore, it is naturally to move forward to design 4D antenna array which corresponds to exploiting space and time domains simultaneously in the future development. However, several practical issues need to be addressed, such as bandwidth expansion due to switching, concern of speed and loss of RF switches, and additional input power consumption for the system, etc.

Massive MIMO systems embedded with high number of antennas in the BSs can improve spectral efficiency and capacity of wireless networks. The hundreds or thousands of antennas can form highly directional beams steering towards specific user terminals (UEs) while minimizing interference from other unintended UEs. This allows for simultaneous multiple communications on the same frequency band. However, the high complexity of antenna array architecture and high total power consumption become hurdles limiting their future development. A potential alternative is to focus on holographic MIMO (HMIMO) designs.

The HMIMO borrows the idea of holography in which it involves reconstructing incident electromagnetic waves from a continuous waveguide-fed metasurface, forming beams at specific directions. The appearance of a HMIMO system is that the frontend is equipped with a metasurface which consists of multiple microstrips and radiating elements. Each microstrip is connected to one RF chain at a side. By utilizing phase accumulation in waveguide modes and adjusting radiating elements appropriately, various beams pointing at different directions can be manipulated. Compared to conventional massive MIMO systems, the complicated phase shifters and number of RF chains used can be eliminated and significantly reduced, respectively. Overall, this allows for a thinner array structure, less power consumption, system building, and maintenance costs reduction.

The unique mechanism of electromagnetic wave manipulation in HMIMO allows radiators to be arranged in sub-half-wavelength spacing. The close spacing provides better spatial control which can increase the data throughput of a system. However, this also results in radiation pattern distortion and radiation efficiency degradation. There will also be a design bound for HMIMO systems. Another concern is about the high Q-factor of conventional microstrip patch radiators following narrowband behaviors. Bandwidth enhancement techniques usually need to be considered during the hardware design process.

References

[1] G.J. Foschini, M.J. Gans, On limits of wireless communications in a fading environment when using multiple antennas, Wireless Personal Communications 6 (3) (Mar. 1998) 311–335.

[2] E. Telatar, Capacity of multi-antenna Gaussian channels, European Transactions on Telecommunications 10 (6) (1999) 585–595, the original version appeared in, Technical Report AT&T Bell Labs, 1995.

[3] C.C. Martin, J.H. Winters, N.R. Sollenberger, Multiple-input multiple-output (MIMO) radio channel measurements, in: 2000 IEEE 52nd Veh. Technol. Conf., Boston, MA, USA, Sep. 2000.

[4] R.S. Blum, J.H. Winters, N.R. Sollenberger, On the capacity of cellular systems with MIMO, in: 2001 IEEE 54th Veh. Technol. Conf., Atlantic City, NJ, USA, Oct. 2001.

[5] W.C. Jakes, A comparison of specific space diversity techniques for reduction of fast fading in UHF mobile radio systems, IEEE Transactions on Vehicular Technology 20 (4) (Nov. 1971) 81–92.

[6] W.C. Jakes, Microwave Mobile Communications, Wiley, New York, 1974.

[7] Lai-U. Choi, R. Murch, A transmit preprocessing technique for multiuser MIMO systems using a decomposition approach, IEEE Transactions on Wireless Communications 3 (1) (Jan. 2004) 20–24.

[8] Q.H. Spencer, A.L. Swindlehurst, M. Haardt, Zero-forcing methods for downlink spatial multiplexing in multiuser MIMO channels, IEEE Transactions on Signal Processing 52 (2) (Feb. 2004) 461–471.

[9] T.L. Marzetta, Noncooperative cellular wireless with unlimited numbers of base station antennas, IEEE Transactions on Wireless Communications 9 (11) (Nov. 2010) 3590–3600.

[10] E.G. Larsson, O. Edfors, F. Tufvesson, T.L. Marzetta, Massive MIMO for next generation wireless systems, IEEE Communications Magazine 52 (2) (Feb. 2014) 186–195.

[11] S.C.K. Ko, R.D. Murch, Compact integrated diversity antenna for wireless communications, IEEE Transactions on Antennas and Propagation 49 (6) (Jun. 2001) 954–960.

[12] C.Y. Chiu, S. Shen, B.K. Lau, R.D. Murch, The design of a trimodal broadside antenna element for compact massive MIMO arrays: utilizing the theory of characteristic modes, IEEE Antennas & Propagation Magazine 62 (Dec. 2020) 46–61.

[13] C.Y. Chiu, C.-H. Cheng, R.D. Murch, C.R. Rowell, Reduction of mutual coupling between closely-packed antenna elements, IEEE Transactions on Antennas and Propagation 55 (6) (Jun. 2007) 1732–1738.

[14] C.Y. Chiu, J.B. Yan, R.D. Murch, 24-port and 36-port antenna cubes suitable for MIMO wireless communications, IEEE Transactions on Antennas and Propagation 56 (4) (Apr. 2008) 1170–1176.

[15] J. Nasir, M.H. Jamaluddin, M. Khalily, M.R. Kamarudin, Irfanullah, N.M. Nor, A dual-port MIMO DRA with high isolation for LTE application, in: 2014 IEEE Asia-Pacific Conf. on Applied Electromagnetics (APACE), Johor Bahru, Malaysia, 2014.
[16] C.Y. Chiu, J.B. Yan, R.D. Murch, J.X. Yun, R.G. Vaughan, Design and implementation of a compact 6-port antenna, IEEE Antennas and Wireless Propagation Letters 8 (2009) 767–770.
[17] K.L. Wong, H.J. Chang, J.Z. Chen, K.Y. Wang, Three wideband monopolar patch antennas in a Y-shape structure for 5G multi-input-multi-output access points, IEEE Antennas and Wireless Propagation Letters 19 (Mar. 2020) 393–397.
[18] C.Y. Chiu, B.K. Lau, R.D. Murch, Bandwidth enhancement technique for broadside tri-modal patch antenna, IEEE Open Journal of Antennas and Propagation 1 (Sep. 2020) 524–533.
[19] F. Hyjazie, H. Boutayeb, Miniaturization of Quad Port Helical Antenna for Wireless 5G Massive MIMO Application, 12th European Conf. on Antennas and Propagat, EuCAP, London, United Kingdom, Apr. 2018.
[20] F. Jiang, C.Y. Chiu, S. Shen, Q.S. Cheng, R.D. Murch, Pixel antenna optimization using N-port characteristic mode analysis, IEEE Transactions on Antennas and Propagation 68 (5) (May 2020) 3336–3347.
[21] S. Soltani, R.D. Murch, A compact planar printed MIMO antenna design, IEEE Transactions on Antennas and Propagation 63 (3) (Mar. 2015) 1140–1149.
[22] C.H. Wu, C.L. Chiu, T.G. Ma, Very compact fully lumped decoupling network for a coupled two-element array, IEEE Antennas and Wireless Propagation Letters 15 (2016) 158–161.
[23] C. Wilson, J. Dean, B.A. Floyd, Mixer-first MIMO receiver with reconfigurable multi-port decoupling and matching, IEEE Journal of Solid-State Circuits 55 (May 2000) 1401–1410.
[24] F. Yang, Y. Rahmat-Samii, Microstrip antennas integrated with electromagnetic band-gap (EBG) structures: a low mutual coupling design for array applications, IEEE Transactions on Antennas and Propagation 51 (10) (Oct. 2003) 2936–2946.
[25] Y.L. Ban, Z.X. Chen, Z. Chen, K. Kang, J.L.W. Li, Decoupled hepta-band antenna array for WWAN/LTE smartphone applications, IEEE Antennas and Wireless Propagation Letters 13 (2014) 999–1002.
[26] A.C.K. Mak, C.R. Rowell, R.D. Murch, Isolation enhancement between two closely packed antennas, IEEE Transactions on Antennas and Propagation 56 (11) (Nov. 2008) 3411–3419.
[27] Quadrifilar Helix Antenna (QHA) —Enabling Highly Efficient Massive MIMO for 5G and Beyond, Communications of Huawei research 2 (Sept 2022), online, https://www-file.huawei.com/-/media/corp2020/pdf/publications/huawei-research/2022/huawei-research-issue2-en.pdf.
[28] F. Hyjazie, H. Boutayeb, Miniaturization of quad port helical antenna for wireless 5G massive MIMO application, in: EuCAP 2018, London, UK, 2018, pp. 1–4.
[29] Q.X. Chu, W. Lin, W.X. Lin, Z.K. Pan, Assembled dual-band broadband quadrifilar helix antennas with compact power divider networks for CNSS application, IEEE Transactions on Antennas and Propagation 61 (2) (Feb. 2013).
[30] R. Garbacz, Modal expansions for resonance scattering phenomena, Proceedings of the IEEE 53 (8) (1965) 856–864.
[31] R. Harrington, J. Mautz, Theory of characteristic modes for conducting bodies, IEEE Transactions on Antennas and Propagation 19 (3) (May 1971) 348–358.
[32] C.Y. Chiu, S. Shen, B.K. Lau, R.D. Murch, The design of a trimodal broadside antenna element for compact massive MIMO arrays: utilizing the theory of characteristic modes, IEEE Antennas & Propagation Magazine 62 (Dec. 2020) 46–61.
[33] F. Jiang, C.Y. Chiu, S. Shen, Q.S. Cheng, R.D. Murch, Pixel antenna optimization using N-port characteristic mode analysis, IEEE Transactions on Antennas and Propagation 68 (May 2020) 3336–3347.
[34] F. Jiang, S. Shen, C.Y. Chiu, Z. Zhang, Q.S. Cheng, R.D. Murch, Pixel antenna optimization based on perturbation sensitivity analysis, IEEE Transactions on Antennas and Propagation 70 (Jan. 2022) 472–486.
[35] F. Jiang, Z. Zhang, M. Li, S. Shen, C.Y. Chiu, Y. Zhang, Q.S. Cheng, R.D. Murch, Multi-port pixel antenna optimization using characteristic mode analysis and sequential feeding port search, IEEE Transactions on Antennas and Propagation 70 (Oct. 2022) 9160–9174.

[36] T. Qiao, F. Jiang, S. Shen, M. Li, C.Y. Chiu, Q.S. Cheng, R.D. Murch, Pixel antenna optimization using the adjoint method and the method of moving asymptote, IEEE Transactions on Antennas and Propagation 71 (Mar. 2023) 2873–2878.
[37] C.Y. Chiu, B.K. Lau, R.D. Murch, Bandwidth enhancement technique for broadside tri-modal patch antenna, IEEE Open Journal of Antennas and Propagation 1 (Sep. 2020) 524–533.
[38] C.Y. Chiu, S. Shen, B.K. Lau, R.D. Murch, The design of a trimodal broadside antenna element for compact massive MIMO arrays: utilizing the theory of characteristic modes, IEEE Antennas & Propagation Magazine 62 (Dec. 2020) 46–61.
[39] C.Y. Chiu, B.K. Lau, R.D. Murch, Co-designed 3+3 port dual-band broadside tri-modal patch antenna, IEEE Open Journal of Antennas and Propagation 2 (Jun. 2021) 767–777.
[40] S. Tang, C.Y. Chiu, R.D. Murch, A millimeter wave tri-modal patch antenna, IEEE Antennas and Wireless Propagation Letters 21 (Feb. 2022) 406–410.
[41] Z. Han, S. Shen, Y. Zhang, S. Tang, C.-Y. Chiu, R. Murch, Using loaded N-port structures to achieve the continuous-space electromagnetic channel capacity bound, IEEE Transactions on Wireless Communications 22 (11) (2023) 7592–7605.
[42] Q. Zhu, G. Jiang, R. Ni, S. Wu, G. Yang, N. Yang, Evaluation of the effective degree of freedom for multiport antenna element, in: 2024 15th GSMM, Hong Kong, 2024, pp. 198–200.
[43] S. Soltani, R.D. Murch, A compact planar printed MIMO antenna design, IEEE Transactions on Antennas and Propagation 63 (3) (March 2015) 1140–1149.
[44] M. Migliore, On the role of the number of degrees of freedom of the field in MIMO channels, IEEE Transactions on Antennas and Propagation 54 (2) (2006) 620–628.
[45] F.K. Gruber, E.A. Marengo, New aspects of electromagnetic information theory for wireless and antenna systems, IEEE Transactions on Antennas and Propagation 56 (11) (2008) 3470–3484.
[46] M.D. Migliore, On electromagnetics and information theory, IEEE Transactions on Antennas and Propagation 56 (10) (2008) 3188–3200.
[47] M.A. Jensen, J.W. Wallace, Capacity of the continuous-space electromagnetic channel, IEEE Transactions on Antennas and Propagation 56 (2) (2008) 524–531.
[48] A. Pizzo, A. de Jesus Torres, L. Sanguinetti, T.L. Marzetta, Nyquist sampling and degrees of freedom of electromagnetic fields, IEEE Transactions on Signal Processing 70 (2022) 3935–3947.
[49] W. Jeon, S.-Y. Chung, Capacity of continuous-space electromagnetic channels with lossy transceivers, IEEE Transactions on Information Theory 64 (3) (2017) 1977–1991.
[50] Z. Wan, J. Zhu, Z. Zhang, L. Dai, C.-B. Chae, Mutual information for electromagnetic information theory based on random fields, IEEE Transactions on Communications 71 (4) (2023) 1982–1996.
[51] R. Li, et al., An electromagnetic information theory-based model for efficient characterization of MIMO systems in complex space, IEEE Transactions on Antennas and Propagation 71 (4) (2023) 3497–3508.
[52] S. Mikki, The Shannon information capacity of an arbitrary radiating surface: an electromagnetic approach, IEEE Transactions on Antennas and Propagation 71 (3) (2023) 2556–2570.
[53] S.S.A. Yuan, X. Chen, C. Huang, W.E.I. Sha, Effects of mutual coupling on degree of freedom and antenna efficiency in holographic MIMO communications, IEEE Open Journal of Antennas and Propagation 4 (2023) 237–244.
[54] M. Gustafsson, J. Lundgren, Degrees of freedom and characteristic modes a perspective and opportunities: estimates for radiating and arbitrarily shaped objects, IEEE Antennas & Propagation Magazine (2024) 2–12.
[55] J. Zhu, Z. Wan, L. Dai, M. Debbah, H.V. Poor, Electromagnetic information theory: fundamentals, modeling, applications, and open problems, IEEE Wireless Communications 31 (3) (2024) 156–162.
[56] C.-X. Wang, Y. Yang, J. Huang, X. Gao, T.J. Cui, L. Hanzo, Electromagnetic information theory: fundamentals and applications for 6G wireless communication systems, IEEE Wireless Communications (2024) 1–8.
[57] S.S.A. Yuan, et al., Breaking the degrees-of-freedom limit of holographic MIMO communications: a 3-D antenna array topology, IEEE Transactions on Vehicular Technology (2024) 1–13.

[58] C. Ehrenborg, M. Gustafsson, M. Capek, Capacity bounds and degrees of freedom for MIMO antennas constrained by Q-factor, IEEE Transactions on Antennas and Propagation 69 (9) (2021) 5388–5400.
[59] C. Ehrenborg, M. Gustafsson, Physical bounds and radiation modes for MIMO antennas, IEEE Transactions on Antennas and Propagation 68 (6) (2020) 4302–4311.
[60] Fundamental Bounds on MIMO Antennas, IEEE Antennas and Wireless Propagation Letters 17 (1) (2018) 21–24.
[61] Z. Han, S. Shen, Y. Zhang, S. Tang, C.-Y. Chiu, R. Murch, Using loaded N-port structures to achieve the continuous-space electromagnetic channel capacity bound, IEEE Transactions on Wireless Communications 22 (11) (2023) 7592–7605.
[62] Z. Han, S. Shen, Y. Zhang, C.-Y. Chiu, R. Murch, A pattern correlation decomposition method for analysis of ESPAR in single-RF MIMO systems, IEEE Transactions on Wireless Communications 21 (7) (July 2022) 4654–4668.
[63] W. Tong, J. Ma, P. Zhu, Y. Chen, AI: the bridge to 6G, Huawei Tech 1 (2024), Online, https://www.huawei.com/en/huaweitech/publication/202401/ai-bridge-to-6g.
[64] S.N. Nallandhigal, K. Wu, Unified and integrated circuit antenna in front end—a proof of concept, IEEE Transactions on Microwave Theory and Techniques 67 (1) (Jan. 2019) 347–364.
[65] 3GPP-TS-38.214, Technical Specification Group Radio Access Network; NR; Physical layer procedures for data (Release 18), Dec. 2023.
[66] M. Katwe, A. Kaushik, L. Mohjazi, M. Abualhayja'a, D. Dardari, K. Singh, M.A. Imran, M.M. Butt, O.A. Dobre, An overview of intelligent meta-surfaces for 6G and beyond: opportunities, trends, and challenges, IEEE Communications Standards Magazine 8 (4) (Dec. 2024) 62–69.
[67] R. Singh, J. An, A. Kaushik, C. Yuen, Y.-J.A. Zhang, O.A. Dobre, Redefining electromagnetic signal and information theory with programmable surfaces: open critical questions and future horizon, https://doi.org/10.36227/techrxiv.175621821.18352632/v1, Aug. 2025.
[68] S. Zhang, B. Di, A. Kaushik, Y.C. Eldar, Holographic-pattern based multi-user beam training in RHS-aided hybrid near-field and far-field communications, IEEE Transactions on Wireless Communications 24 (10) (Oct. 2025) 8892–8907.

CHAPTER 6

Application of ESIT: multishape radio and its experimental studies using OAM, Airy, and Bessel beams

Doohwan Lee, Hirofumi Sasaki, Yasunori Yagi, Kosuke Suzuoki, and Riichi Kudo

NTT Network Innovation Laboratories, NTT Corporation, Yokosuka, Japan

6.1 Introduction

The 6G networks aim to deliver new value beyond the communication domain, such as autonomy, safety, and sustainability, while also enhancing existing network capabilities, including reduced latency and improved transmission rates. To achieve this, research is underway to develop key 6G technologies. These efforts span the advancement of technologies, such as evolved multiinput and multioutput (MIMO), new antenna designs, and new waveform designs, as well as the exploration of novel technologies such as integrated sensing and communication, nonterrestrial networks, and holographic communications [1–3].

Electromagnetic signal and information theory (ESIT) is an integrated framework designed to explore the fundamental potential of wireless communications by combining deterministic electromagnetic theory with statistical information theory. With the ESIT framework, researchers have developed a channel model that integrates circuit theory and communication theory. This model has revealed insights into the degrees of freedom and communication capacity, taking into account the physical characteristics of electromagnetic waves. ESIT is anticipated to play a crucial role in identifying the ultimate limits of wireless communications by leveraging the physical properties of radio waves, alongside emerging technologies such as reconfigurable intelligent surfaces (RIS), continuous-aperture MIMO, and holographic communications [4–7].

Radio waves in subterahertz (sub-THz) frequency bands have short wavelengths, enabling the miniaturization of antenna sizes. Even with a reasonably sized antenna, the near-field can extend hundreds of meters. Therefore, research is actively underway to explore near-field communication using these bands, which had been underutilized. In the 6G era, it is expected that the use of the physical properties of radio waves, which have not been actively

Electromagnetic Signal and Information Theory
https://doi.org/10.1016/B978-0-44-336354-2.00013-7
Copyright © 2027 Elsevier Inc. All rights are reserved, including those for text and data mining, AI training, and similar technologies.

leveraged in wireless communication, will become feasible. The highly sophisticated use of electromagnetic waves, based on ESIT, along with more customized beam control and recently developed device technologies such as programmable metamaterials and RIS, will open new frontiers for 6G research [8–10].

6.2 Multishape radio

To fully use the unexploited physical properties of electromagnetic waves, we introduce new approaches to electromagnetic-wave manipulation and refinement that leverage these properties more effectively. To develop a wide range of electromagnetic-wave-manipulation technologies that go beyond current capabilities, we are exploring ways to integrate various advanced technologies. In this context, we introduce a novel concept called "multishape radio," which fully exploits previously untapped physical properties of electromagnetic waves.

Multishape radio is a concept that uses the physical characteristics of radio waves that had not been actively used, thus expanding the potential applications of new radio-wave technologies. We term this concept multishape radio on the basis of the observation that radio waves with unique physical properties exhibit distinct shapes [11]. Table 6.1 lists the representative physical properties of radio waves and their potential applications from the perspective of multishape radio.

Table 6.1: Physical properties of electromagnetic waves and their applications to multishape radio.

Physical Property	Example	Usage
Curved radio	Airy beam	Low-interference wireless transmission, obstacle avoidance
Radio with helical phase front	OAM beam	High-capacity wireless transmission, wireless sensing, physical security
Nondivergent radio	Bessel beam	Wireless power transfer, high-capacity wireless transmission, obstacle avoidance

The first example of multishape radio is a radio with helical phase fronts, known as an orbital angular momentum (OAM) beam. OAM is a physical property of electromagnetic waves. An OAM-carrying beam exhibits a unique intensity profile and phase front in which equal-phase spots propagate in a helical pattern. Electromagnetic waves with different OAM modes are orthogonal to each other, enabling the transmission of different signals using distinct OAM modes without aliasing [12]. Therefore, OAM can be used to achieve high-capacity wireless transmission by increasing the spatial multiplexing order. Significant research has been conducted on OAM in the field of wireless communications. For instance, Yan et al. achieved a data rate of 32 Gb/s using 28- and 60-GHz frequency bands by transmitting four OAM modes over a 2.5-meter distance [13]. In our previous study, we successfully demonstrated a data rate of 1.58 Tb/s using eight OAM modes and two polarizations in the sub-THz band over a 1-m transmission distance [14].

The second example of multishape radio is a curved beam, the main lobe of which exhibits a curved power distribution along its propagation path. Unlike conventional radio beams, in which the main lobe travels straight, curved beams enable a variety of applications, such as obstacle avoidance. An Airy beam is a typical example of a curved beam and has been studied extensively in the field of optical communications [15]. Research in wireless communications has begun to explore the unique characteristics of curved power distribution [16]. Notably, Airy beams possess an additional unique property, namely an asymmetric sidelobe distribution, which can be leveraged for low-interference wireless transmission of multiple data streams [17].

The third example of multishape radio is the Bessel beam (BB), which is a nondiffractive solution to the Helmholtz equation. Its shape remains unchanged along the propagation path up to a certain distance. Due to its nondiffractive properties, the beam does not diverge, making it suitable for wireless power transfer [18]. It can also be used for high-capacity transmission through enhanced antenna gain [19]. Joseph et al. successfully demonstrated the use of a conical lens to generate a BB and transmit it in the sub-THz band [20]. Yu et al. also showed the feasibility of communication using BBs in the sub-THz band [21]. Both research teams demonstrated the self-healing properties of BBs, which enable wireless transmission even if the wireless link between the transmitter (Tx) and receiver (Rx) is partially blocked by obstacles. BBs also offer various potentials for use in wireless communications. Further research will continue to explore these possibilities.

These are just a few examples of multishape radio; many physical properties of electromagnetic waves remain unexplored, and additional examples of multishape radio are likely to emerge.

6.3 Multiplexing using OAM beams

6.3.1 Background

OAM is a physical property of electromagnetic waves characterized by a helical phase front along the propagation direction. This characteristic can be used to create multiple orthogonal channels, enhancing radio-spectrum efficiency in wireless communication. By transmitting signals with different OAM modes, multiple data streams can be sent simultaneously.

Beams with distinct OAM modes propagate coaxially and remain orthogonal, allowing for the multiplexing of signals within these beams. The OAM modes are defined by their spatial

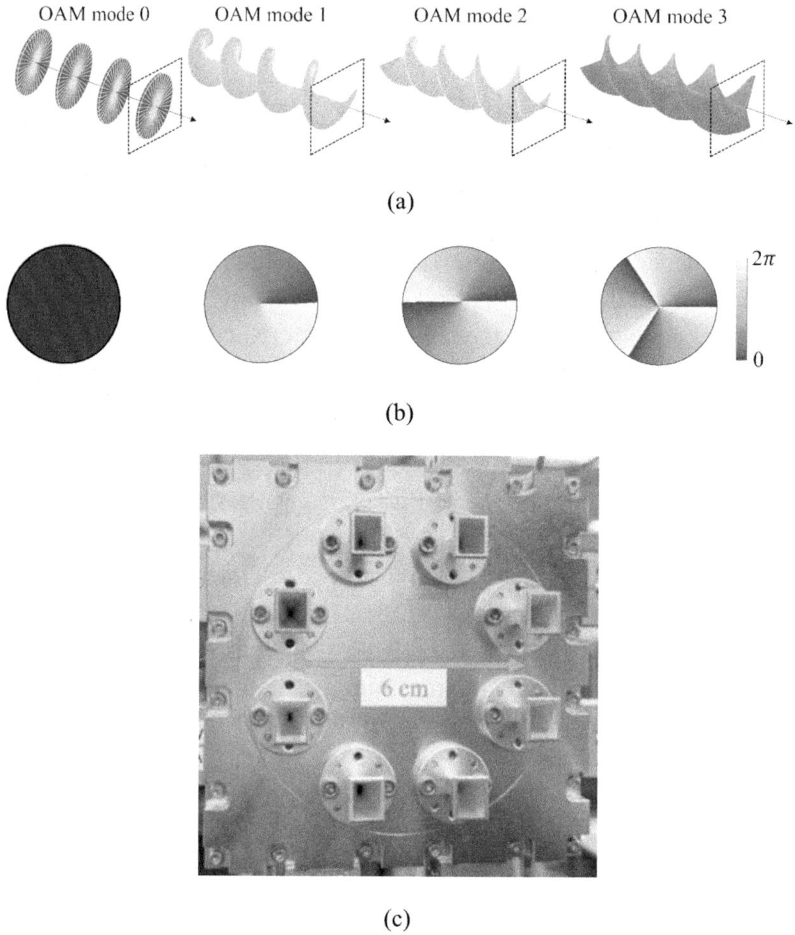

Figure 6.1:
Illustration of OAM Beams: (a) wavefront shapes of OAM modes, (b) spatial-phase distribution on plane perpendicular to propagation axis, and (c) developed UCA integrated with sub-THz-band Butler matrix.

phase distribution. Fig. 6.1(a) illustrates the wavefronts of the OAM modes. OAM mode 0 is equivalent to a plane wave, with a flat phase front, while the other OAM modes exhibit helical wavefronts. When observing the phase distribution on a plane perpendicular to the propagation direction, as shown in Fig. 6.1(b), the phase rotates linearly, with the number of rotations being an integer equal to the OAM mode state. This rotation ensures a periodic structure in the circumferential direction, preserving the OAM states throughout propagation. Consequently, a key advantage of OAM multiplexing is that the orthogonality between OAM modes remains unaffected by the propagation distance, providing flexibility in deployment.

Much work has been dedicated to antenna design for OAM-beam generation using spiral phase plates (SPPs) or holographic plates (HPs). However, SPP- or HP-based OAM multiplexing systems appear challenging to deploy, as each OAM mode requires a separate antenna at a different location. In contrast, OAM-beam generation using a uniform circular array (UCA) is more practical, as multiple OAM modes can be generated simultaneously with a single UCA.

UCA-based OAM-beam generation is achieved by applying a constant phase change, $\varphi = 2\pi l/n$ (where l is the OAM mode and n is the number of antenna elements), to each antenna element. Beams with different OAM modes can be approximately generated or separated simultaneously through discrete Fourier transform (DFT) processing, and they remain orthogonal as long as they propagate coaxially. When the number of antenna elements in a UCA is defined as n, the circulant matrix is diagonalized by the $n \times n$ DFT matrix, with

$$\boldsymbol{D}_n(x, y) = \exp\left(-j\left(\frac{2\pi(x-1)}{n}\right)l(y)\right)/\sqrt{n},$$

as

$$\boldsymbol{H} = \boldsymbol{D}_n \boldsymbol{\Sigma} \boldsymbol{D}_n^H, \boldsymbol{\Lambda} = \boldsymbol{\Sigma} = \left\{\begin{matrix} \lambda_{l(1)} & & & \mathbf{0} \\ & \lambda_{l(2)} & & \\ & & \ddots & \\ \mathbf{0} & & & \lambda_{l(n)} \end{matrix}\right\},$$

where l denotes the OAM mode corresponding to the yth eigenvector, H and Σ represent the channel matrix between the Rx and Tx UCAs and the diagonal matrix, respectively, and Λ denotes the equivalent channel matrix incorporating DFT processing. Mathematically, the channel between UCAs forms a circulant matrix, enabling it to be diagonalized using the DFT matrix, similar to the singular-value decomposition method. This indicates that oppositely arranged Tx and Rx UCAs can establish multiple isolated channels, independent of the propagation distance.

6.3.2 Experimental study

We designed and fabricated a UCA with a 6-cm diameter for operation in the sub-THz frequency bands, as shown in Fig. 6.1(c). The designed UCA is integrated with a Butler matrix, analog circuit that executes a DFT process using a phase shifter and a hybrid coupler.

Fig. 6.2(a) illustrates the experimental setup for OAM multiplexing using our implemented parabolic reflector antenna, designed for 20-m transmission. Table 6.2 summarizes the experimental parameters. Transmission signals for each OAM mode are independently

Table 6.2: Experimental parameters.

Parameter	Values	Parameter	Values
OAM modes	−4, −3, −2, −1, 0, 1, 2, 3	Total baud rate	16 + 16 = 32 GBd
Diameter of reflector	12 cm (Sub)/35 cm (Main)	Frame length	32,400 symbols
Diameter of UCA	6 cm	Forward error correction	BCH (Outer)/LDPC (Inner)
Num. of antennas	8 elements	Equalization	SC-FDE
Transmission distance	20 m	Modulation	16QAM, 64QAM
Frequencies	136–152, 152–168 GHz	Coding rates	1/2, 3/5, 2/3, 3/4, 4/5, 5/6

Table 6.3: Experimental results.

Frequency Band	Transmission rate (1 m)	Transmission rate (20 m)
Lower (136–152 GHz)	331 Gb/s	335 Gb/s
Higher (152–168 GHz)	390 Gb/s	371.73 Gb/s
Total (136–168 GHz)	721 Gb/s	707.73 Gb/s

processed before and after the 8 × 8 Butler matrix at both the Tx and Rx. Thus, equalization between different signals in different OAM modes is unnecessary. Fig. 6.2(b) presents the representative constellations of transmitted signals in the 152–168-GHz band. Experimental results confirm the successful transmission of eight distinct signals using various modulation levels and channel coding rates. The bit error rates for all transmitted signals remained below 10^{-7}.

Table 6.3 summarizes the experimental results. For reference, a 1-m transmission experiment was also conducted using only UCAs without parabolic reflectors. In both the 1- and 20-m experiments, similar transmission rates exceeding 700 Gb/s over a 32-GHz bandwidth were achieved. Further enhancement is possible by implementing polarization multiplexing, which could potentially double the transmission capacity to exceed 1 Tb/s. In our experiments, a total capacity of 1.4 Tb/s was achieved using dual polarization at both 1 m without a parabolic reflector and 20 m with parabolic reflectors. These results confirm the feasibility of using OAM beams for wireless multiplexing transmission.

6.4 Low-interference transmission using airy beams

6.4.1 Background

An Airy wave is derived using the paraxial approximation of the wave equation [15]. The paraxial Helmholtz equation is expressed as

$$i\frac{\partial \Psi}{\partial z} + \frac{1}{2}\frac{\partial^2 \Psi}{\partial s^2} = 0, \tag{6.1}$$

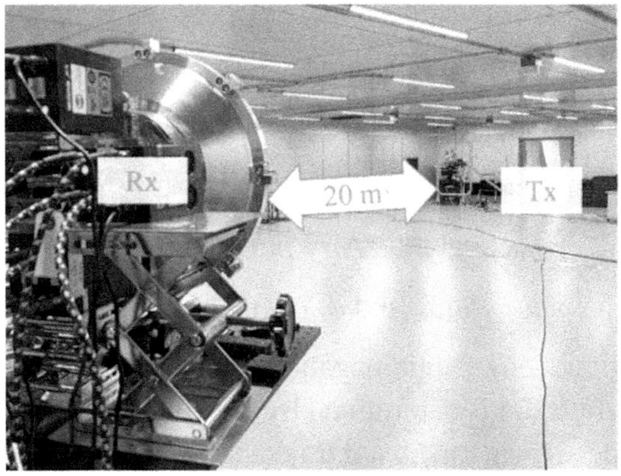

(a)

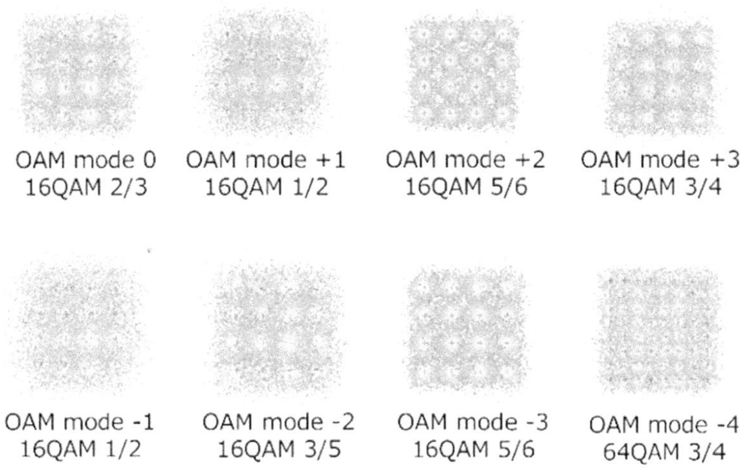

(b)

Figure 6.2:
(a) Experimental environment for wireless OAM multiplexing transmission at distance of 20 m in shielded room using a prototype of OAM transceiver equipped with UCA and parabolic reflector antenna; (b) Representative constellations of transmitted signals in 152–168-GHz band.

where Ψ, z, and s, respectively, represent the intensity of the electromagnetic wave, transverse coordinates, longitude-coordinate distance, and wavenumber. The ideal Airy beam is obtained as a solution to the paraxial Helmholtz equation when the initial condition at $z = 0$ is given by an Airy function, i.e., $\Psi_0(s, 0) = Ai(\gamma s)$. This solution exhibits a curved power-distribution trajectory,

$$\Psi_{ideal}(s,z) = Ai\left(\gamma s - \frac{\gamma^4 z^2}{4k_0^2}\right) e^{i\phi_{ideal}(s,z)}, \tag{6.2}$$

where $Ai(\cdot)$, k_0, and γ are respectively the Airy function, wavenumber, and scale parameter inversely proportional to the pulse width, and $\phi_{ideal}(s,z)$ equals $\left(\frac{\gamma^2 z}{2k_0}\right)\left(\gamma s - \frac{\gamma^4 z^2}{6k_0^2}\right)$, indicating the phase variation along the propagation direction. Note that the intensity distribution of the ideal Airy beam is determined by the Airy function and remains invariant along the propagation while showing a curved trajectory in the (s,z)-domain according to the equation $s = \frac{\gamma^3 z^2}{4k_0^2}$.

Generating the ideal Airy beam is practically impossible because it requires infinite power ($\int Ai\ (s)^2 ds = \infty$). Therefore, in practical cases, a truncated Airy beam, which consumes finite power, is considered. We consider a truncated Airy beam with an exponential truncation window. This truncated Airy beam is obtained as a solution to the paraxial Helmholtz equation when the initial condition at $z = 0$ is given by an Airy function modified with a truncation window, i.e., $\Psi_{truncated}(s,0) = Ai(\gamma s)e^{\alpha s}$, where α is the truncation factor of the exponential truncation function. The resulting solution maintains a curved power-distribution trajectory,

$$\Psi_{truncated}(s,z) = Ai\left(\gamma s - \frac{\gamma^4 z^2}{4k_0^2} + i\frac{\alpha\gamma z}{k_0}\right) e^{i\phi_{truncated}(s,z)+\alpha\left(s-\frac{\gamma^3 z^2}{2k_0^2}\right)}, \tag{6.3}$$

where $\phi_{truncated}(s,z)$ equals $\left(\frac{\gamma^2 z}{2k_0}\right)\left(\gamma s - \frac{\gamma^4 z^2}{6k_0^2} + \frac{\alpha^2}{\gamma^2}\right)$ and indicates the phase variation along the propagation. Note that the intensity distribution of the truncated finite-energy Airy beam is also determined by the Airy function and remains invariant along the propagation. Similar to the ideal Airy beam, it follows a curved trajectory in the (s,z)-domain described by the equation $s = \frac{\gamma^3 z^2}{4k_0^2}$.

In addition to these properties, an Airy beam possesses important characteristics that make it effective for wireless transmission. As shown in Fig. 6.3(a), an Airy beam exhibits an asymmetric sidelobe distribution. By carefully positioning the sidelobe-free region, multiple data streams can be transmitted without significant interference. Taking advantage of the fact that the sidelobes of an Airy beam appear only in a single quadrant, four Airy beams can be transmitted simultaneously without causing harmful interference.

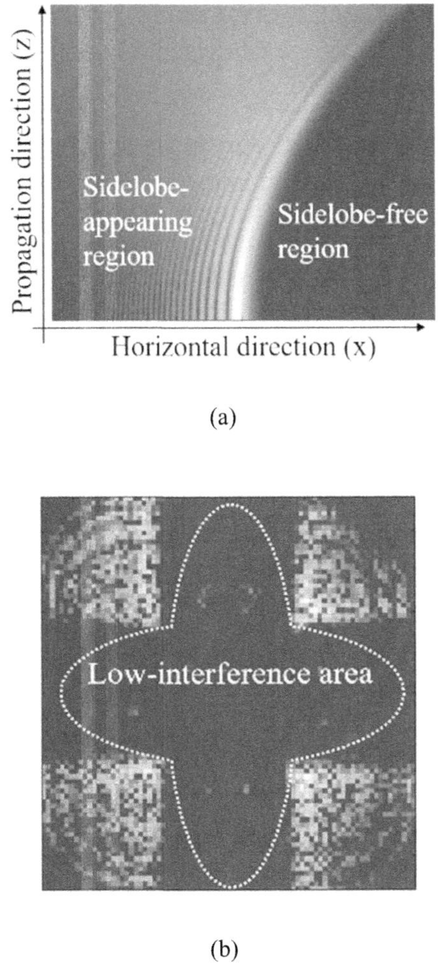

Figure 6.3:
(a) Propagation of Airy beam and its sidelobe characteristics; (b) Experimental result of low-interference area generation using four Airy beams.

6.4.2 Experimental study

Hereafter, we refer to a truncated Airy beam simply as an Airy beam. An Airy beam is generated through the following process. First, a truncated Gaussian beam is produced using a physical antenna such as a horn antenna. This Gaussian beam then passes through a phase modulation lens (PML), which imposes a cubic phase modulation onto the beam, generating an Airy beam in the k-space. Next, the beam is transmitted through a Fourier transform lens (FTL), resulting in the formation of an Airy beam in real space. This process involves only phase manipulation without altering intensity, making Airy-beam generation feasible in the sub-THz band. The PML and FTL are fabricated using three dimensional (3D) printing techniques.

In the previous section, we described a 1D Airy beam for simplicity. In this section, we extend the concept to a 2D Airy beam for practical implementation.

First, we consider PML design. We use a 2D Airy beam in the k-space. It can be obtained with simple calculus by setting coordinate s as (k_x, k_y), where $\psi(x, y, z)$ equals $\psi(x, z)\ \psi(y, z)$ because the x- and y-dimensions are independent. We then add a 2D phase term, $e^{\frac{-ik_0\ (k_x^2+k_y^2)}{2f}}$, into the PML to reduce the transmitter size including antenna and lenses. The normal system that generates an Airy beam has a system length of $2f$; however, we can reduce this into f by integrating the above phase term into the PML [22]. The final phase profile of the PML is given as

$$\phi_{\text{PML}}\left(k_x, k_y\right) = \beta\left((k_x^3 + k_y^3\right) - k_0(k_x^2 + k_y^2)/(2f). \tag{6.4}$$

Second, we consider the design of the FTL. The FTL is also generated in the k-space with a phase profile given as

$$\phi_{\text{FTL}}\left(k_x, k_y\right) = -k_0(k_x^2 + k_y^2)/(2f). \tag{6.5}$$

Fig. 6.4 presents the fabricated PMLs and FTL. To validate curved-trajectory control, we designed and fabricated two types of PMLs using a 3D printer. The two leftmost figures in Fig. 6.4 correspond to the PMLs, where β was set to 562500 and 1125000, respectively. The parameter α, which serves as the truncation factor and acts as a power-limit coefficient, was set to 12 and 6 accordingly. A smaller β results in a more curved trajectory. The bottom-right figure in Fig. 6.4 represents the FTL, where α was adjusted on the basis of the varying values of β.

The focal length (f) was set to 10 cm. The phase modulation of all PMLs and the FTL was achieved by controlling the depth of the lens, using the phase difference between electromagnetic waves propagating through the air and antenna. Modulo arithmetic (e.g., 370 degrees being equivalent to 10 degrees) was also applied to reduce the overall volume. The center frequency was set at 142 GHz, and the diameters of all lenses were 15 cm.

Table 6.2 and Fig. 6.5 present the experimental parameters and experimental setup for four-Airy-beam transmission, respectively. Each Airy beam was generated by a transmitted beam passing through two lenses: a PML and an FTL. These lenses were fabricated using a 3D printer. Four PMLs were positioned such that the sidelobes of the four Airy beams appeared

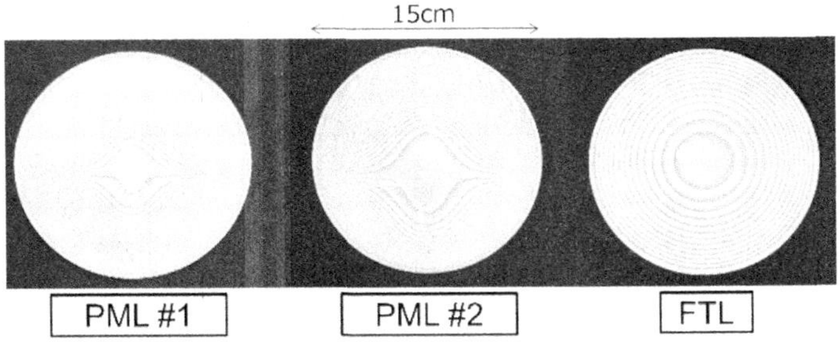

Figure 6.4:
Fabricated PMLs, FTL, and design parameters.

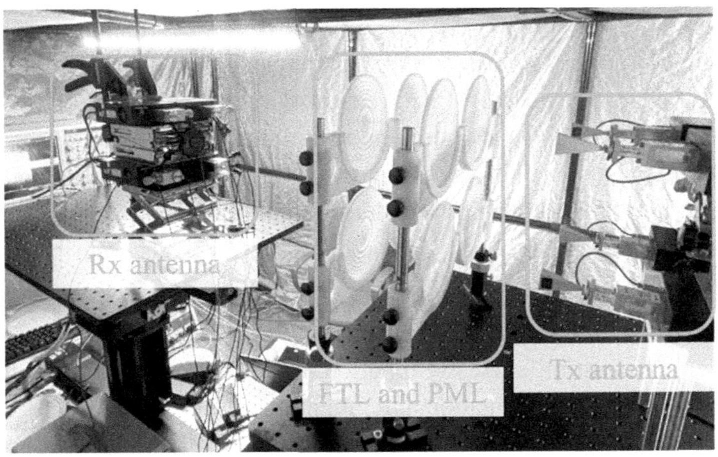

Figure 6.5:
Experimental setup for 4-Airy-beam transmission.

in different quadrant areas, minimizing interference. The distances between the Rx antennas and FTLs, FTLs and PMLs, and PMLs and Tx antennas were 40, 10, and 15 cm, respectively.

Fig. 6.3(b) presents the experimental results of low-interference area generation using four Airy beams. By carefully positioning the PMLs, a low-interference area was created, enabling four-stream transmission without mutual interference. Table 6.4 summarizes the wireless transmission results for the four streams. Due to the functional limitations of signal generators, the experiments were conducted in two 16-GHz sub-THz frequency bands. Using the low-interference areas, four data streams were successfully transmitted without causing significant interference. MIMO-channel estimation and equalization processing could also be abbreviated, providing a substantial reduction in computational cost.

Table 6.4: Experimental results of Airy beams.

Antenna Number	Transmission Rate (132–152 GHz)	Antenna Number	Transmission Rate (152–168 GHz)
1	42.67 Gb/s	1	38.4 Gb/s
2	38.4 Gb/s	2	42.67 Gb/s
3	53.33 Gb/s	3	48 Gb/s
4	48 Gb/s	4	51.2 Gb/s
Sum: 182.4 Gb/s		Sum: 180.27 Gb/s	

6.5 Signal-to-noise-ratio enhancement using bessel beams

6.5.1 Background

To mitigate the high propagation losses in higher frequency bands, more efficient beamforming methods are required. BBs have gained significant attention due to their ability to enhance received power [23]. A BB continuously improves received power up to a certain distance from the beam source. This section provides an overview of BBs and their generation method. A BB represents a nondiffractive solution to the Helmholtz equation, meaning its shape remains unchanged throughout propagation. An ideal BB is described as

$$E(r, z) = A_0 \exp(ik_z)\, J_n(k_r r) \exp(\pm in\theta), \tag{6.6}$$

where J_n represents the nth-order Bessel function of the first kind. The coordinates z and r, respectively, represent the propagation and radial directions, θ represents the azimuthal angle, k_z and k_r represent wavenumbers in z and r, respectively, satisfying $k_0 = 2\pi/\lambda = \sqrt{k_z^2 + k_r^2}$ (λ denotes wavelength), and A_0 represents the amplitude-scale constant. From this equation, an ideal BB exhibits an axisymmetric shape, with its intensity distribution remaining unchanged regardless of the propagation distance. A 0th-order BB was considered, where the peak intensity of the beam along the axis is the highest. This follows the shape of the Bessel function of the first kind for each order [24].

An ideal BB-intensity distribution $I(z, r)$ at $n = 0$ is represented as $A_0^2 |J_0(k_r r)|^2$. This indicates that generating an ideal BB requires an infinite amount of power. In practical applications, a pseudo-BB, which can be generated with finite power, is used for this reason [23]. However, the key property of an ideal BB—its invariant distribution during propagation—is subject to distance limitations.

There are two primary methods for generating a pseudo-BB: using an axicon lens or using a ring slit in combination with an FTL. For this study, we used the axicon-lens method, as it is more suitable for communication applications in terms of power efficiency [24]. An overview of the pseudo-BB generation process under these assumptions is illustrated in Fig. 6.6. The

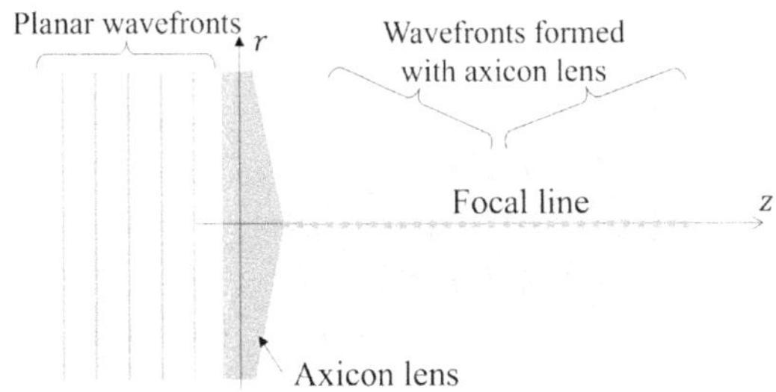

Figure 6.6:
Overview of pseudo-BB generation.

intensity distribution of a pseudo-BB, generated by injecting a plane-wave Gaussian beam with a beam waist of w0w_0 into an axicon lens with a transmission-phase characteristic of $\exp(-ik_r r)$ is derived from scalar diffraction theory, as given by the following equation [25]:

$$(r, z) = \frac{4Pk_0 z \sin^2 \alpha_0}{w_0^2 \cos \alpha_0} \exp\left(-\frac{2z^2 \tan^2 \alpha_0}{w_0^2}\right) J_0^2 (k_r r) . \tag{6.7}$$

Here, P represents the total power of the Gaussian beam incident on the lens, and $\alpha_0 = \arcsin(k_r / k_0)$. This equation reveals that the intensity distribution of a pseudo-BB along the z-axis ($r = 0$) differs from that of an ideal BB. Specifically, the pseudo-BB exhibits a dependence on z, with $I = 0$ as $z = \infty$. However, similar to the ideal BB, the beam null and beam width depend only on k_r. Therefore, the radius of the initial null of the beam can be approximated as $2.405/k_r$ on the basis of the properties of the 0th-order Bessel function of the first kind.

6.5.2 Experimental study

Fig. 6.7 provides an overview of the experimental system, which includes a signal generator, a six-times multiplication upconverter, and an arbitrary waveform generator. After generating a 141-GHz continuous wave, spatial-phase modulation was applied using a programmed control system. A 15-cm dielectric axicon lens was used, with $\alpha = \pi/5\lambda$. A Gaussian beam was assumed as the input to the axicon lens, and the distance between the horn antenna and lens was set to 20 cm to ensure that the intensity distribution remained within the lens, approximately 3σ. To compensate for the effects of spherical waves caused by the close proximity between the antenna and lenses, the phase characteristics of a spherical lens with a focal length of 20 cm were incorporated into the axicon lens, in addition to the transmission-phase characteristics described in the previous section. As a result, a BB was successfully generated using a single lens.

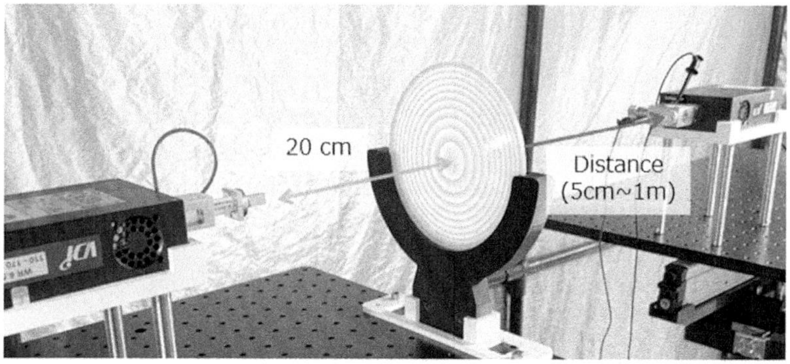

Figure 6.7:
Experimental setup of BB.

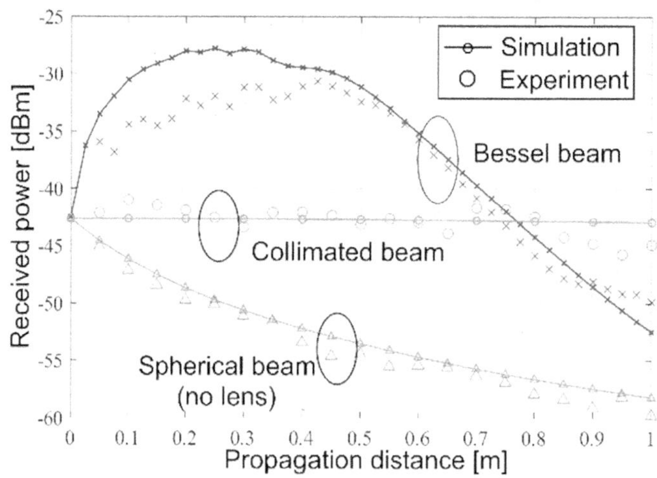

Figure 6.8:
Experimental results of received power on lens center axis at each beam when using dielectric lenses.

On the receiving side, the planar received power distribution was measured at each propagation distance with a resolution of 2 mm in both vertical and horizontal directions, and the peak power was obtained for each propagation distance. When measuring the received power, the time-domain waveform was first acquired, followed by a fast Fourier transform to extract only the 141-GHz component, which was then treated as the received power.

For comparison with conventional technology, we also conducted a test without a collimated beam and lens, where only the effects of spherical waves were compensated using a spherical lens with a focal length of 20 cm. Fig. 6.8 presents a comparison between the experimental results and simulation results obtained using the angular spectrum method. Although the received power of the BB degraded in the region close to the lens, the overall trend remained consistent. It was confirmed that the received power improved by 10 dB at a propagation distance of 42.5 cm.

6.6 Conclusion

ESIT is an integrated framework designed to explore the fundamental potential of wireless communications by combining deterministic electromagnetic theory with statistical information theory. ESIT is expected to help identify the ultimate limits of wireless communications by leveraging the physical properties of radio waves. We introduced "multishape radio" as a novel concept that exploits previously underutilized physical properties of electromagnetic waves. This concept expands the application of new radio waveforms within the ESIT framework.

We explored three representative examples of multishape radio: OAM, Airy, and BBs. We first demonstrated that OAM beams can be used for wireless multiplexing to achieve high-capacity transmission. In our experiments, we achieved a wireless-transmission rate exceeding 700 Gb/s over a 32-GHz bandwidth using OAM beams at a distance of 20 m. These results confirm the feasibility of OAM beams for wireless multiplexing transmission. We then showed that Airy beams can be used for low-interference wireless transmission. By precisely controlling the low-interference regions of Airy beams, we successfully achieved low-interference transmission at 360 Gb/s using four Airy beams within a 32-GHz bandwidth in the sub-THz band. Finally, we demonstrated that BBs can enhance the Rx signal-to-noise ratio (SNR). By comparing the performance of conventional Gaussian beams, collimated beams, and BBs, we experimentally confirmed that BBs improve the Rx SNR by over 10 dB. These examples illustrate the potential of multishape radio. Many physical properties of electromagnetic waves remain unexplored, suggesting the possibility of even more advanced multishape-radio technologies in the future.

To further position this study within the context of the latest research, several recent works have expanded the theoretical and experimental foundations of electromagnetic signal and information theory. Emerging perspectives on programmable surfaces and the integration of deterministic and statistical domains have been presented in [26], while the synergistic role of cmWave and sub-THz bands in enabling high-capacity 6G communications has been emphasized in [27]. The evolving landscape of intelligent metasurfaces and their standardization trends has been comprehensively reviewed in [28]. Furthermore, [29] provides an overview of key innovations and challenges driving the evolution toward 6G, which is consistent with the ESIT framework's vision of unifying physical-layer wave properties and information-theoretic principles.

References

[1] W. Saad, M. Bennis, M. Chen, A vision of 6G wireless systems: applications, trends, technologies, and open research problems, IEEE Network 34 (3) (2020) 134–142.

[2] W. Jiang, B. Han, M.A. Habibi, H.D. Schotten, The road towards 6G: a comprehensive survey, IEEE Open Journal of the Communications Society 2 (2021) 334–366.

[3] Z. Wei, et al., Integrated sensing and communication signals toward 5G-A and 6G: a survey, IEEE Internet of Things Journal 10 (13) (2023) 11068–11092.

[4] Z. Wu, L. Dai, Multiple access for near-field communications: SDMA or LDMA?, IEEE Journal on Selected Areas in Communications 41 (6) (2023) 1918–1935.

[5] J. Zhu, Z. Wan, L. Dai, M. Debbah, H.V. Poor, Electromagnetic information theory: fundamentals, modeling, applications, and open problems, IEEE Wireless Communications 31 (3) (2024) 156–162.

[6] C.-X. Wang, et al., Electromagnetic Information Theory: Fundamentals and Applications for 6G Wireless Communication Systems, IEEE Wireless Communications, 2025 (Early access).

[7] C. Huang, et al., Holographic MIMO surfaces for 6G wireless networks: opportunities, challenges, and trends, IEEE Wireless Communications 27 (5) (2020) 118–125.

[8] T. Gong, et al., Holographic MIMO communications: theoretical foundations, enabling technologies, and future directions, IEEE Communications Surveys and Tutorials 26 (1) (2024) 196–257.

[9] H. Sun, L. Zhu, W. Mei, R. Zhang, Power measurement-based channel estimation for IRS-enhanced wireless coverage, IEEE Transactions on Wireless Communications 23 (12) (2024) 19183–19198.

[10] Z. Chen, et al., Terahertz wireless communications for 2030 and beyond: a cutting-edge frontier, IEEE Communications Magazine 59 (11) (2021) 66–72.

[11] D. Lee, Y. Yasunori, H. Shiba, Multishape radio: new approach to utilizing the physical properties of electromagnetic waves, IEICE Communications Express 11 (9) (2022) 571–576.

[12] Y. Shen, et al., Optical vortices 30 years on: OAM manipulation from topological charge to multiple singularities, Light: Science & Applications 8 (Oct 2019).

[13] Y. Yan, et al., High-capacity millimeter-wave communications with orbital angular momentum multiplexing, Nature Communications 5 (2014) 4876.

[14] H. Sasaki, Y. Yagi, R. Kudo, D. Lee, 1.58 Tbps OAM multiplexing wireless transmission with wideband butler matrix for sub-THz band, IEEE Journal on Selected Areas in Communications 42 (6) (2024) 1613–1625.

[15] N.K. Efremidis, Z. Chen, M. Segev, D.N. Christodoulides, Airy beams and accelerating waves: an overview of recent advances, Optica 6 (2019) 686–701.

[16] H. Guerboukha, et al., Curving THz wireless data links around obstacles, Communications Engineering 3 (2024) 58.

[17] D. Lee, Y. Yagi, K. Suzuoki, R. Kudo, Experimental demonstration of wireless transmission using Airy beams in sub-THz band, IEEE Open Journal of the Communications Society (2025) (Early Access).

[18] S. Paković, et al., Bessel–Gauss beam launchers for wireless power transfer, IEEE Open Journal of Antennas and Propagation 2 (2021) 654–663.

[19] A. Singh, I.V.A.K. Reddy, D. Bodet, J.M. Jornet, Bessel beams for 6G – a performance analysis, in: 56th Asilomar Conference on Signals, Systems, and Computers, Pacific Grove, CA, USA, 2022.

[20] I.V. Reddy, D. Bodet, A. Singh, et al., Ultrabroadband terahertz-band communications with self-healing Bessel beams, Communications Engineering 2 (2023) 70.

[21] Y. Katsuue, et al., Obstacle-tolerant terahertz wireless link using self-healing Bessel beams, Applied Physics Letters 123 (25) (Dec 2023).

[22] C. Liu, L. Niu, K. Wang, J. Liu, 3D-printed diffractive elements induced accelerating terahertz Airy beam, Optics Express 24 (25) (Dec 2016).
[23] J. Durnin, et al., Physical Review Letters 58 (1987) 1449–1501.
[24] D. McGloin, K. Dholakia, Bessel beams: diffraction in a new light, Contemporary Physics 46 (1) (2005) 15–28.
[25] O. Brzobohatý, T. Cižmár, P. Zemánek, High quality quasi-Bessel beam generated by round-tip axicon, Optics Express 16 (17) (2008) 12688.
[26] R. Singh, J. An, A. Kaushik, C. Yuen, Y.-J.A. Zhang, O.A. Dobre, Redefining electromagnetic signal and information theory with programmable surfaces: open critical questions and future horizon, TechRxiv, https://doi.org/10.36227/techrxiv.175621821.18352632/v1, August 2025.
[27] M.V. Katwe, A. Kaushik, K. Singh, M. Di Renzo, S. Sun, D. Lee, A.G. Armada, Y.C. Eldar, O.A. Dobre, T.S. Rappaport, CmWave and sub-THz: key radio enablers and complementary spectrum for 6G, IEEE Wireless Communications (2025), in press (May).
[28] M. Katwe, A. Kaushik, L. Mohjazi, M. Abualhayja'a, D. Dardari, K. Singh, M.A. Imran, M.M. Butt, O.A. Dobre, An overview of intelligent meta-surfaces for 6G and beyond: opportunities, trends, and challenges, IEEE Communications Standards Magazine 8 (4) (2024) 62–69.
[29] R. Singh, A. Kaushik, W. Shin, M. Di Renzo, V. Sciancalepore, D. Lee, H. Sasaki, A. Shojaeifard, O.A. Dobre, Towards 6G evolution: three enhancements, three innovations, and three major challenges, IEEE Network (2025) Early access (May).

CHAPTER 7

Fairness-aware beamforming for cell-free continuous aperture array systems

Mohammadali Mohammadi[a], Deepak Mishra[b], Hien Quoc Ngo[a], and Michail Matthaiou[a]

[a]*School of Electronics, Electrical Engineering and Computer Science, Queen's University Belfast, Belfast, United Kingdom* [b]*School of Electrical Engineering and Telecommunications, University of New South Wales (UNSW) Sydney, Sydney, NSW, Australia*

7.1 Introduction

The evolution of wireless communication networks has been driven by remarkable advancements in multiple-input multiple-output (MIMO) technology. Since the introduction of third-generation (3G) networks, MIMO has become a foundational element of wireless communication standards, boosting network capacity by leveraging spatial degrees of freedom (DoF). By equipping transceivers with multiple antennas, MIMO systems enable the simultaneous transmission and reception of multiple data streams, significantly enhancing spectral efficiency (SE). To push the limits of spatial DoFs and capacity, integrating a large number of antennas within a limited surface area became a pivotal approach, paving the way for massive MIMO (mMIMO) in fifth-generation (5G) networks. Characterized by a large array of antennas at the base stations, mMIMO delivers exceptional improvements in SE and energy efficiency (EE). It has become a cornerstone of 5G systems, facilitating advanced beamforming, interference suppression, and spatial multiplexing techniques [1]. These capabilities have translated into higher user data rates and increased network capacity, surpassing the performance of previous fourth-generation (4G) networks. The commercialization of mMIMO technology has been a notable success. By the end of 2023, mobile network operators had extended mMIMO-powered 5G coverage to 45% of the global population, demonstrating its widespread adoption and transformative impact on modern wireless communications [2].

The shift from mMIMO in 5G toward the envisioned extremely large-scale MIMO (XL-MIMO) – also known as Gigantic MIMO – in 6G has sparked an ongoing wave of advanced antenna array designs. This wave, which began even before 6G's conceptualization, focuses on integrating hundreds or even thousands of antennas into transceivers, pushing the

Electromagnetic Signal and Information Theory
https://doi.org/10.1016/B978-0-44-336354-2.00014-9
Copyright © 2027 Elsevier Inc. All rights are reserved, including those for text and data mining, AI training, and similar technologies.

limits of network performance. Among the most discussed and promising designs are holographic MIMO [3], large intelligent surfaces (LISs) [4], and dynamic metasurface antennas (DMAs) [5], each offering groundbreaking improvements in SE and EE to address the ever-growing demands of next-generation wireless systems. A central objective of these emerging architectures is to create an approximately continuous electromagnetic (EM) aperture, known as the *continuous aperture array (CAPA)*.[1] More specifically, a CAPA can be viewed as a spatially discrete array (SPDA) composed of an infinite number of antennas with infinitesimally small spacing [6]. The goal of a CAPA is to overcome the physical limitations inherent in traditional SPDA deployments, which restrict spatial DoF and, consequently, limit the achievable SE. By replacing discrete antenna arrays with (near-)continuous apertures, a CAPA represents a significant advancement, enabling more flexible, high-capacity, and energy-efficient wireless networks. This is crucial for realizing the ambitious objectives of 6G and beyond.

In this context, Katwe et al. [7] provided an in-depth exploration of the integration of intelligent metasurfaces into 6G and beyond networks, outlining their potential usage scenarios and emphasizing the transformative role of this technology in shaping the future of wireless communications. Furthermore, Singh et al. [8] demonstrated that programmable metasurfaces can function as dynamic boundaries capable of actively reshaping wave propagation, thereby unlocking new opportunities in spatial multiplexing, wavefront engineering, and adaptive communications. Furthermore, the authors discussed fundamental challenges in linking EM field behavior with information-theoretic metrics and in designing efficient sensing–modulation schemes under practical system constraints.

7.1.1 Recent results

To evaluate the performance of CAPAs, a novel framework known as electromagnetic information theory (EMIT) has gained significant traction in recent research. This approach models the system from an EM perspective, tackling the complexities introduced by CAPA's continuous EM radiating surface. In this model, the transmit signal is characterized as information-bearing sinusoidal source currents distributed seamlessly across the aperture, generating EM waves that propagate toward communication users. Upon reception, users decode the transmitted information from the arriving EM waves. The continuous radiating surface of CAPA presents a dual nature – introducing intricate challenges in system modeling and analysis, while also unlocking new possibilities for enhanced performance and system optimization. As a result, EMIT has become a powerful tool for understanding and improving CAPA-based systems, driving ongoing advancements in this emerging field.

Existing studies on CAPA systems can be primarily categorized into two distinct classes: *(i)* single-transceiver CAPA systems, which focus on characterizing the system's fundamental

[1] Henceforth, we will be using the terms CAPA-MIMO, CAPA, and continuous aperture array interchangeably.

limits, including the DoF and channel capacity, by modeling the continuous aperture as a spatially distributed source of EM waves; and *(ii)* multiuser CAPA systems, which extend the analysis to multiuser scenarios, emphasizing performance evaluation and the development of optimization frameworks for system design, accounting for interference management, resource allocation, and spatial multiplexing gains.

7.1.1.1 Single-transceiver CAPA systems

Two major lines of research have emerged in this context. The first line focuses on investigating the fundamental DoFs in single-transceiver CAPA systems – a topic with roots tracing back to the 1980s. In early pioneering work Bucci et al. [9] analyzed the DoFs of scattered EM fields, leveraging the Nyquist number over a given observation interval to characterize the system's spatial resolution. Then, Miller [10] established a more rigorous link between the DoFs and the physical dimensions of CAPA transmitters and receivers using eigenfunction analysis, providing a theoretical framework to quantify the spatial information-carrying capacity of continuous apertures. Subsequent advancements refined these models, enhancing their practical applicability. Notably, Dardari in [4] derived precise analytical expressions for communication DoFs, revealing that CAPA systems can achieve DoFs greater than one – even under line-of-sight (LoS) conditions, where conventional systems are typically constrained. This breakthrough laid the groundwork for further exploration into near-field scenarios, where CAPA's continuous radiating surface can exploit spatial variations more effectively. In particular, Decarli et al. [11] introduced a multifocusing technique aimed at approximating optimal DoFs for large-scale CAPAs, demonstrating improved performance in focusing energy across multiple spatial regions. Complementing this, Ding et al. [12] proposed a spatial bandwidth-based approach to derive closed-form DoF approximations for linear CAPAs along three orthogonal axes, providing a structured analysis of the spatial resolution capabilities in different directions. Further, Xie et al. [13] extended this analysis by examining the concept of effective DoFs, offering deeper insights into how practical factors – such as aperture size, geometry, and propagation conditions – influence the achievable DoFs in realistic deployment scenarios. This line of research continues to evolve, offering a more comprehensive understanding of the spatial information capacity of CAPA systems and laying the theoretical groundwork for advanced communication and sensing applications.

A second major line of research focuses on channel capacity – a pivotal aspect of CAPA systems. Building on an eigenfunction-based formulation, Jensen et al. [14] established a comprehensive EM framework for CAPA capacity analysis. This work underscored how finite transmit power inherently limits channel capacity, while also proposing practical strategies to mitigate supergain effects, which arise from extreme focusing of radiated power. Following this, Jeon et al. [15] extended the analysis by incorporating physical losses into the model, exploring their impact on both the DoF and the achievable capacity. This marked an important step toward bridging theoretical capacity bounds with realistic propagation environments. Several subsequent studies have introduced innovative methodologies to refine capacity evaluation and optimize transmission strategies. More specifically, Migliore [16] leveraged Kolmogorov's information theory to provide a statistical interpretation of CAPA capacity, offering deeper insight into the relationship between spatial information density and power constraints. Meanwhile, Wan et al. [17] presented a comparative analysis between CAPA systems and conventional discrete MIMO architectures. They demonstrated that while MIMO systems can asymptotically approach CAPA performance, this requires an impractical deployment of infinitely many antennas, all without mutual coupling – highlighting the distinct advantages of CAPA structures in achieving compact, high-capacity designs. More recent advancements have continued to push the boundaries of CAPA capacity modeling. Wan et al. [18] proposed a novel numerical computation scheme tailored for analyzing the capacity of random EM fields, enhancing the ability to evaluate performance under complex, nondeterministic scenarios. In parallel, Mikki [19] introduced a generalized capacity computation approach applicable to arbitrary CAPA surface geometries, broadening the framework's applicability across diverse antenna configurations. To further improve practical transmission efficiency, Sanguinetti et al. [20] conceived a wavenumber-division multiplexing (WDM) scheme. This technique efficiently exploits spatial frequency components, achieving near-optimal capacity under LoS conditions by maximizing spectral separation across the wave domain. Expanding upon this foundation, Ouyang et al. [21] explored the wavenumber-domain channel response to characterize ergodic channel capacity and the diversity-multiplexing trade-off in CAPA-based fading environments – providing a more nuanced understanding of performance variations under real-world fading conditions.

7.1.1.2 Multiuser CAPA systems

More recent advancements have shifted focus to multiuser scenarios. For uplink systems operating under LoS conditions, Hu et al. [22] explored the uplink capacity using matched-filtering techniques, laying the groundwork for further capacity analysis in more complex environments. Following this foundational work, Sun et al. [23] optimized the signal-to-interference-plus-noise ratio (SINR) and proposed an adaptive interference mitigation strategy aimed at enhancing the overall performance of multiuser uplink systems. Expanding the scope of multiuser scenarios, Zhao et al. [24] investigated both uplink and downlink capacities by leveraging uplink–downlink duality principles, providing a more comprehensive framework for analyzing the system's performance across different communication directions. From an optimization perspective, Zhang et al. [25] developed pattern-division multiplexing (PDM) to design the patterns for CAPA-MIMO and introduced a Fourier-based beamforming approach designed to maximize the downlink sum rate, highlighting CAPA's potential for significant performance improvements over traditional MIMO systems. This Fourier-based optimization technique was subsequently adapted for uplink beamforming in multiuser scenarios by Qian et al. [26], where both the base station and the users were equipped with holographic planar arrays. They formulated the problem of maximizing the SE for the considered system, for which an iterative water-filling algorithm was proposed for the optimal power allocation matrices. Ouyang et al. [27] analyzed the received signal-to-noise ratio (SNR) achieved by aperture selection in an uplink CAPA-based system. Wan et al. [28] studied beamforming optimization in CAPA-based multiuser communications and proposed two low-complexity approaches to solve the nonconvex integral-based functional programming problem, namely 1) calculus of variations (CoV)-based approach that derives closed-form optimal continuous source patterns and leads to an iterative algorithm, and 2) correlation-based zero-forcing (Corr-ZF) approach that uses closed-form ZF source current patterns to eliminate interference, transforming the problem into a simpler power allocation task solvable by the water-filling method. Zhang et al. [29] proposed a holographic pattern-based multiuser beam training scheme for hybrid near- and far-field communications. The method jointly performs angle and distance estimation for all users, using a holography-inspired codebook for simultaneous angular search and real-time distance-adaptive codewords exploiting the additive property of holographic patterns.

The scalability of CAPA-MIMO systems is governed by the spatial DoF, which grows with aperture size and operating frequency. CAPAs offer moderate DoF and strong penetration at cmWave frequencies, making them ideal for indoor and cell-free deployments. At sub-THz bands, they enable massive DoFs and ultra-high spatial resolution, supporting extreme capacity and sensing. Practical challenges, such as hardware discretization and metasurface quantization are effectively addressed using PDM and wavenumber-domain frameworks, which preserve key spatial modes and maintain performance. These features establish CAPAs as a promising enabler for 6G, outperforming conventional SPDAs in array gain and diversity–multiplexing trade-offs, especially in high-frequency, energy-constrained scenarios.

Another promising research direction is distributed MIMO networks, which, in their practical implementation, evolve into cell-free massive MIMO (CF-mMIMO) networks. In a CF-mMIMO network, a large number of distributed access points collaboratively serve all users by utilizing a central processing unit (CPU) through fronthaul connections. This approach eliminates the need for centralized base stations, thereby enhancing coverage, improving reliability, and increasing system capacity. By distributing antennas across a wide area, CF-mMIMO networks enable more efficient resource allocation, seamless handovers, and an improved overall user experience, particularly in densely populated urban environments and high-mobility areas. In this chapter, we introduce the integration of CAPA into CF-mMIMO networks, termed cell-free CAPA (CF-CAPA), and address the challenge of beam-pattern design to ensure user fairness within the network.

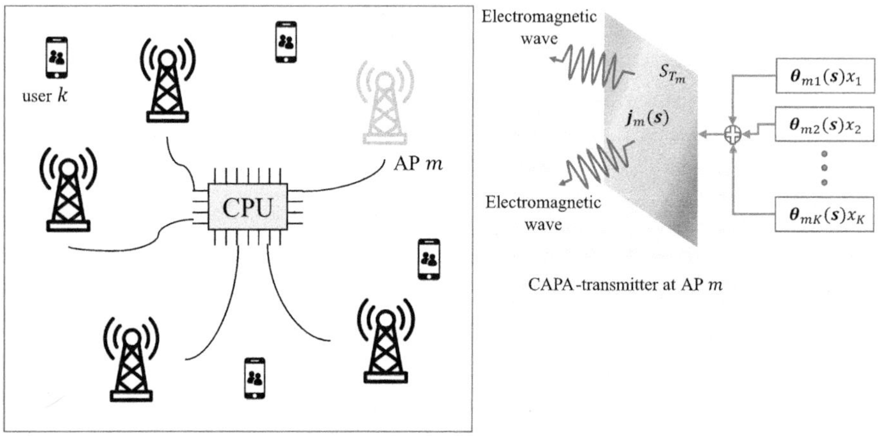

Figure 7.1:
Illustration of distributed CAPA-based multiuser downlink communication system.

Notation: Scalars, vectors, and matrices are denoted by lower-case, bold-face lower-case, and bold-face upper-case letters, respectively; $\mathbb{C}$, $\mathbb{R}$, $\mathbb{R}_+$, and $\mathbb{Z}$ denote the set of complex, real, positive real, and integer numbers; $(\cdot)^{-1}$ $(\cdot)^T$, $(\cdot)^*$, and $(\cdot)^H$ denote the inverse, transpose, conjugate, and conjugate-transpose, respectively; $|\cdot|$ denotes the absolute value of a complex scalar; $\|\cdot\|$ denotes the Euclidean norm of its argument; $\|\cdot\|_F$ denotes the Frobenius norm of its argument; $(\cdot)^{-1}$ denotes the inverse operation; $\Re\{\cdot\}$ denotes the real part of its argument; $\lceil\cdot\rceil$ is the ceil function which represents the smallest integer greater than or equal to its argument. Finally, the big-O notation $\mathcal{O}(\cdot)$ asymptotically describes the order of computational complexity.

7.2 Cell-free/distributed continuous-aperture MIMO

We study a distributed CAPA-based multiuser downlink communication system, as shown in Fig. 7.1, where M APs equipped with CAPA transmitters simultaneously serve K users over the same frequency band. For notational simplicity, we define the sets $\mathcal{K} \triangleq \{1, \ldots, K\}$ and $\mathcal{M} \triangleq \{1, \ldots, M\}$ to collect the indices of the users and transmitters, respectively. We model this system from the EM theory perspective and optimize its performance from the fairness point of view.

Consider CAPA m with a continuous aperture $\mathcal{S}_{\mathrm{T}m}$ with an area of $\mathcal{A}_{\mathrm{T}m} = |\mathcal{S}_{\mathrm{T}m}|$ operating in a 3D homogeneous medium. This surface contains sinusoidal source currents to emit EM waves for wireless communication. We use $\mathbf{s} \triangleq [s_x, s_y, s_z]^T$ in the Cartesian coordinate system to denote the position of point $\mathbf{s} \in \mathcal{A}_{\mathrm{T}m}$. Let $\mathbf{x} = [x_1, \ldots, x_K]^T \in \mathbb{C}^{K\times 1}$ denote the symbols transmitted to K users, satisfying $\mathbb{E}\{\mathbf{x}\mathbf{x}^H\} = \mathbf{I}_K$. At the mth transmitter, these symbols are modulated over K different CAPA-MIMO patterns $\{\boldsymbol{\theta}_{mk}(\mathbf{s})\}$, $\forall k \in \mathcal{K}$, which aims to make these symbols orthogonal at different users as much as possible, such that high channel capacity can be achieved [25].

7.2.1 Preliminaries

Let $\mathbf{j}_m(\mathbf{s}, t) \in \mathbb{R}^3$ denote the monochromatic current density of the mth transmitter at a generic location $\mathbf{s}$ and time t. The current density $\mathbf{j}_m(\mathbf{s}, t) = \Re\{\mathbf{j}_m(\mathbf{s})e^{j\omega t}\}$ can be generated at any position $\mathbf{s} \in \mathcal{S}_{\mathrm{T}m}$ and time t with ω being the angular frequency of the current. Simplifying the communication system to operate in narrowband, aligning with the commonly employed time-harmonic assumption in EM analysis, we can disregard the time-related component $e^{j\omega t}$ [30,31].

Let $\mathbf{r}_k \in \mathbb{R}^3$ denote the three-dimensional (3D) location of the user k. Based on Maxwell's equations, the current distribution $\mathbf{j}(\mathbf{r}_k)$ and the electric field electric $\mathbf{E}(\mathbf{r}_k)$ satisfy the following inhomogeneous Helmholtz wave equation:

$$\nabla \times \nabla \times \mathbf{E}(\mathbf{r}_k) - \kappa_0^2 \mathbf{E}(\mathbf{r}_k) = \mathrm{j}\kappa_0 \eta \mathbf{j}(\mathbf{r}_k), \tag{7.1}$$

where $\eta \triangleq \mu_0 c = 120\pi\,\Omega$ denotes the intrinsic impedance, while in free space $\mu_0 = 4\pi \times 10^{-7}$ H/m and $c = 3 \times 10^8$ m/s is the speed of light.

In a homogeneous medium, the electric fields $\mathbf{E}(\mathbf{r}_k) \in \mathbb{C}^{3\times 1}$ at point $\mathbf{r}_k$, induced by the current distribution $\mathbf{j}_m(\mathbf{s})$ on the mth continuous MIMO aperture, can be expressed as [32]

$$\mathbf{E}(\mathbf{r}_k) = \int_{\mathcal{S}_{\mathrm{T}m}} \mathbf{G}(\mathbf{r}_k, \mathbf{s})\mathbf{j}_m(\mathbf{s})d\mathbf{s}, \tag{7.2}$$

where $\mathbf{s}$ represents the point located in $\mathcal{S}_{\mathrm{T}m}$ and for line-of-sight scenarios, the integral kernel $\mathbf{G}(\mathbf{r}_k, \mathbf{s}) \in \mathbb{C}^{3\times 1}$ is often referred to as the Green's function and determined by the characteristics of the surrounding wireless environment. By considering the radiation field (including the Fresnel and Fraunhofer regions), we can ignore the higher order terms of the $\mathbf{G}(\mathbf{r}_k, \mathbf{s})$ with respect to the distance. Thus, after retaining only the lower-order terms in the region where the EM field has stabilized into normal radiation, we have

$$\mathbf{G}(\mathbf{r}_k, \mathbf{s}) = -\frac{j\eta e^{j\frac{2\pi}{\lambda}\|\mathbf{r}_k - \mathbf{s}\|}}{2\lambda\|\mathbf{r}_k - \mathbf{s}\|}\left(\mathbf{I}_3 - \frac{(\mathbf{r}_k - \mathbf{s})(\mathbf{r}_k - \mathbf{s})^T}{\|\mathbf{r}_k - \mathbf{s}\|^2}\right), \tag{7.3}$$

where λ is the signal wavelength. The approximation in (7.3) holds accurately when the transmitter is positioned outside the reactive near-field of the receiving antenna. From a signal processing standpoint, it is important to highlight that $\mathbf{G}(\mathbf{r}_k, \mathbf{s})$ can be seen as the 3D impulse response of the propagation medium, linking the surface currents at location $\mathbf{s}$ to the resulting fields at location $\mathbf{r}_k$ [33]. In uniform media, the dyadic Green's function $\mathbf{G}(\mathbf{r}_k, \mathbf{s})$ is influenced by both the separation distance between the source and receiver points, as well as their specific positions. This representation will later be simplified using the Fourier series in the frequency domain.

Throughout the remainder of this chapter, we use the LoS channel as an example. However, it is important to emphasize that our proposed approach is not restricted to any specific channel type.

7.2.2 Problem formulation

Following a similar approach to traditional MIMO beamforming [34], the symbols x_k intended for transmission from the mth CAPA-MIMO transmitter are modulated onto K distinct CAPA-MIMO patterns, denoted as $\{\boldsymbol{\theta}_{mk}(\mathbf{s})\}_{k\in\mathcal{K}}$. The objective is to maximize the orthogonality of these symbols for different users, thereby enhancing channel capacity.
For simplicity, we assume that CAPA-MIMO m utilizes a linear superposition to integrate multiple information-bearing patterns $\{\boldsymbol{\theta}_{mk}(\mathbf{s})\}_{k\in\mathcal{K}}$ for coherent transmission. Consequently, the resulting current distribution $\mathbf{j}_m(\mathbf{s})$ on the mth continuous MIMO aperture can be modeled as

$$\mathbf{j}_m(\mathbf{s}) = \sum\nolimits_{k\in\mathcal{K}} \boldsymbol{\theta}_{mk}(\mathbf{s}) x_k, \quad \forall m \in \mathcal{M}, \mathbf{s} \in \mathcal{S}_{\mathrm{T}m}. \tag{7.4}$$

Note that $\boldsymbol{\theta}_{mk}(\mathbf{s})$ is the component of current density at the transmitter m that carries the symbol x_k. We assume that all K EM wave receivers, representing K users, are positioned within the far-field region of the CAPA-MIMO transmitters. Each user is equipped with a tripolarization antenna that is idealized and has an effective area of $\mathcal{A}_{\mathrm{r}} = \frac{\lambda^2}{4\pi}$, ensuring that $\mathcal{A}_{\mathrm{r}} \ll \mathcal{A}_{\mathrm{T}m}$, for all $m \in \mathcal{M}$ [22]. Under these conditions, each user can be effectively modeled as a discrete point in three-dimensional space, similar to a conventional discrete antenna setup.

By invoking (7.2) and (7.4), the received EM wave at user k can be expressed as

$$\begin{aligned}\mathbf{y}_k &= \mathbf{E}(\mathbf{r}_k) + \mathbf{n}_k \\ &= \sum\nolimits_{m\in\mathcal{M}} x_k \int_{\mathcal{S}_{\mathrm{T}m}} \mathbf{G}_{mk}(\mathbf{s}) \boldsymbol{\theta}_{mk}(\mathbf{s})\, \mathrm{d}\mathbf{s} \\ &\quad + \sum\nolimits_{m\in\mathcal{M}} \sum\nolimits_{j\in\mathcal{K}\setminus k} x_j \int_{\mathcal{S}_{\mathrm{T}m}} \mathbf{G}_{mk}(\mathbf{s}) \boldsymbol{\theta}_{mj}(\mathbf{s})\, \mathrm{d}\mathbf{s} + \mathbf{n}_k,\end{aligned} \tag{7.5}$$

where $\mathbf{G}_{mk}(\mathbf{s}) \triangleq \mathbf{G}(\mathbf{r}_k, \mathbf{s}), \mathbf{s} \in \mathcal{S}_{\mathrm{T}m}$ and $\mathbf{n}_k \in \mathbb{C}^3$ is the EM noise at user k. In (7.5), the first term is the desired signal received from all transmitters, and the second term is the interuser interference caused by transmission towards other users. For simplicity, we follow the isotropic propagation assumption, hence $\mathbf{n}_k$, for all users can be modeled as mutually independent additive white Gaussian noise (AWGN) with zero mean and variance $\sigma^2\mathbf{I}_3$ [25].

In the ideal case, user k with a tripolarization antenna can sense the three polarizations of the EM waves reaching the receiver. Then, the three polarization components are combined to decode symbol x_k with a polarization combiner $\boldsymbol{\psi}_k \in \mathbb{C}^3$. There are two possible methods to achieve an adjustable combiner $\boldsymbol{\psi}_k$: (i) the receiver antenna can dynamically adjust its polarization direction to combine polarization components in the analog domain, utilizing polarization-adjustable reconfigurable antennas [35]; (ii) the receiver can digitize the polarization components and subsequently combine them in the digital domain. Considering adjustable combining polarization, the decoded signal at user k is given by $\hat{x}_k = \boldsymbol{\psi}_k^H \mathbf{y}_k$. Let E_k denote the mean-square error (MSE) of the decoded symbol, defined as

$$\begin{aligned} E_k(\boldsymbol{\Psi}, \boldsymbol{\Theta}(\mathbf{s})) &= \mathbb{E}\left\{|\hat{x}_k - x_k|^2\right\} \\ &= \left|1 - \sum_{m\in\mathcal{M}} \int_{\mathcal{S}_{\mathrm{T}m}} \boldsymbol{\psi}_k^H \mathbf{G}_{mk}(\mathbf{s})\boldsymbol{\theta}_{mk}(\mathbf{s})\,\mathrm{d}\mathbf{s}\right|^2 \\ &\quad + \sum\nolimits_{m\in\mathcal{M}} \sum\nolimits_{j\in\mathcal{K}\setminus k} \left|\int_{\mathcal{S}_{\mathrm{T}m}} \boldsymbol{\psi}_k^H \mathbf{G}_{mk}(\mathbf{s})\boldsymbol{\theta}_{mj}(\mathbf{s})\,\mathrm{d}\mathbf{s}\right|^2 + \sigma_n^2 \|\boldsymbol{\psi}_k\|^2, \end{aligned} \tag{7.6}$$

where $\boldsymbol{\Psi} \triangleq \{\boldsymbol{\psi}_k\}_{k\in\mathcal{K}}$ denotes the combining vector for all K users and $\boldsymbol{\Theta}(\mathbf{s}) \triangleq \{\boldsymbol{\theta}_{mk}\}_{k\in\mathcal{K}}^{m\in\mathcal{M}}$ contains the beam-pattern vectors between all APs and all users.

Moreover, the SINR for decoding the desired information symbol at user k can be obtained by using (7.5) as

$$\mathrm{SINR}_k = \frac{\sum_{m\in\mathcal{M}} \left|\int_{\mathcal{S}_{\mathrm{T}m}} \mathbf{G}_{mk}(\mathbf{s})\boldsymbol{\theta}_{mk}(\mathbf{s})\,\mathrm{d}\mathbf{s}\right|^2}{\sum_{m\in\mathcal{M}} \sum_{j\in\mathcal{K}\setminus k} \left|\int_{\mathcal{S}_{\mathrm{T}m}} \mathbf{G}_{mk}(\mathbf{s})\boldsymbol{\theta}_{mj}(\mathbf{s})\mathrm{d}\mathbf{s}\right|^2 + \sigma^2}. \tag{7.7}$$

To investigate the optimal structure of general CAPA beamforming, we formulate a general optimization problem that maximizes system performance using arbitrary monotonic utility functions, expressed as

$$(\mathcal{P}1) \quad \max_{\boldsymbol{\Theta}(\mathbf{s})} \quad U(\mathrm{SINR}) \tag{7.8a}$$

$$\text{s.t.} \quad \sum\nolimits_{k\in\mathcal{K}} \int_{\mathcal{S}_{\mathrm{T}m}} \|\boldsymbol{\theta}_{mk}(\mathbf{s})\|^2\,\mathrm{d}\mathbf{s} \le P_{\mathrm{T}}, \forall m \in \mathcal{M}, \tag{7.8b}$$

where $\text{SINR} = [\text{SINR}_1, \ldots, \text{SINR}_K]^T$ and function $U(\cdot)$ is the system utility function that is strictly monotonically increasing in each argument SINR_k. Constraint (7.8b) guarantees that the total transmitted power stays within the allocated budget P_{T} [25,28]. Note that P_{T} can be viewed as the allowable maximum "transmit power" of the mth CAPA-MIMO, which is implicitly associated with the physical radiation power.

While similar optimization problems in conventional SPDA systems have been widely explored in previous research [34], solving the problem ($\mathcal{P}1$) remains challenging because it belongs to the category of functional programming problems. Unlike conventional optimization problems where the decision variables are discrete vectors or matrices, this problem involves continuous functions as variables.

Communication performance is typically assessed through various metrics, many of which are monotonic functions of the SINR. These include achievable data rate, mean squared error, and bit error rate. Furthermore, commonly used system utility functions—such as the weighted arithmetic mean, weighted geometric mean, and max-min criterion—also exhibit monotonicity with respect to these performance metrics. In CF-mMIMO systems, ensuring equitable service across a network is important; as such, fairness-oriented optimization problems are crucial in resource allocation. Max–min fairness seeks to maximize the minimum user performance metric, ensuring uniform coverage and eliminating service gaps. Proportional fairness balances efficiency and fairness by maximizing the sum of logarithmic data rates, making it well suited for balancing system throughput and user experience. Weighted fairness assigns priority-based resource allocation, benefiting applications with diverse quality-of-service (QoS) requirements. The choice of fairness criterion depends on network goals, influencing system performance, user satisfaction, and coverage consistency.

In this chapter, we focus on the max–min optimization problem. This optimization approach focuses on maximizing the minimum SE across all users in the system, which is especially important in scenarios where users are geographically dispersed and face varying levels of interference and channel conditions. By addressing these disparities, max–min optimization helps to maintain a balanced QoS for all users, even in the presence of limited network resources. Mathematically, the max–min optimization problem can be formulated as

$$(\mathcal{P}2) \quad \max_{\boldsymbol{\Theta}(\mathbf{s})} \quad \min_{k\in\mathcal{K}} \mathcal{R}_k \tag{7.9a}$$

$$\text{s.t.} \quad \sum\nolimits_{k\in\mathcal{K}} \int_{\mathcal{S}_{\text{T}m}} \|\boldsymbol{\theta}_{mk}(\mathbf{s})\|^2 \, d\mathbf{s} \leq P_{\text{T}}, \forall m = 1, \ldots, M, \tag{7.9b}$$

where $\mathcal{R}_k$ is the achievable SE of user k, which can be expressed as

$$\mathcal{R}_k = \log_2(1 + \text{SINR}_k), \ k \in \mathcal{K}. \tag{7.10}$$

To deal with the complex form of $\mathcal{R}_k$ in (7.10), we employ [36, Theorem 1] to transform it to a more tractable form. To this end, we introduce the auxiliary variable $\boldsymbol{\rho} = [\rho_1, \ldots, \rho_K]^T \in \mathbb{R}_+^K$. Then, introducing the slack variable $t \geq 0$, optimization problem $(\mathcal{P}1)$ is equivalent to

$$(\mathcal{P}3) \quad \max_{\boldsymbol{\rho}, \boldsymbol{\Psi}, \boldsymbol{\Theta}(\mathbf{s}), t} \quad t \tag{7.11a}$$

$$\text{s.t.} \quad \log_2 \rho_k - \frac{1}{\ln 2}(\rho_k E_k(\boldsymbol{\Psi}, \boldsymbol{\Theta}(\mathbf{s})) - 1) \geq t, \quad \forall \in \mathcal{K}, \tag{7.11b}$$

$$\sum\nolimits_{k \in \mathcal{K}} \int_{\mathcal{S}_{\mathrm{T}m}} \|\boldsymbol{\theta}_{mk}(\mathbf{s})\|^2 \, \mathrm{d}\mathbf{s} \leq P_{\mathrm{T}}, \quad \forall m \in \mathcal{M}. \tag{7.11c}$$

The problem $(\mathcal{P}3)$ is a complicated nonconvex optimization problem due to the strong coupling among the optimization variables. To address this issue, we can optimize the slack variables $\boldsymbol{\rho}$, combiners $\boldsymbol{\Psi}$, and continuous pattern functions $\boldsymbol{\Theta}(\mathbf{s})$ alternatively until the convergence of the objective function. Now, we proceed to provide the update process for the optimization parameters.

7.2.2.1 Optimizing $\boldsymbol{\rho}$ with given $\boldsymbol{\Psi}$ and $\boldsymbol{\Theta}(\mathbf{s})$

To optimize $\boldsymbol{\rho}$, when the combiners $\boldsymbol{\Psi}$ and pattern functions $\boldsymbol{\Theta}(\mathbf{s})$ are fixed, we observe that only constraint (7.11b) depends on $\boldsymbol{\rho}$. Therefore, in this case we need to maximize the left-hand side of (7.11b), to ensure that the objective function reaches its maximum in the next steps of the optimization process. To this end, by removing the unrelated component in the optimization problem $(\mathcal{P}2)$, the subproblem for optimizing the auxiliary variable $\boldsymbol{\rho}$ is reduced to

$$(\mathcal{P}3.1) \quad \max_{\rho_k} \quad \log_2 \rho_k - \frac{1}{\ln 2}(\rho_k E_k(\boldsymbol{\Psi}, \boldsymbol{\Theta}(\mathbf{s})) - 1), \quad \forall k \in \mathcal{K}. \tag{7.12a}$$

By setting the first derivative of the objective function of the unconstrained optimization problem $(\mathcal{P}3.1)$ with respect to ρ_k to zero, the solution to this problem can be obtained as

$$\rho_k^\star = E_k^{-1}(\boldsymbol{\Psi}, \boldsymbol{\Theta}(\mathbf{s})), \qquad \forall k \in \mathcal{K}. \tag{7.13}$$

7.2.2.2 Optimizing $\boldsymbol{\psi}_k$ with given $\boldsymbol{\rho}$ and $\boldsymbol{\Theta}(\mathbf{s})$

For given auxiliary variables $\boldsymbol{\rho}$ and pattern functions $\boldsymbol{\Theta}(\mathbf{s})$ for CAPA-MIMO transmitters, $\boldsymbol{\psi}_k$ only appears in (7.11b) inside $E_k(\boldsymbol{\Psi}, \boldsymbol{\Theta}(\mathbf{s}))$. Therefore, by maximizing this term on the left-hand side of (7.11b), we ensure that the objective function attains its maximum in subsequent steps of the optimization process. After removing the unrelated components in problem $(\mathcal{P}3)$, we arrive at the following optimization problem over $\boldsymbol{\psi}_k$ for each user:

$$(\mathcal{P}3.2) \quad \max_{\boldsymbol{\psi}_k} \quad -\rho_k \boldsymbol{\psi}_k^H \mathbf{A}_k \boldsymbol{\psi}_k + 2\rho_k \Re\{\boldsymbol{\psi}_k^H \boldsymbol{\beta}_k\}, \quad \forall k \in \mathcal{K}, \tag{7.14a}$$

where

$$\mathbf{A}_k = \sum\nolimits_{m\in\mathcal{M}} \sum\nolimits_{j\in\mathcal{K}} \int_{\mathcal{S}_{\mathrm{T}m}} \mathbf{G}_{mk}(\mathbf{s})\boldsymbol{\theta}_{mj}(\mathbf{s})\mathrm{d}\mathbf{s}\left(\int_{\mathcal{S}_{\mathrm{T}m}} \mathbf{G}_{mk}(\mathbf{s}')\boldsymbol{\theta}_{mj}(\mathbf{s}')\mathrm{d}\mathbf{s}'\right)^H + \sigma^2\mathbf{I}_3, \quad (7.15a)$$

$$\boldsymbol{\beta}_k = \sum\nolimits_{m\in\mathcal{M}} \int_{\mathcal{S}_{\mathrm{T}m}} \mathbf{G}_{mk}(\mathbf{s})\boldsymbol{\theta}_{mk}(\mathbf{s})\,\mathrm{d}\mathbf{s}. \quad (7.15b)$$

Noticing that ($\mathcal{P}3.2$) is a standard convex quadratic programming (QP), the optimized combiner $\boldsymbol{\psi}_k$ can be obtained as

$$\boldsymbol{\psi}_k^\star = \mathbf{A}_k^{-1}\boldsymbol{\beta}_k, \qquad \forall k \in \mathcal{K}. \quad (7.16)$$

7.2.2.3 Optimizing $\boldsymbol{\Theta}(\mathrm{s})$ for given $\boldsymbol{\rho}$ and $\boldsymbol{\Psi}$

Now, given fixed auxiliary variables $\boldsymbol{\rho}$ and user combiners $\boldsymbol{\Psi}$, the subproblem of optimizing the continuous pattern function can be expressed as

$$(\mathcal{P}3.3) \quad \max_{\boldsymbol{\Theta}(\mathbf{s}),t} \quad t \quad (7.17a)$$

$$\text{s.t.} \quad \sum\nolimits_{m\in\mathcal{M}} \sum\nolimits_{j\in\mathcal{K}} \left| \int_{\mathcal{S}_{\mathrm{T}m}} \boldsymbol{\psi}_k^H \mathbf{G}_{mk}(\mathbf{s})\boldsymbol{\theta}_{mj}(\mathbf{s})\mathrm{d}\mathbf{s} \right|^2$$

$$- 2\Re\left\{ \sum\nolimits_{m\in\mathcal{M}} \int_{\mathcal{S}_{\mathrm{T}m}} \boldsymbol{\psi}_k^H \mathbf{G}_{mk}(\mathbf{s})\boldsymbol{\theta}_{mk}(\mathbf{s})\mathrm{d}\mathbf{s} \right\} \geq t, \quad \forall k \in \mathcal{K}, \quad (7.17b)$$

$$\sum\nolimits_{k\in\mathcal{K}} \int_{\mathcal{S}_{\mathrm{T}m}} \|\boldsymbol{\theta}_{mk}(\mathbf{s})\|^2\,\mathrm{d}\mathbf{s} \leq P_{\mathrm{T}}, \quad \forall m \in \mathcal{M}. \quad (7.17c)$$

Since the information-bearing patterns $\boldsymbol{\theta}_{mk}(\mathbf{s})$ and the Green function $\mathbf{G}_{mk}(\mathbf{s})$, for all $m \in \mathcal{M}$ and $k \in \mathcal{K}$, are all continuous physical quantities, their analysis or optimization in general can become computationally demanding. This is because continuous quantities entail an infinite number of parameters, which can lead to a significant computational load when assessing or enhancing system performance. A common approach to making these problems more manageable is to project these continuous physical quantities onto a space defined by a complete set of orthogonal functions. Consequently, the design of these functions is transformed into the design of their projection lengths in the orthogonal basis space, making the optimization of continuous functions feasible. Approximations can then be derived by limiting the problem to a finite-dimensional space through a suitable truncation.

Inspired by the pattern design methods in antenna theory, the continuous pattern functions $\{\boldsymbol{\theta}_{mk}(\mathbf{s})\}$ can be projected onto an orthogonal space [30] to efficiently solve the optimization problem (7.12). By invoking Fourier bases to expand the spatial continuous pattern functions, for an arbitrary continuous and absolutely integrable pattern function $\boldsymbol{\theta}_{mk}(\mathbf{s})$, can be expressed as

$$\boldsymbol{\theta}_{mk}(\mathbf{s}) = \sum_{n_x=-\infty}^{\infty} \sum_{n_y=-\infty}^{\infty} \sum_{n_z=-\infty}^{\infty} \mathbf{w}_{mk,\mathbf{n}} \Psi_{\mathbf{n}}(\mathbf{s}), \quad \mathbf{s} \in \mathcal{S}_{\mathrm{T}m}, \tag{7.18}$$

where $\mathbf{n} = (n_x, n_y, n_z) \in \mathbb{Z}^3$, $\mathbf{w}_{mk,\mathbf{n}} \in \mathbb{C}^3$ is the projection length in the wavenumber domain, and $\Psi_{\mathbf{n}}(\mathbf{s}) \in \mathbb{C}$ is the Fourier base function. More specifically, we have

$$\begin{aligned} \mathbf{w}_{mk,\mathbf{n}} &= \frac{1}{\sqrt{A_{Tm}}} \int_{\mathbf{s} \in \mathcal{S}_{\mathrm{T}m}} \boldsymbol{\theta}_{mk}(\mathbf{s}) \Psi_{\mathbf{n}}^*(\mathbf{s}) d\mathbf{s}, \\ \Psi_{\mathbf{n}}(\mathbf{s}) &= \frac{1}{\sqrt{A_{Tm}}} e^{2\pi j \left(\frac{n_x}{L_x}\left(s_x - \frac{L_x}{2}\right) + \frac{n_y}{L_y}\left(s_y - \frac{L_y}{2}\right) + \frac{n_z}{L_z}\left(s_z - \frac{L_z}{2}\right) \right)}, \quad \mathbf{s} \in \mathcal{S}_{\mathrm{T}m}, \end{aligned} \tag{7.19}$$

where L_x, L_y, and L_z denote the maximum projection lengths of volume $\mathcal{S}_{\mathrm{T}m}$ on the x-, y-, and z-axis of 3D coordinate system, respectively. The Fourier bases $\Psi_{\mathbf{n}}(\mathbf{s})$ satisfy the 3D orthogonality, i.e.,

$$\int_{\mathbf{s} \in \mathcal{S}_{\mathrm{T}m}} \Psi_{\mathbf{n}}(\mathbf{s}) \Psi_{\mathbf{n}'}^*(\mathbf{s}) d\mathbf{s} = \begin{cases} 1 & \mathbf{n} = \mathbf{n}', \\ 0 & \mathbf{n} \neq \mathbf{n}'. \end{cases} \tag{7.20}$$

Using this series expansion, the continuous pattern functions $\boldsymbol{\Theta}(\mathbf{s})$ in problem ($\mathcal{P}3.3$) can be equivalently replaced by their discrete projection lengths $\mathbf{w}_{mk,\mathbf{n}}$, $\forall m \in \mathcal{M}$, $k \in \mathcal{K}$, in the wavenumber space. To this end, we replace $\int_{\mathcal{S}_{\mathrm{T}m}} \mathbf{G}_{mk}(\mathbf{s}) \boldsymbol{\theta}_{mj}(\mathbf{s}) d\mathbf{s}$ by its discrete form, given by

$$\int_{\mathcal{S}_{\mathrm{T}m}} \mathbf{G}_{mk}(\mathbf{s}) \boldsymbol{\theta}_{mj}(\mathbf{s}) d\mathbf{s} = \sum_{n_x=-\infty}^{\infty} \sum_{n_x=-\infty}^{\infty} \sum_{n_x=-\infty}^{\infty} \boldsymbol{\Omega}_{mk,\mathbf{n}} \mathbf{w}_{mj,\mathbf{n}}, \forall m \in \mathcal{M}, \ \forall k, j \in \mathcal{K}, \tag{7.21}$$

where

$$\boldsymbol{\Omega}_{mk,\mathbf{n}} = \int_{\mathbf{s} \in \mathcal{S}_{\mathrm{T}m}} \mathbf{G}_{mk}(\mathbf{s}) \Psi_{\mathbf{n}}(\mathbf{s}) d\mathbf{s} \tag{7.22}$$

is the Fourier transform of function $\mathbf{G}_{mk}(\mathbf{s})\mathrm{rect}(\mathbf{s} \in \mathcal{S}_{\mathrm{T}m})$ over the CAPA-MIMO aperture $\mathcal{S}_{\mathrm{T}m}$ at wavenumber $\boldsymbol{\kappa}_{\mathbf{n}} = 2\pi[\frac{n_x}{L_x}, \frac{n_y}{L_y}, \frac{n_z}{L_y}]$, while $\mathrm{rect}(\cdot)$ is the generalized rectangular function whose value takes one/zero when the condition in its argument is true/false. More specifically, according to Parseval's theorem, we have

$$\int_{\mathbf{s} \in \mathcal{S}_{\mathrm{T}m}} \|\mathbf{G}_{mk}(\mathbf{s})\|_F^2 d\mathbf{s} = \sum_{n_x=-\infty}^{\infty} \sum_{n_x=-\infty}^{\infty} \sum_{n_x=-\infty}^{\infty} \|\boldsymbol{\Omega}_{mk,\mathbf{n}}\|_F^2, \tag{7.23}$$

which indicates that $\|\mathbf{G}_{mk}(\mathbf{s})\|_F^2$ can be viewed as the channel gain in the spatial domain at position $\mathbf{r}_k$, while $\|\boldsymbol{\Omega}_{mk,\mathbf{n}}\|_F^2$ can be viewed as the channel gain in the wavenumber domain at wavenumber $\boldsymbol{\kappa}_{\mathbf{n}}$. Finally, the constraint (7.17c) can be replaced by

$$\int_{\mathcal{S}_{\mathrm{T}m}} \|\boldsymbol{\theta}_{mk}(\mathbf{s})\|^2 \, \mathrm{d}\mathbf{s} = \sum_{n_x=-\infty}^{\infty} \sum_{n_x=-\infty}^{\infty} \sum_{n_x=-\infty}^{\infty} \|\boldsymbol{\Omega}_{mk,\mathbf{n}} \mathbf{w}_{mk,\mathbf{n}}\|^2. \tag{7.24}$$

Therefore, by invoking (7.23) and (7.24) the functional programming ($\mathcal{P}3.3$) can be reformulated as a common digital signal processing problem. However, the infinite number of expansion items results in the optimization of an unacceptable number of projection lengths for practical computing devices. To address this challenge, some recent works [18,25], have shown that the value of the channel gain in the wavenumber domain, i.e., $\|\boldsymbol{\Omega}_{mk,\mathbf{n}}\|_F^2$, is high in the low-wavenumber band and negligible in the high-wavenumber band in most cases. For example, the authors in [18], have analytically proved that, the wavenumber-domain channel gain within the band of $[-\kappa_0, \kappa_0]$ dominates in the whole wavenumber domain. Inspiring by this property, we use the following results from [25, Proposition 1], to approximate the electric field and transmit power as

$$\int_{\mathcal{S}_{\mathrm{T}m}} \mathbf{G}_{mk}(\mathbf{s})\boldsymbol{\theta}_{mj}(\mathbf{s})\mathrm{d}\mathbf{s} \approx \sum_{n_x=-N_x}^{N_x} \sum_{n_x=-N_y}^{N_y} \sum_{n_x=-N_z}^{N_z} \boldsymbol{\Omega}_{mk,\mathbf{n}} \mathbf{w}_{mj,\mathbf{n}}, \forall m \in \mathcal{M}, \ \forall k, j \in \mathcal{K}, \tag{7.25a}$$

$$\int_{\mathcal{S}_{\mathrm{T}m}} \|\boldsymbol{\theta}_{mk}(\mathbf{s})\|^2 \mathrm{d}\mathbf{s} \approx \sum_{n_x=-N_x}^{N_x} \sum_{n_x=-N_y}^{N_y} \sum_{n_x=-N_z}^{N_z} \|\mathbf{w}_{mk,\mathbf{n}}\|^2, \ \forall m \in \mathcal{M}, \forall k \in \mathcal{K}, \tag{7.25b}$$

where the numbers of reserved expansion items are given by

$$N_x = \left\lceil \frac{\kappa_0 L_x}{2\pi} \right\rceil, \ N_y = \left\lceil \frac{\kappa_0 L_y}{2\pi} \right\rceil, \ N_z = \left\lceil \frac{\kappa_0 L_z}{2\pi} \right\rceil. \tag{7.26}$$

By exploiting the above results, the total number of expansion items becomes finite, which is acceptable for practical computing devices.

Accordingly, we employ the continuous–discrete transforms with finite-item approximation in (7.25) to reformulate ($\mathcal{P}3.3$) to the following tractable form

$$(\mathcal{P}3.4) \quad \max_{\mathbf{v},t} \quad t \tag{7.27a}$$

$$\text{s.t.} \quad \sum_{m \in \mathcal{M}} \sum_{j \in \mathcal{K}} \left| \mathbf{h}_{mk}^H \mathbf{v}_{mj} \right|^2 - 2\Re\left\{ \sum_{m \in \mathcal{M}} \mathbf{h}_{mk}^H \mathbf{v}_{mk} \right\} \geq t, \quad \forall \in \mathcal{K}, \tag{7.27b}$$

$$\sum_{k \in \mathcal{K}} \|\mathbf{v}_{mk}\|^2 \leq P_{\mathrm{T}}, \quad \forall m \in \mathcal{M}, \tag{7.27c}$$

where $\mathbf{h}_{mk} \in \mathbb{C}^{N_F \times 1}$ and $\mathbf{v}_{mk} \in \mathbb{C}^{N_F \times 1}$ are the vectorized set of $\mathbf{h}_{mk,\mathbf{n}} \triangleq \boldsymbol{\psi}_k^H \boldsymbol{\Omega}_{mk,\mathbf{n}}$ and $\mathbf{w}_{mk,\mathbf{n}}$ for all $\mathbf{n} = (n_x, n_y, n_z) \in \{\{-N_x, \ldots, N_x\}, \{-N_y, \ldots, N_y\}, \{-N_z, \ldots, N_z\}\}$ with $N_F = (2N_x + 1)(2N_y + 1)(2N_z + 1)$. The optimization problem ($\mathcal{P}3.4$) is a standard quadratically constrained quadratic programming (QCQP) and can be efficiently solved via standard convex problem solvers such as CVX [37]. Then, the optimal projection length $\mathbf{w}^\star_{mk,\mathbf{n}}$, for all $m \in \mathcal{M}$ and $k \in \mathcal{K}$ can be used to obtain the continuous pattern functions $\{\boldsymbol{\theta}_{mk}(\mathbf{s})\}_{k\in\mathcal{K}}^{m\in\mathcal{M}}$ as follows:

$$\boldsymbol{\theta}^\star_{mk}(\mathbf{s}) = \sum_{n_x=-N_x}^{N_x} \sum_{n_x=-N_y}^{N_y} \sum_{n_x=-N_z}^{N_z} \mathbf{w}^\star_{mk,\mathbf{n}} \Psi_{\mathbf{n}}(\mathbf{s}), \quad \mathbf{s} \in \mathcal{S}_{\mathrm{T}m}. \tag{7.28}$$

Remark 1. The proposed PDM framework for CAPA-MIMO inherently balances energy-efficiency and SE. Increasing transmit power can enhance the achievable minimum SE or sum-SE, but it also leads to higher energy consumption, as captured in the power constraint formulation. PDM addresses this challenge by concentrating energy in low-wavenumber components, which contribute more to channel gain, thereby minimizing energy dispersion across less impactful high-wavenumber components. Moreover, the finite-term approximation of the Fourier series in (7.25) ensures that only the most influential basis functions are retained, optimizing performance while controlling energy costs. Future developments could explore energy-aware or energy-efficiency-driven optimization objectives, as well as adaptive truncation strategies that dynamically adjust to rate and energy demands—making the framework well suited for energy-constrained environments such as Internet of Things and edge computing deployments.

7.2.3 Overall algorithm and complexity analysis

Based on the proposed approach, we can alternatively optimize the auxiliary variable, the polarization combiner at each user, and continuous pattern functions at transmitters until convergence. The overall algorithm is summarized as **Algorithm 1**, where ϵ is the predefined threshold to claim the convergence. All pattern functions $\{\boldsymbol{\theta}_{mk}(\mathbf{s})\}_{k\in\mathcal{K}}^{m\in\mathcal{M}}$ and user combiners $\boldsymbol{\Psi} = \{\boldsymbol{\psi}_k\}_{k\in\mathcal{K}}$ are randomly initialized by the standard complex-Gaussian stochastic processes and variables, respectively.

The primary source of computational complexity in the proposed pattern design scheme arises from the update processes for $\boldsymbol{\rho}$, $\boldsymbol{\psi}_k$, $\mathbf{w}$, and the pattern function $\boldsymbol{\Theta}(\mathbf{s})$. The computational complexity of updating $\boldsymbol{\rho}$ is mainly caused by calculating the MSE E_k and is $\mathcal{O}(12K^2 I_s + K^2 + 4K)$, where I_s is the sampling number of the continuous integral operation [25]. The computational complexity of $\boldsymbol{\psi}_k$ stems from the matrix inversion in (7.16), which involves

Algorithm 1 Proposed Pattern Design Algorithm for Solving (7.12).

Input: Channel functions $\mathbf{G}_{mk}(\mathbf{s})$ with respect to $\mathbf{s} \in \mathcal{S}_{\mathrm{T}m}$ and for all $m \in \mathcal{M}$ and $k \in \mathcal{K}$.
Initialize: Set $i = 0$. Initialize $\boldsymbol{\Psi} = \{\boldsymbol{\psi}_k^{(0)}\}$ and $\boldsymbol{\Theta}(\mathbf{s}) = \{\boldsymbol{\theta}_{mk}^{(0)}(\mathbf{s})\}$ to a feasible value and obtain and $\mathcal{R}_k^{(0)}$.
Repeat
    Update $\rho_k^{(i)}$, $\forall k \in \mathcal{K}$, by using (7.13) for given $\boldsymbol{\Psi} = \{\boldsymbol{\psi}_k^{(i)}\}$ and $\boldsymbol{\Theta}(\mathbf{s}) = \{\boldsymbol{\theta}_{mk}^{(i)}(\mathbf{s})\}$.
    Update $\boldsymbol{\psi}_k^{(i+1)}$, $\forall k \in \mathcal{K}$, by using (7.16) for given $\boldsymbol{\rho} = \{\rho_k^{(i)}\}$ and $\boldsymbol{\Theta}(\mathbf{s}) = \{\boldsymbol{\theta}_{mk}^{(i)}(\mathbf{s})\}$.
    Update $\mathbf{w}_{mk,\mathbf{n}}^{(i+1)}$, $\forall k \in \mathcal{K}, m \in \mathcal{M}$, by solving ($\mathcal{P}$2.4) for given $\boldsymbol{\rho} = \{\rho_k^{(i)}\}$ and $\boldsymbol{\Psi} = \{\boldsymbol{\psi}_k^{(i+1)}\}$.
    Update $\boldsymbol{\theta}_{mk}^{(i+1)}(\mathbf{s})$, $\forall k \in \mathcal{K}, m \in \mathcal{M}$, by using (7.28) with $\mathbf{w}_{mk,\mathbf{n}}^{(i+1)}$.
    Update $\mathcal{R}_k^{(i+1)}$, $\forall k \in \mathcal{K}$, by using (7.10) and $\boldsymbol{\theta}_{mk}^{(i+1)}(\mathbf{s})$.
    Update the objective value as $\min_{k=1,\ldots,K} \mathcal{R}_k^{(i+1)}$.
    Set $i = i + 1$.
Until: The fractional reduction of the objective function value falls below a predefined threshold ϵ or prespecified maximum number of iterations, $i_{\max}$, is reached.
Return: Optimized $\boldsymbol{\theta}_{mk}^{\star}(\mathbf{s})$ and $\mathcal{R}_k^{\star}$.

operations of order $\mathcal{O}(10K^2 I_s + K^2 + 21K)$. Updating $\mathbf{w}$ involves solving the QCQP problem in (7.27) using CVX, with a computational complexity of $\mathcal{O}((K+M)((K+M)N_F)^3)$. Finally, the complexity of updating $\boldsymbol{\theta}_{mk}(\mathbf{s})$, caused by the discrete–continuous transform in (7.28), is $\mathcal{O}(I_s N_F)$. Since the aperture of continuous MIMO transmitters are nearly continuous, so it is reasonable to assume $I_s \gg K$ and $N_F \gg K$. Therefore, the overall computational complexity of **Algorithm 1** can be approximated by $\mathcal{O}(I_m(I_s K^2 + (K+M)^4 N_F^3))$ where I_m is the iteration number required by the convergence of the minimum SE among all users.

7.2.4 Numerical results and discussion

In this subsection, we present numerical results from Monte Carlo simulations to validate the effectiveness of the PDM design for joint CAPA beamforming, power allocation, and combiner design in the CF-CAPA system. Unless otherwise specified, the simulation parameters are set as follows [25]. The continuous MIMO aperture has a square shape with the area of $\mathcal{A}_{\mathrm{T}m} = 0.25$ m^2. It is assumed that the transmit CAPA m is deployed within the x–y-plane and centered at the origin of the coordinate system, where

$$\mathcal{S}_{T_m} = \left\{ \mathbf{s} = [s_x, s_y, s_z]^T \,\big|\, |s_x| \le \frac{L_x}{2}, s_y| \le \frac{L_y}{2} \right\}, \tag{7.29}$$

with $L_x = L_y = \sqrt{\mathcal{A}_{\mathrm{T}m}}$. We assume that $M = 10$ CAPA-MIMO transmitters are deployed in an area of 100×100 m^2 to serve $K = 4$ users. The frequency of information-carrying current density $\mathbf{j}_m(\mathbf{s})$ and electric field $\mathbf{E}(\mathbf{r}_k)$ is set to $f = 2.4$ GHz. The maximum transmit power of each CAPA-MIMO is set to $P_T = 50$ mA^2, and the noise is set to $\sigma^2 = 5.6 \times 10^{-3}\mathrm{V}^2/\mathrm{m}^2$. The sampling number of integral $\int_{\mathcal{S}_{\mathrm{T}m}} \mathrm{d}\mathbf{s}$ is set to $I_s = 1024$.

As a benchmark for comparison, we consider the WDM scheme. The core idea of the WDM design is to directly utilize multiple orthogonal Fourier basis functions to independently generate beam patterns for different users. For the given WDM-based beam patterns, we evaluate two scenarios: full-power transmission and a fairness-oriented configuration. Additionally, to highlight the performance gain achieved by the distributed network, we include a reference case where a single CAPA-MIMO transmitter is placed at the center of the communication area. This CAPA-MIMO transmitter transmits with the same power as each distributed CAPA-MIMO transmitter, and the total area is set to $M\mathcal{A}_{\mathrm{T}m}$ m^2.

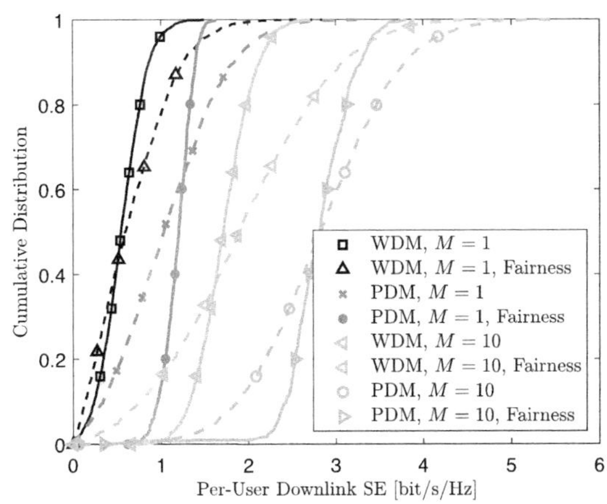

Figure 7.2:
CDF of the per-user achievable SE for different pattern designs and network setups ($K = 4$, $N_F = 81$).

Fig. 7.2 shows the cumulative distribution function (CDF) of the minimum achievable SE among users, comparing the performance of the PDM and WDM designs under both optimized and full-power transmission schemes. The results clearly demonstrate the performance gains achieved by the distributed CAPA-MIMO design over the single-transmitter setup, as well as the additional benefits provided by transmission power optimization. More specifically, the distributed design with PDM and transmission power optimization offers up to an 86% gain over its WDM counterpart and up to a 145% gain compared to the single CAPA-MIMO setup in terms of the average minimum SE. We would like to highlight that the wavenumber-domain channel gain of CAPA-MIMO is primarily concentrated within the low-wavenumber band $[-\kappa_0, \kappa_0]$. With $N_F = 81$ ($N_x = N_y = 3$ and $N_z = 0$), the number of functional DoFs becomes sufficiently large to enable effective design of continuous pattern functions across different CAPA-MIMO transmitters. As a result, the contribution of high-wavenumber components in the Fourier expansion is negligible, and their exclusion has minimal impact on pattern design. Therefore, increasing N_F beyond this point only adds complexity without yielding substantial performance gains.

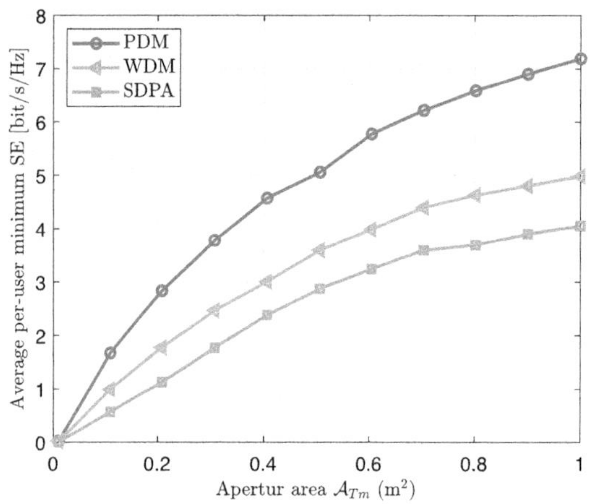

Figure 7.3:
Average per-user minimum SE versus the aperture area ($M = 10$, $K = 4$, $N_F = 81$).

Fig. 7.3 illustrates the impact of the aperture area on system performance under different pattern designs. The figure includes the conventional SPDA design, where the surface is populated with discrete antenna elements, each having an effective area of $\frac{\lambda^2}{4\pi}$ and spaced at $\frac{\lambda}{2}$ intervals. Beamforming for the SPDA can be achieved using conventional optimization techniques or suboptimal linear designs, such as ZF and maximum ratio transmission (MRT). In this study, we employ ZF beamforming to suppress interuser interference. The results demonstrate that the proposed CAPA design, based on the PDM approach, consistently outperforms both the WDM-based design and the SPDA across the considered aperture area range. Specifically, at an aperture size of $\mathcal{A}_{Tm} = 0.3\ \text{m}^2$, CAPA achieves performance improvements of 54% and 110% compared to the WDM and SPDA designs, respectively. Moreover, it is observed that as the aperture size increases, the channels between different users tend to become more orthogonal, thereby amplifying the performance gain of the PDM-based design over the benchmark schemes.

7.3 Future research direction and conclusion

This chapter examined the integration of CAP into CF-mMIMO networks, with a focus on fairness-aware beam pattern design. The proposed approach ensures equitable service across users while leveraging the spatial richness of CAPA. A particularly promising direction for future research is the application of the EMIT framework to better understand near-field dynamics. By embedding electromagnetic principles into system modeling, EMIT enables a more accurate representation of wave interactions, which can significantly enhance the design and performance of near-field wireless systems. Despite its potential, EMIT currently faces several challenges in wireless channel modeling. These include the difficulty of reconciling continuous and discrete modeling approaches, and the complexity of simulating metasurface-based channels, especially in near-field regimes. While EMIT captures rich spatial and dimensional information, further research is needed to fully exploit space-domain characteristics for enhanced multiplexing and diversity gains. Integrating probabilistic and deterministic models within EMIT remains a complex but necessary step toward optimizing system-level performance. Energy sustainability also emerges as a critical concern, particularly given the high power demands and hardware costs associated with advanced technologies like CF-mMIMO and millimeter-wave communications. EMIT aims to address these challenges by improving system efficiency and capacity, although practical implementation and validation are still in early stages.

Looking ahead, the proposed EMIT-based analysis and optimization methodology can be extended to evaluate key performance indicators such as cell throughput, spectral efficiency, and block error rate (BLER) in CF-mMIMO systems. Beyond fairness-aware design, the beam pattern optimization framework can also be adapted to maximize energy and spectral efficiency under various deployment scenarios. Moreover, the fusion of CAPA with advanced multiplexing and beamforming techniques unlocks a wide range of transformative applications for next-generation wireless networks. CAPA systems support high-capacity cell-free deployments with uniform coverage, making it ideal for dense indoor environments. Its compatibility with integrated sensing and communication (ISAC) could enable precise localization, gesture recognition, and environmental awareness—key features for smart infrastructure and autonomous systems. In immersive media, CAPAs can underpin holographic beamforming for extended reality (XR) and Metaverse platforms, offering real-time spatial rendering and interaction. For industrial applications, energy-efficient PDM designs support low-latency and high-reliability communication, making CAPA suitable for Industrial IoT, factory automation, and mission-critical control systems. Furthermore, CAPA's high beamforming gain and spatial precision make it well suited for wireless energy transfer and simultaneous wireless information and power transfer (SWIPT). In SWIPT scenarios, a CAPA enables unprecedented concentration of electromagnetic energy, enhancing both wireless information transfer (WIT) and wireless power transfer (WPT) [38]. These capabilities collectively position CAPA as a foundational technology for 6G, capable of meeting the stringent demands of emerging wireless ecosystems.

Another interesting future direction is the application of the proposed EMIT analysis to evaluate the secrecy capacity of wireless systems. This involves designing the beam pattern for physical layer security-based solutions, including artificial noise injection, to assess how well the system can protect information from eavesdroppers while maintaining reliable communication. Finally, noting that EMIT is based on continuous domain analysis, another promising direction could be investigating EMIT signal design for wireless power transfer or SWIPT. One promising application could be integrated sensing, communications, and powering for energy and spectrally efficient usage of EM signals. These directions highlight the versatility of EMIT analysis in improving the performance, security, and efficiency of advanced wireless systems.

References

[1] J. Zhang, E. Björnson, M. Matthaiou, D.W.K. Ng, H. Yang, D.J. Love, Prospective multiple antenna technologies for beyond 5G, IEEE Journal on Selected Areas in Communications 38 (8) (Aug. 2020) 1637–1660.

[2] Ericsson, Ericsson mobility report, Online, June 2024, [Online]. Available, http://www.ericsson.com/mobility-report. (Accessed 24 March 2025).

[3] T. Gong, et al., Holographic MIMO communications: theoretical foundations, enabling technologies, and future directions, IEEE Communications Surveys and Tutorials 26 (1) (2024) 196–257, First quarter.

[4] D. Dardari, Communicating with large intelligent surfaces: fundamental limits and models, IEEE Journal on Selected Areas in Communications 38 (11) (Nov. 2020) 2526–2537.

[5] N. Shlezinger, et al., Dynamic metasurface antennas for 6G extreme massive MIMO communications, IEEE Wireless Communications 28 (2) (Apr. 2021) 106–113.

[6] A. Sayeed, N. Behdad, Continuous aperture phased MIMO: basic theory and applications, in: Proc. IEEE Allerton, Sept. 2010, pp. 1196–1203.

[7] M.V. Katwe, A. Kaushik, L. Mohjazi, M. Abualhayja'a, D. Dardari, K. Singh, M.A. Imran, M.M. Butt, O.A. Dobre, An overview of intelligent meta-surfaces for 6G and beyond: opportunities, trends, and challenges, IEEE Communications Standards Magazine 8 (4) (Dec. 2024) 62–69.

[8] R. Singh, J. An, A. Kaushik, C. Yuen, Y.-J.A. Zhang, O.A. Dobre, Redefining electromagnetic signal and information theory with programmable surfaces: open critical questions and future horizon, TechRxiv, preprint. Online, https://doi.org/10.36227/techrxiv.175621821.18352632/v1, Aug. 2025.

[9] O.M. Bucci, G. Franceschetti, On the degrees of freedom of scattered fields, IEEE Transactions on Antennas and Propagation 37 (7) (Jul. 1989) 918–926.

[10] D.A. Miller, Communicating with waves between volumes: evaluating orthogonal spatial channels and limits on coupling strengths, Applied Optics 39 (11) (2000) 1681–1699.

[11] N. Decarli, D. Dardari, Communication modes with large intelligent surfaces in the near field, IEEE Access 9 (2021) 165648–165666.

[12] L. Ding, E.G. Ström, J. Zhang, Degrees of freedom in 3d linear large-scale antenna array communications—a spatial bandwidth approach, IEEE Journal on Selected Areas in Communications 40 (10) (Oct. 2022) 2805–2822.

[13] Z. Xie, Y. Liu, J. Xu, X. Wu, A. Nallanathan, Performance analysis for near-field MIMO: discrete and continuous aperture antennas, IEEE Wireless Communications Letters 12 (12) (Dec. 2023) 2258–2262.
[14] M.A. Jensen, J.W. Wallace, Capacity of the continuous-space electromagnetic channel, IEEE Transactions on Antennas and Propagation 56 (2) (Feb. 2008) 524–531.
[15] W. Jeon, S.-Y. Chung, Capacity of continuous-space electromagnetic channels with lossy transceivers, IEEE Transactions on Information Theory 64 (3) (Mar. 2018) 1977–1991.
[16] M.D. Migliore, Horse (electromagnetics) is more important than horseman (information) for wireless transmission, IEEE Transactions on Antennas and Propagation 67 (4) (Apr. 2019) 2046–2055.
[17] Z. Wan, J. Zhu, L. Dai, Can continuous aperture MIMO obtain more mutual information than discrete MIMO?, IEEE Communications Letters 27 (12) (Dec. 2023) 3185–3189.
[18] Z. Wan, J. Zhu, Z. Zhang, L. Dai, C.-B. Chae, Mutual information for electromagnetic information theory based on random fields, IEEE Transactions on Communications 71 (4) (Apr. 2023) 1982–1996.
[19] S. Mikki, The Shannon information capacity of an arbitrary radiating surface: an electromagnetic approach, IEEE Transactions on Antennas and Propagation 71 (3) (Mar. 2023) 2556–2570.
[20] L. Sanguinetti, A.A. D'Amico, M. Debbah, Wavenumber-division multiplexing in line-of-sight holographic MIMO communications, IEEE Transactions on Wireless Communications 22 (4) (Apr. 2023) 2186–2201.
[21] C. Ouyang, Z. Wang, X. Zhang, Y. Liu, Diversity and multiplexing for continuous-aperture array (CAPA)-based communications, IEEE Transactions on Wireless Communications 24 (10) (Oct. 2025) 8662–8680.
[22] S. Hu, F. Rusek, O. Edfors, Beyond massive MIMO: the potential of data transmission with large intelligent surfaces, IEEE Transactions on Signal Processing 66 (10) (2018) 2746–2758.
[23] Z. Sun, Y. Jing, X. Yu, SINR analysis and interference mitigation for continuous-aperture holographic MIMO uplink, IEEE Transactions on Vehicular Technology 74 (1) (Jan. 2025) 1747–1752.
[24] B. Zhao, C. Ouyang, X. Zhang, Y. Liu, Continuous aperture array (CAPA)-based wireless communications: capacity characterization, arXiv preprint, arXiv:2406.15056, 2024.
[25] Z. Zhang, Z. Dai, Pattern-division multiplexing for multi-user continuous-aperture MIMO, IEEE Journal on Selected Areas in Communications 41 (8) (Aug. 2023) 2350–2366.
[26] M. Qian, L. You, X.-G. Xia, X. Gao, On the spectral efficiency of multi-user holographic MIMO uplink transmission, IEEE Transactions on Wireless Communications 23 (10) (Oct 2024) 15421–15434.
[27] C. Ouyang, Y. Liu, X. Zhang, Aperture selection for CAP arrays (CAPAs), in: Proc. IEEE GLOBECOM, Dec. 2024, pp. 1263–1268.
[28] Z. Wang, C. Ouyang, Y. Liu, Beamforming optimization for continuous aperture array (CAPA)-based communications, IEEE Transactions on Wireless Communications (2025) 1.
[29] S. Zhang, B. Di, A. Kaushik, Y.C. Eldar, Holographic-pattern-based multiuser beam training in RHS-aided hybrid near-field and far-field communications, IEEE Transactions on Wireless Communications 24 (10) (Oct. 2025) 8892–8907.
[30] F.K. Gruber, E.A. Marengo, New aspects of electromagnetic information theory for wireless and antenna systems, IEEE Transactions on Antennas and Propagation 56 (11) (Nov. 2008) 3470–3484.
[31] A.A. D'Amico, A.d.J. Torres, L. Sanguinetti, M. Win, Cramér–Rao bounds for holographic positioning, IEEE Transactions on Signal Processing 70 (Nov. 2022) 5518–5532.
[32] W.C. Chew, Waves and Fields in Inhomogenous Media, vol. 16, Wiley, Hoboken, NJ, USA, Feb. 1999.
[33] D. Tse, P. Viswanath, Fundamentals of Wireless Communication, Cambridge University Press, Cambridge, U.K., 2005.
[34] M. Mohammadi, Z. Mobini, H.Q. Ngo, M. Matthaiou, Ten years of research advances in full-duplex massive MIMO, IEEE Transactions on Communications 73 (3) (Mar. 2025) 1756–1786.
[35] H. Li, Z. Zhou, Y. He, W. Sun, Y. Li, I. Liberal, N. Engheta, Geometry-independent antenna based on epsilon-near-zero medium, Nature Communications 13 (1) (2022) 3568.
[36] Q. Shi, M. Razaviyayn, Z.-Q. Luo, C. He, An iteratively weighted MMSE approach to distributed sum-utility maximization for a MIMO interfering broadcast channel, IEEE Transactions on Signal Processing 59 (9) (Sept. 2011) 4331–4340.

[37] M. Grant, S. Boyd, CVX: Matlab software for disciplined convex programming, version 2.1 [online], Available, http://cvxr.com/cvx, 2014.
[38] Q. Huang, J. Hu, Y. Zhao, K. Yang, Holographic integrated data and energy transfer, IEEE Transactions on Wireless Communications 23 (12) (Dec. 2024) 18987–19002.

CHAPTER 8

FAS-RIS empowered 6G wireless networks

Tuo Wu[a], Kangda Zhi[b], Junteng Yao[c], Xiazhi Lai[d], Jie Tang[a], and Kai-Kit Wong[e]

[a]South China University of Technology, Guangdong, China [b]School of Electrical Engineering and Computer Science, Technical University of Berlin, Berlin, Germany [c]Faculty of Electrical Engineering and Computer Science, Ningbo University, Zhejiang, China [d]School of Computer Science, Guangdong University of Education, Guangdong, China [e]Department of Electronic and Electrical Engineering, University College London, United Kingdom

8.1 Introduction and background

The rapid evolution of wireless communications is driven by the ever-growing demand for higher data rates, extensive connectivity, and ultra-low latency [1], highlighting the need for innovative solutions to overcome physical and technical limitations in current communication systems. The emergence of the Internet of Things (IoT) and the imminent deployment of sixth-generation (6G) networks have intensified these requirements, creating a pressing need for novel communication paradigms that can support increasingly diverse applications with unprecedented performance metrics.

As the wireless communications landscape continues to evolve, global anticipation for sixth-generation (6G) networks is rapidly increasing [2]. Building on the achievements of fifth-generation (5G) systems, 6G is envisioned to deliver unprecedented data rates, enhanced reliability, broader coverage, and support for massive connectivity, all of which will empower a host of disruptive applications such as immersive experiences [1], ultrareliable low-latency communications (URLLC), pervasive connectivity, artificial intelligent (AI)-driven services, and integrated sensing and communication (ISAC) [2]. These demanding requirements present significant challenges for existing wireless technologies, necessitating fundamental innovations at both the physical and network layers.

8.1.1 Challenges in current wireless systems

Multiple-input multiple-output (MIMO) technology has historically enhanced wireless communication by leveraging spatial diversity. However, MIMO systems require antennas

Electromagnetic Signal and Information Theory
https://doi.org/10.1016/B978-0-44-336354-2.00015-0
Copyright © 2027 Elsevier Inc. All rights are reserved, including those for text and data mining, AI training, and similar technologies.

to be spaced at least half a wavelength apart to achieve significant diversity gains. This spatial requirement poses a major challenge for compact devices such as smartphones and IoT sensors, which are inherently space-constrained and cannot accommodate multiple antennas without compromising design and usability [3]. As device miniaturization continues to be a driving force in consumer electronics, this spatial limitation becomes increasingly problematic.

A critical cornerstone of current 5G networks is massive multiple-input multiple-output (MIMO), which relies on a large number of antennas to harness spatial diversity [4]. However, the performance of massive MIMO is inherently constrained by the number of available radio-frequency (RF) chains at the terminals. In striving to meet 6G's ambitious goals, one prevailing approach is to further increase the number of antennas at the base station, giving rise to extra-large MIMO (XL-MIMO). However, escalating hardware costs, heightened power consumption, and increased operational complexity present significant barriers to scaling up conventional MIMO systems [2]. These challenges have prompted researchers to explore alternative approaches that can deliver comparable or superior performance without the associated drawbacks of traditional MIMO systems.

8.1.2 Fluid antenna systems: a promising solution

To address these bottlenecks, fluid antenna systems (FAS) have emerged as a promising solution. FAS utilizes liquid-based or reconfigurable pixel-based antennas that can dynamically adjust their position within a confined space to optimize signal reception. This flexibility effectively exploits spatial degrees of freedom, enabling significant diversity gains even in devices with limited physical space. By dynamically moving to the optimal position, a fluid antenna can adapt to changing signal environments, enhancing communication performance without the need for multiple fixed antennas.

FAS have emerged as a promising solution to address these challenges [5–11]. Unlike traditional fixed-position antennas (FPAs) that offer only a single point of transmission, FAS can dynamically reconfigure their shape and position to adapt to rapidly changing radio environments [8]. This flexibility introduces a new degree of freedom (DoF) at the physical layer, one that surpasses the limitations imposed by the number of RF chains. By enabling terminals to access finely defined spatial regions for signal processing, FAS can fully exploit the available spatial diversity, potentially transforming antenna design and performance. In fact, by dynamically switching among multiple ports arranged along a flexible medium, FAS can harness the rich spatial variations of the wireless channel even within a compact form factor. However, the close proximity of these ports inevitably leads to significant spatial correlation among the channel coefficients, which must be accurately modeled to assess system performance and guide practical design [2,12,13].

Recent advancements in FAS technology have demonstrated impressive performance gains in various wireless communication scenarios. These systems leverage different implementation approaches, including liquid metal antennas, mechanically reconfigurable structures, and pixel-based designs, each offering unique advantages in terms of flexibility, response time, and integration capabilities. The ability of FAS to adapt to changing channel conditions in real-time makes it particularly valuable for mobile and dynamic environments, where traditional fixed antenna systems often struggle to maintain optimal performance.

8.1.3 Reconfigurable intelligent surfaces: environmental control

Concurrently, reconfigurable intelligent surfaces (RIS) have gained significant attention as a cost-effective means to customize wireless propagation environments [14,15]. RIS comprises a large array of passive reflective elements whose electromagnetic properties and physical positions can be dynamically adjusted. By intelligently controlling the phase shifts and positions of these elements, RIS can reflect incoming signals in desired directions, effectively enhancing signal strength at target receivers, extending coverage, and improving reliability [16,17]. RIS offers the ability to reconfigure the radio environment, making it particularly suitable for overcoming obstacles, mitigating signal blockages, and enhancing communication in challenging scenarios [18].

The passive nature of RIS elements makes them energy-efficient and cost-effective compared to traditional active relays and repeaters. Furthermore, RIS can be deployed on existing infrastructure such as building facades, interior walls, and roadside structures, requiring minimal additional space and installation effort. This flexibility in deployment, combined with their ability to reshape the electromagnetic environment, positions RIS as a key enabling technology for future wireless networks, particularly in complex urban environments where signal propagation is often hindered by numerous obstacles. Recent advances in RIS technology have explored new paradigms such as indexed multiple access [19] and intelligent metasurfaces for 6G and beyond [20], while emerging work on RIS-aided fluid antenna systems has demonstrated promising performance gains in nonterrestrial networks [21]. Furthermore, fundamental questions regarding electromagnetic signal and information theory with programmable surfaces are being actively investigated [22], opening new research directions for future wireless systems.

8.1.4 FAS-RIS integration: a synergistic approach

The integration of FAS and RIS into a unified FAS-RIS system presents a novel paradigm for wireless communications, especially for 6G networks [8]. While each technology offers significant advantages individually, their combination creates a powerful synergy that addresses multiple challenges simultaneously. FAS provides adaptability at the device level through dynamic antenna positioning, while RIS enables control over the propagation environment through intelligent reflection. Together, they offer unprecedented flexibility in optimizing wireless communication links, particularly in challenging scenarios where traditional approaches fall short.

However, several critical challenges remain to be addressed in the development and deployment of FAS-RIS systems. First, the dynamic nature of both FAS and RIS elements introduces complex channel modeling and estimation problems. Second, the joint optimization of FAS port selection and RIS phase shifts requires sophisticated algorithms that can adapt to rapidly changing environments. Third, the practical implementation of FAS-RIS systems demands innovative materials and control mechanisms that can ensure reliable operation across diverse scenarios.

8.1.5 Chapter contributions and organization

Motivated by these challenges and opportunities, this chapter makes the following key contributions:

- We present a comprehensive framework for FAS-RIS system design and implementation, incorporating various material choices and architectural considerations for different application scenarios.
- We develop detailed models for the channel characteristics and system behavior of FAS-RIS, taking into account both the spatial correlation among FAS ports and the phase-dependent reflection properties of RIS elements.
- We propose and analyze novel optimization strategies for FAS-RIS systems, considering both CSI-based and CSI-free schemes to balance performance with practical implementation constraints.
- We provide extensive performance analysis through theoretical derivations and numerical simulations, demonstrating the advantages of FAS-RIS systems in terms of outage probability, throughput, and energy efficiency.
- We identify and discuss key research directions and potential applications of FAS-RIS technology in future wireless networks, particularly focusing on emerging 6G applications.

Compared to existing works on FAS systems [7] and FAS-RIS integration [8], this chapter makes several distinctive contributions: (1) we provide a unified architectural framework that systematically categorizes FAS-RIS implementations into four types (base station FAS, user equipment FAS, FAS-enabled RIS, and fluid-RIS), offering comprehensive design guidelines for different deployment scenarios; (2) we develop a block-correlation matrix approximation (BCMA) model specifically tailored for FAS-RIS systems, which significantly reduces computational complexity from $O(N^3)$ to approximately $O(\sum_{b=1}^{B} L_b^2)$ while maintaining analytical accuracy; (3) we propose joint optimization strategies for both CSI-based and CSI-free scenarios, addressing practical implementation constraints such as channel estimation overhead; and (4) we provide comprehensive performance analysis including outage probability, throughput optimization, and computational complexity, demonstrating substantial performance gains over traditional fixed-position antenna systems.

By combining the adaptive positioning capabilities of FAS with the environmental customization offered by RIS, FAS-RIS systems can overcome limitations that neither technology could address alone. This synergy allows for more precise and flexible beamforming, improved signal coverage, and enhanced system performance, particularly in environments where direct line-of-sight communication is obstructed or where devices are highly dynamic [7].

In FAS-RIS systems, RIS can be enhanced by incorporating fluid antenna elements, allowing the reflective units to not only adjust their electromagnetic properties but also their physical positions. This FAS-enabled RIS can dynamically adapt to rapidly changing wireless environments, optimizing signal reflection paths and directions in real time. Such adaptability is particularly beneficial in dense urban areas, dynamic scenarios like drone or high-speed train communications, and in millimeter-wave and terahertz communication systems where signal blockage and path loss are significant challenges.

The remainder of this chapter is organized as follows: Section 8.2 presents the fundamentals and architecture of FAS-RIS systems, including various implementation approaches and design considerations. Section 8.3 provides a detailed case study of a practical FAS-RIS system, including system modeling, performance analysis, and optimization techniques. Section 8.4 discusses future research directions and potential applications of FAS-RIS technology in various domains. Finally, Section 8.5 concludes the chapter with a summary of key findings and insights.

8.2 Fundamentals and architecture of FAS-RIS systems

This section introduces the general hardware architecture of FAS-RIS and elaborates on the benefits of FAS-RIS communication systems. The integration of FAS and RIS technologies creates a powerful synergy that enhances wireless communication capabilities through dynamic antenna positioning and intelligent environmental control.

Based on the deployment and configuration, FAS-RIS systems can be categorized into several types as shown in Table 8.1 and illustrated in Fig. 8.1. Each type offers unique

Table 8.1: FAS-RIS System Components: Characteristics and Applications.

Component Type	Application	Material Implementation	Key Features/Advantages
Base station FAS	High-speed backhaul communication	Liquid metal (LM), metallic pixels, nonmetallic liquids (e.g., water-based)	High conductivity, rapid reconfiguration, environmentally friendly options, efficient signal transmission
	Industrial automation	Metallic pixels, LM with pattern control	Precise beam control, microsecond-level switching, adaptive to dynamic environments
	Smart city IoT	Hybrid materials, programmable pixels	Wide coverage, flexible deployment, cost-effective solutions
User equipment FAS	Wearable devices	Stretchable LM, flexible composites	Conformable to body movement, comfortable wear, reliable connection
	Mobile handsets	Planar LM, compact pixel arrays	Low profile, easy integration, efficient space utilization
	IoT sensors	3D liquid structures, hybrid materials	Enhanced coverage, adaptable form factor, energy efficiency
FAS-enabled RIS	Smart surfaces	Metallic pixel arrays, LM coating	High-precision phase control, continuous state adjustment
	Dynamic reflectors	LM-based elements, composite materials	Adaptive reflection, real-time optimization, robust performance
	Multifunctional surfaces	Hybrid materials, programmable elements	Combined functionality, versatile applications, enhanced control
Fluid-RIS	Indoor positioning	MEMS-based arrays, smart materials	Fast response, high precision, reliable tracking
	Outdoor coverage	Weather-resistant LM, robust composites	Environmental adaptability, stable performance
	Harsh environment deployment	Advanced composites, protected LM systems	Enhanced durability, reliable operation, wide condition tolerance

advantages and is suited for different application scenarios. Each of these designs addresses specific challenges in wireless communications and can be selected based on the particular requirements of the deployment scenario.

8.2.1 FAS-RIS system design and implementation

This subsection details the design and implementation aspects of various FAS-RIS system components as categorized in Table 8.1. Each component plays a crucial role in the overall system architecture and contributes to the enhanced performance of FAS-RIS systems.

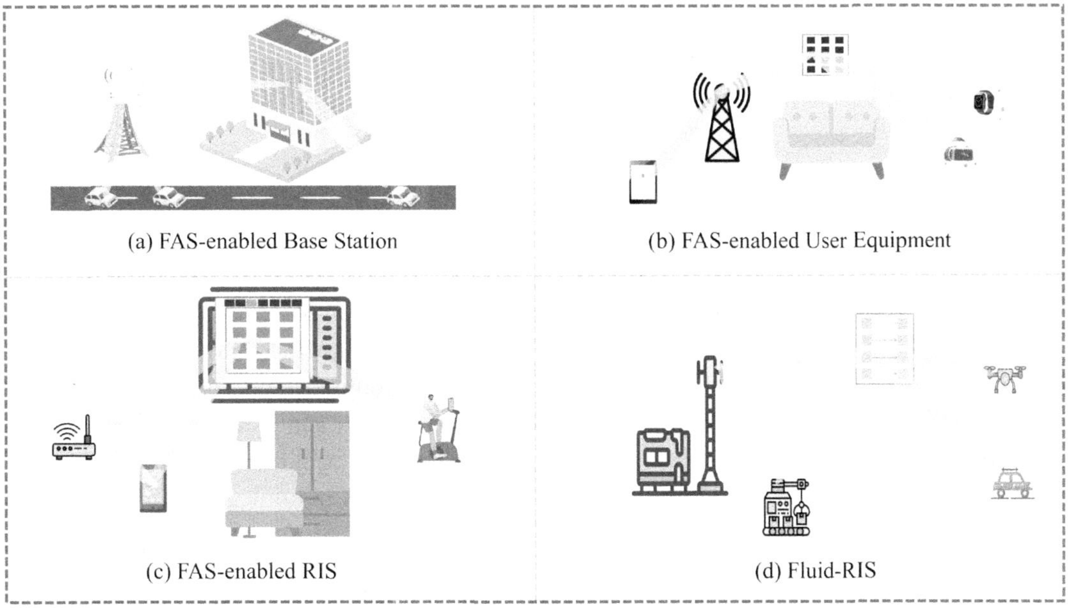

Figure 8.1:
FAS-RIS system design.

FAS-enabled base station design

As shown in Fig. 8.1(a), deploying fluid antenna systems (FAS) with multiple activated ports at the base station (BS) significantly enhances the beamforming gain of FAS-RIS systems. This configuration allows the BS to flexibly concentrate and steer signal strength, effectively compensating for the double-path loss inherent in cascaded channels. Consequently, signal coverage and strength are improved, and a high signal-to-noise ratio (SNR) is maintained even in complex propagation environments, thereby enhancing the quality and stability of communication links.

Moreover, the fluid antennas within the FAS possess dynamic reconfigurability; by altering the shape, position, or composition of the fluid, they can adjust the radiation characteristics and direction of the antenna in real time. This flexibility enables the BS to rapidly adapt to changing channel conditions, maintaining optimal signal transmission performance in various environments.

For instance, in smart city IoT communications, FAS can adjust signal paths in real time to overcome building obstructions. In industrial automation, it adapts to changes in equipment layout. In Vehicle-to-Everything (V2X) applications on high-speed railways and highways, FAS dynamically adjusts signal direction in response to fast-moving vehicles, ensuring stable connections. By combining the dynamic reconfigurability of FAS with FAS-RIS, the system's service quality and communication reliability are enhanced, providing flexible and efficient solutions for various applications.

FAS-enabled user equipment design

As illustrated in Fig. 8.1(b), deploying FAS on user equipment (UE) brings significant advantages to FAS-RIS systems. UEs equipped with FAS can dynamically adjust their antenna's radiation characteristics in real time—such as radiation patterns and directivity—enabling the UE to more effectively align with signals reflected by the RIS. This enhances signal reception and improves link quality. The dynamic reconfigurability of FAS at the user end allows rapid adaptation to changing channel conditions, maintaining optimal SNR even in complex propagation environments like urban canyons or indoor settings with multiple obstructions, thus providing more reliable and stable communication links crucial for applications requiring continuous connectivity.

Moreover, UEs with FAS can actively participate in advanced communication technologies such as spatial multiplexing and user-centric beamforming. By adjusting antenna parameters, they can reduce interference with neighboring devices and improve spectral efficiency. The collaboration between UEs and FAS-RIS enhances overall network capacity and performance. In practical applications like mobile augmented reality (AR), virtual reality (VR), and high-definition video streaming, UEs equipped with FAS can adapt to varying channel conditions, ensuring higher data rates and lower latency. By integrating FAS at the user end, the system becomes more robust and flexible, meeting the high demands of next-generation wireless networks. Combining the dynamic capabilities of UEs with FAS and FAS-RIS not only improves service quality and communication reliability but also provides scalable and efficient solutions for various application scenarios in 6G networks.

FAS-enabled RIS

As depicted in Fig. 8.1(c), integrating FAS into RIS can significantly enhance system performance and flexibility. By allowing the reflective elements of the RIS to be fluid, FAS-enabled RIS can dynamically adjust the physical positions and electromagnetic properties of the reflective units, achieving more precise and flexible beamforming. Specifically, FAS can be implemented in RIS through two approaches:

Firstly, enabling the reflective elements in the RIS to move, granting positional flexibility to the reflective units. Movable reflective elements can dynamically adjust their physical positions to adapt to changes in the wireless environment, optimizing signal reflection paths and directions. This mechanism allows the RIS to reconfigure itself in real time to maximize channel capacity and signal quality, which is particularly beneficial in rapidly changing communication scenarios.

Secondly, embedding fluid antenna systems into the RIS panel, making each reflective unit possess dynamically reconfigurable electromagnetic properties. By altering the shape, position, or composition of the fluid antenna, the reflective units can adjust their reflection coefficients, phase, and amplitude in real time. This embedded design endows the RIS with higher control precision and flexibility, enabling more complex beamforming and multifunctionality.

This dynamic reconfigurability enables FAS-enabled RIS to adapt to rapidly changing wireless environments, optimizing the reflection direction and strength of signals in real time, thereby improving channel capacity and signal quality. FAS-enabled RIS has immense potential in various application scenarios. For example, in dense urban environments, dynamically adjustable RIS can realign in real time to avoid obstacles and optimize signal paths; in high-mobility scenarios such as drone communications or high-speed train communications, FAS-enabled RIS can swiftly respond to the position changes of moving nodes, ensuring continuous and stable connections; in millimeter-wave and terahertz communications, FAS-enabled RIS can enhance signal coverage by dynamically adjusting reflection properties, improving system reliability.

By integrating FAS into RIS, the system gains greater controllability and adaptability, providing efficient and flexible solutions for future wireless communication networks. This innovative combination not only enhances communication performance but also lays the foundation for constructing intelligent wireless environments.

Fluid-RIS design

The concept of fluid-RIS (FRIS), integrating the adaptability of fluid antennas with the environmental control of RIS, has emerged as a promising direction. As shown in Fig. 8.1(d), initial conceptualizations, sometimes referred to as FRIS, envisioned RIS elements equipped with the capability to dynamically adjust their physical positions. This approach, potentially using micro-actuator systems like piezoelectric actuators or MEMS, aimed to provide precise control over the spatial location of each element for enhanced beamforming and adaptation. Centralized control units managing element positions, along with dedicated power and communication links, were considered part of this architecture.

However, from a practical implementation viewpoint, physically moving the elements raises significant concerns regarding hardware complexity, slow response times, positioning precision difficulties, increased power consumption, and mechanical reliability and maintenance costs, casting doubt on the feasibility of large-scale, physically movable RIS deployments.

A more recent and practically oriented interpretation of FRIS redefines the "fluid" aspect not as physical movement, but as the ability to dynamically switch the active reflection point (port) among many closely spaced candidates within sub-areas of the RIS surface [23]. This port-switching approach avoids the mechanical challenges while still offering enhanced flexibility. Results have shown substantial performance gains compared to conventional RIS by exploiting this finer-grained control over the reflection points [23]. Further refining this concept, recent work models FRIS elements with discrete on–off responses and discrete phase shifts, addressing the idealized assumption of perfect phase shifters and jointly optimizing element selection and beamforming for practical implementation [24]. This model typically involves a BS with a fixed-position antenna (FPA), a single-FPA user, and the FRIS, where elements can be individually switched "on" (actively modifying the signal) or "off" (connected to a matched load, not modulating the signal).

Therefore, the modern fluid-RIS design focuses more on electronic reconfigurability at a sub-element level rather than mechanical movement. This approach offers several key advantages: dynamically selecting reflection points within subareas allows for enhanced beamforming precision and finer control over the reflected wavefront compared to conventional RIS with fixed element centers. Furthermore, the system gains improved adaptability to rapid channel variations and user mobility by electronically switching ports instead of physically moving elements. This electronic reconfigurability also leads to reduced complexity and cost, as eliminating mechanical components significantly lowers hardware complexity, power consumption, and potential maintenance issues. Importantly, this FRIS concept demonstrates compatibility with practical implementations using discrete phase shifts and on–off element selection, as explored in recent studies [24]. This electronically reconfigurable Fluid-RIS paradigm maintains the core benefit of adaptability associated with "fluid" systems while offering a more viable path towards practical implementation in future wireless networks.

8.2.2 Application scenarios and use cases

The FAS-RIS system finds applications across various scenarios, each with unique requirements and implementation approaches:

High-speed communication

High-speed backhaul communication represents a critical application of base station FAS. Using liquid metal and metallic pixel implementations, these systems achieve superior signal transmission efficiency. The high conductivity of liquid metal enables rapid reconfiguration capabilities, while the integration of nonmetallic liquids offers environmentally friendly alternatives. This application is particularly vital in scenarios requiring high-bandwidth data transfer between network nodes.

Industrial and smart city applications

In industrial automation settings, FAS-RIS systems employ metallic pixels and pattern-controlled liquid metal to achieve precise beam control. The microsecond-level switching capability enables real-time adaptation to dynamic industrial environments. For smart city IoT deployments, hybrid materials and programmable pixels provide cost-effective solutions with wide coverage capabilities, supporting the diverse connectivity needs of urban infrastructure.

Mobile and wearable applications

The implementation of FAS in mobile and wearable devices showcases the versatility of the technology. Wearable applications utilize stretchable liquid metal and flexible composites, ensuring comfort while maintaining reliable connectivity. In mobile handsets, planar liquid metal and compact pixel arrays enable efficient space utilization without compromising performance. IoT sensor applications benefit from 3D liquid structures and hybrid materials, offering enhanced coverage and energy efficiency.

Smart surface applications

FAS-enabled RIS technology finds extensive use in smart surface applications. Using metallic pixel arrays and liquid metal coating, these surfaces provide high-precision phase control and continuous state adjustment capabilities. Dynamic reflectors, implemented with liquid metal-based elements and composite materials, enable real-time optimization of signal reflection. Multifunctional surfaces combine various materials to support diverse applications with enhanced control capabilities.

Environmental-specific deployments

Fluid-RIS systems are adapted for various environmental conditions. Indoor positioning applications utilize MEMS-based arrays and smart materials for precise tracking. Outdoor deployments employ weather-resistant liquid metal and robust composites to ensure stable performance under varying conditions. For harsh environment applications, advanced composites and protected liquid metal systems provide the necessary durability and reliability.

8.2.3 Technical principle of FAS-RIS

The integration of FAS and RIS creates a novel wireless communication paradigm that leverages the complementary advantages of both technologies. FAS technology enables the receiver to dynamically select the optimal reception point within a confined area, while RIS technology facilitates intelligent reflection and manipulation of electromagnetic waves to enhance signal propagation.

From a signal processing perspective, the FAS-RIS system operates as an integrated communication chain. The process begins with the transmitter (typically a BS) emitting a signal that propagates toward the RIS. Upon reaching the RIS, its elements, which are configured with specific phase shifts, reflect the incident electromagnetic waves toward the receiver. At the receiving end, the FAS dynamically selects the optimal port position to maximize the received signal quality, completing the signal transmission process. This coordinated operation between the RIS's intelligent reflection and the FAS's dynamic reception optimization creates a highly efficient and adaptable communication system.

8.2.4 Channel modeling

The channel modeling for FAS-RIS systems requires careful consideration of both the BS-RIS link and the RIS-user link, along with the correlation properties of the fluid antenna ports.

BS-RIS channel

The channel between the BS and the RIS can be modeled as

$$\mathbf{h} = [h_1, h_2, \ldots, h_M]^T, \tag{8.1}$$

where h_m represents the complex channel coefficient between the BS and the mth RIS element. Depending on the deployment scenario, this channel can follow various statistical distributions:

- In rich scattering environments without line-of-sight (LoS), Rayleigh fading is appropriate, $h_m \sim \mathcal{CN}(0, \epsilon_1)$.
- In scenarios with both LoS and non-LoS components, Rician fading is more suitable, $h_m = \sqrt{\frac{K}{K+1}}\bar{h}_m + \sqrt{\frac{1}{K+1}}\tilde{h}_m$, where K is the Rician factor, $\bar{h}_m$ is the deterministic LoS component, and $\tilde{h}_m \sim \mathcal{CN}(0, \epsilon_1)$ is the random non-LoS component.

RIS-FAS channel

The channel between the RIS and the FAS can be represented as

$$\mathbf{V} = \begin{bmatrix} v_{1,1} & v_{1,2} & \cdots & v_{1,N} \\ v_{2,1} & v_{2,2} & \cdots & v_{2,N} \\ \vdots & \vdots & \ddots & \vdots \\ v_{M,1} & v_{M,2} & \cdots & v_{M,N} \end{bmatrix}, \tag{8.2}$$

where $v_{m,k}$ denotes the channel coefficient between the mth RIS element and the kth FAS port.

The spatial correlation among different FAS ports is a critical aspect of the system modeling. As previously mentioned, the correlation coefficient between ports k and l can be modeled using the 3D Clarke's model

$$g_{k,l} = \text{sinc}\left(\frac{2\pi(k-l)W}{N-1}\right). \tag{8.3}$$

End-to-end channel

The end-to-end channel from the BS to the kth FAS port via the RIS can be expressed as

$$\rho_k = \mathbf{v}_k^T \mathbf{\Phi} \mathbf{h}, \tag{8.4}$$

where $\mathbf{v}_k = [v_{1,k}, v_{2,k}, \ldots, v_{M,k}]^T$ is the channel vector from all RIS elements to the kth FAS port, and $\mathbf{\Phi} = \text{diag}(e^{j\phi_1}, e^{j\phi_2}, \ldots, e^{j\phi_M})$ is the diagonal reflection coefficient matrix of the RIS, with ϕ_m being the phase shift introduced by the mth RIS element.

8.2.5 Received signal-to-noise ratio

For a given transmit signal s with power P_S, the received signal at the kth FAS port can be written as

$$y_k = \sqrt{P_S}\rho_k s + n_k, \tag{8.5}$$

where $n_k \sim \mathcal{CN}(0, \sigma_n^2)$ is the additive white Gaussian noise.

The resulting signal-to-noise ratio (SNR) at the kth port is

$$\text{SNR}_k = \frac{P_S|\rho_k|^2}{\sigma_n^2}. \tag{8.6}$$

With FAS technology, the receiver selects the port with the maximum SNR,

$$\mathrm{SNR}^* = \max_{k\in\{1,2,\ldots,N\}} \mathrm{SNR}_k = \frac{P_S|\rho^*|^2}{\sigma_n^2}, \tag{8.7}$$

where $\rho^* = \max_k |\rho_k|$.

8.2.6 RIS phase optimization

The performance of the FAS-RIS system heavily depends on the phase configuration of the RIS elements. Two primary approaches for phase optimization are:

Maximize instantaneous SNR

When full channel state information (CSI) is available, the RIS phases can be optimized to maximize the instantaneous SNR at the best FAS port,

$$\mathbf{\Phi}^* = \arg\max_{\mathbf{\Phi}} \max_k \frac{P_S|\mathbf{v}_k^T \mathbf{\Phi}\mathbf{h}|^2}{\sigma_n^2}. \tag{8.8}$$

This optimization problem is challenging due to its nonconvex nature and the coupling between the RIS phase shifts and the FAS port selection. Alternating optimization methods can be employed, where the RIS phase shifts are optimized for each potential FAS port, and then the best port is selected.

Statistical optimization

In scenarios where instantaneous CSI is difficult to acquire, the RIS phases can be optimized based on statistical channel information,

$$\mathbf{\Phi}^* = \arg\max_{\mathbf{\Phi}} \mathbb{E}\left[\max_k \frac{P_S|\mathbf{v}_k^T \mathbf{\Phi}\mathbf{h}|^2}{\sigma_n^2}\right]. \tag{8.9}$$

This approach requires knowledge of the statistical properties of the channels, such as the correlation matrices and mean values.

8.2.7 Performance metrics

Several key metrics are used to evaluate the performance of FAS-RIS systems:

Outage probability

The outage probability is defined as the probability that the instantaneous SNR falls below a threshold γ_{th},

$$P_{\text{out}} = \Pr(\text{SNR}^* < \gamma_{\text{th}}). \tag{8.10}$$

Ergodic capacity

The ergodic capacity represents the average data rate achievable by the system,

$$C = \mathbb{E}[\log_2(1 + \text{SNR}^*)]. \tag{8.11}$$

Energy efficiency

Energy efficiency measures the data rate per unit of energy consumed,

$$\eta = \frac{C}{P_{\text{total}}}, \tag{8.12}$$

where P_{total} includes the transmit power, as well as the power consumed by the RIS and FAS components.[1]

These theoretical models and performance metrics form the foundation for analyzing and optimizing FAS-RIS integrated systems. In subsequent sections, we will explore specific case studies and application scenarios to illustrate the practical benefits of this technology integration.

8.3 Case study

In this case, we consider an FAS-RIS downlink communication system. The system comprises a BS equipped with a single fixed-position antenna, an RIS with M reflecting elements, and a mobile user (MU) equipped with a single fluid antenna. The direct link between the BS and the MU is assumed to be blocked by obstacles such as buildings or natural barriers, necessitating the use of the RIS to facilitate communication. The BS, RIS, and MU are located at coordinates $(0, 0, 5)$ m, $(15, 15, 5)$ m, and $(50, 0, 0)$ m, respectively. The channel between the BS and the RIS is characterized by a distance-dependent flat Rician fading model, with a large-scale path loss exponent of 2.2 and a Rician factor of 1.

[1] The detailed analysis of energy efficiency for FAS-RIS systems, including the trade-offs between performance gains and power consumption, will be investigated in our future work.

8.3.1 System model

The communication system consists of a base station (BS) equipped with a single fixed-position antenna, an RIS with M reflecting elements, and a receiver equipped with a fluid antenna capable of switching among N ports within a linear space of $W\lambda$, where W is the normalized size, and λ is the carrier wavelength. The direct link between the BS and the receiver is assumed broken due to blockages such as massive buildings or trees, and therefore the RIS is employed to help the transmission of signal from the BS to the receiver. It is also assumed that the time delay for port switching is negligibly small.

Some other kinds of smart antenna techniques, such as extremely large-scale (XL) multiple-input multiple-output (MIMO) and movable antenna (MA), can also be applied to improve the system performance. However, the FAS is more suitable for space-limited devices, such as receivers in vehicular application scenarios. Hence we consider the FAS for receiver in this study.

8.3.2 Communication model

In the communication process, the BS emits the RF signal $s \sim \mathcal{CN}(0, 1)$, and then the RIS reflects the transmitted signal to the receiver. In this work, the Rayleigh fading channel is considered, which is suitable for non-line-of-sight scenarios, such as BS-to-vehicle scenarios and vehicle-to-vehicle scenarios.

Given the absence of the direct link between the BS and the receiver, the receiver relies on the signals solely from the reflections of the RIS. Consequentially, at the kth FAS port of the user, the received signal is given by

$$y_k = \sqrt{P_S} \sum_{m=1}^{M} h_m v_{m,k} e^{-2i\pi\theta_m} s + n_k, \tag{8.13}$$

where $h_m \sim \mathcal{CN}(0, \epsilon_1)$ denotes the complex channel coefficient from the BS to the mth RIS element, and $v_{m,k} \sim \mathcal{CN}(0, \epsilon_2)$ denotes the channel from the mth RIS element to the kth FAS port; ϵ_1 and ϵ_2 are the average channel gains of the BS-to-RIS link and RIS-to-receiver link, respectively; P_S is the transmit power, $n_k \sim \mathcal{CN}(0, \sigma_n^2)$ is the additional white Gaussian noise, θ_m denotes the reflecting phase of the mth RIS element, and the RIS is assumed to have full knowledge of the phases of h_m and $v_{m,k}$; thus can adjust θ_m to maximize the channel gain, i.e., $\theta_m = -\arg(h_m v_{m,k})$. Following this, the combined channel at the kth FAS port can be expressed as

$$\gamma_k = \sum_{m=1}^{M} |h_m||v_{m,k}|. \tag{8.14}$$

It can be readily observed that both $|h_m|$ and $|v_{m,k}|$ follow Rayleigh distribution, with

$$\mathbf{E}(|h_m||v_{m,k}|) = \mathbf{E}(|h_m|)\mathbf{E}(|v_{m,k}|) = \pi\sqrt{\epsilon_1\epsilon_2}/4, \quad \text{and}$$
$$\mathbf{Var}(|h_m||v_{m,k}|) = \mathbf{E}(|h_m|^2|v_{m,k}|^2) - (\mathbf{E}(|h_m||v_{m,k}|))^2 = \epsilon_1\epsilon_2(1-\pi^2/16).$$

At the FAS, the port with the maximum channel gain will be selected for receiving the signal. Consequentially, the resulting channel at the optimal port is written as

$$\gamma^* = \max_{k\in\{1,2,\ldots,N\}} \gamma_k. \tag{8.15}$$

Remark 1. The signal processing method in FAS resembles the selection combining technique in multiple-antenna systems. However, our distinctive contribution is to analyze the system performance including the correlation among the ports.

8.3.3 FAS channel correlation model

For the correlation between two different ports of the FAS, the 3D Clarke's model is adopted. Specifically, the correlation coefficient between ports k and l is modeled by

$$g_{k,l} = \text{sinc}\left(\frac{2\pi(k-l)W}{N-1}\right), \tag{8.16}$$

where $\text{sinc}(x) = \frac{\sin(x)}{x}$. Note that $\text{sinc}(x)$ is an odd function, which indicates that the correlation coefficient matrix of $\mathbf{v}_m = [v_{m,1},\ldots,v_{m,N}]$ is a Toeplitz matrix as

$$\begin{aligned}\mathbf{\Sigma}\in\mathbb{R}^{N\times N} &= \mathbf{topelitz}(g_{1,1}, g_{1,2},\ldots,g_{1,N})\\ &= \begin{pmatrix} g_{1,1} & g_{1,2} & \cdots & g_{1,N}\\ g_{1,2} & g_{1,1} & \cdots & g_{1,N-1}\\ \vdots & \vdots & \ddots & \vdots\\ g_{1,N} & g_{1,N-1} & \cdots & g_{1,1}\end{pmatrix}.\end{aligned} \tag{8.17}$$

8.3.4 Outage performance analysis

In our analysis, we evaluate the outage performance of the FAS-RIS system. Let us denote by R the target transmission rate, and then the outage probability is accordingly given as

$$P_{\text{out}} = F_{\gamma^*}(\Lambda_{\text{th}}) = \Pr\left(\log_2\left(1+\frac{P_S(\gamma^*)^2}{\sigma_n^2}\right)\le R\right), \tag{8.18}$$

where $\Lambda_{\text{th}} = ((2^R - 1)\sigma_n^2/P_S)^{1/2}$.

To proceed with the outage probability analysis, it is necessary to derive the CDF of γ^*, i.e., $F_{\gamma^*}(y)$. Given the fact that the exact expression for $F_{\gamma^*}(y)$ is intractable, we develop several approximations for $F_{\gamma^*}(y)$. The results presented are developed based on the central limit theorem (CLT), and a multivariate Gaussian distribution is assumed. Consequently, the mean and variance of γ_k, as well as the correlation coefficient between ports k and l, must be computed to facilitate further derivations.

CLT approximation

In practice, the number of RIS elements, M, is supposed to be very large, and thus the CLT can be applied. Consequentially, $\mathbf{\Gamma} = [\gamma_1, \ldots, \gamma_N]$ can be approximated by $\bar{\mathbf{\Gamma}} = [\bar{\gamma}_1, \ldots, \bar{\gamma}_N]$, which follows a multivariate Gaussian distribution with identical correlation coefficient matrix and mean vector to those of $\mathbf{\Gamma}$. The mean and variance of $\bar{\gamma}_k$ are identical to those of γ_k, which can be computed as

$$E_\gamma = M\mathbf{E}(|h_m||v_{m,k}|) = \frac{M\pi\sqrt{\epsilon_1\epsilon_2}}{4}, \tag{8.19}$$

$$V_\gamma = M\mathbf{Var}(|h_m||v_{m,k}|) = M\epsilon_1\epsilon_2\big(1 - \frac{\pi^2}{16}\big). \tag{8.20}$$

Block-correlation channel approximation

To overcome the computational complexity of high-dimensional integrals in the CLT approach, we resort to a block-correlation approximation model. It is important to note that as the dimension N increases, the correlation coefficient matrix becomes predominantly influenced by a small fraction of eigenvalues. To efficiently capture the significant eigenvalues, we employ a block matrix, which simplifies the representation of the original matrix.

Specifically, we employ a block-diagonal correlation matrix to describe the spatial correlation structure of the FAS N ports, which can be expressed as

$$\hat{\mathbf{\Sigma}} \in \mathbb{R}^{N\times N} = \text{Blkdiag}(\mathbf{C}_1, \ldots, \mathbf{C}_B), \tag{8.21}$$

where $\text{Blkdiag}(\cdot)$ returns a block diagonal matrix, with each submatrix $\mathbf{C}_b$ being a constant correlation matrix of size L_b and having the correlation parameter μ_b^2 as entries, i.e.,

$$\mathbf{C}_b \in \mathbb{R}^{L_b\times L_b} = \text{topelitz}(1, \mu_b^2, \ldots, \mu_b^2), \tag{8.22}$$

where $\sum_{b=1}^{B} L_b = N$, and L_b is found by the algorithm described in [25].

For L_b ports in the bth submatrix, the channels $\mathbf{h}_k$, for $k \in \{1, \ldots, L_b\}$, are inherently correlated. The mathematical expression can be given by

$$\mathbf{h}_k = \sqrt{\frac{\alpha K}{K+1}}\bar{\mathbf{h}}_k + \mu_b \tilde{\mathbf{h}}_b + \sqrt{1-\mu_b^2}\mathbf{e}_k, \tag{8.23}$$

where $\tilde{\mathbf{h}}_b$ and $\mathbf{e}_k$ are i.i.d. RVs, following the distribution $\mathcal{CN}\left(\mathbf{0}, \frac{\alpha}{K+1}\mathbf{I}_M\right)$ in which $\mathbf{I}_M$ is an identity matrix.

Using the CLT, the signal-to-noise ratio (SNR) at the kth port of the user can be expressed as

$$\gamma_k = \frac{PA_k^2}{\sigma^2}, \tag{8.24}$$

where $A_k = |\mathbf{h}_k \boldsymbol{\Phi} \mathbf{g}^H|$, P represents the transmission power of the BS, and σ^2 denotes the noise power at the user. The maximum SNR across all ports is given by

$$\gamma = \frac{P|A_{\max}|^2}{\sigma^2}, \tag{8.25}$$

where $A_{\max} = \max\{A_1, A_2, \ldots, A_N\}$.

Based on this model, the outage probability can be derived as

$$\begin{aligned}\mathbb{P}^{\text{out}} = \prod_{b=1}^{B} \int_0^\infty \frac{2r_b}{\bar{\sigma}^2\mu^2} e^{-\frac{r_b^2+|\eta|^2}{\bar{\sigma}^2\mu^2}} I_0\left(\frac{2r_b|\eta|}{\bar{\sigma}^2\mu^2}\right) \\ \times \left[1 - Q_1\left(\sqrt{\frac{2}{\bar{\sigma}^2(1-\mu^2)}}r_b, \sqrt{\frac{2\gamma_{\text{th}}}{\bar{\sigma}^2(1-\mu^2)}}\right)\right]^{L_b} dr_b,\end{aligned} \tag{8.26}$$

where $\gamma_{\text{th}} = (2^R - 1)\sigma^2/P$, $Q_1(\cdot,\cdot)$ is the first-order Marcum Q-function, $\eta = \sqrt{\frac{\alpha K}{K+1}}\bar{\mathbf{h}}_k \boldsymbol{\Phi} \mathbf{g}^H$ represents the deterministic component of the channel, and $\bar{\sigma}^2 = \frac{M\alpha\beta}{K+1}$ is the variance of the random component.

This block-correlation model dramatically reduces the computational complexity of analysis while maintaining accuracy, making it particularly suitable for analyzing FAS-RIS systems with a large number of ports.

Performance bounds and asymptotic analysis

To gain deeper insights into the system performance, we can derive the upper bound, lower bound, and asymptotic expressions for the outage probability.

The upper bound is derived using the independent antennas equivalent (IAE) model, where $\mu_b = 1$. This simplification leads to

$$\check{\mathbb{P}}^{\text{out}} = \left[1 - Q_1\left(\sqrt{\frac{2}{\bar{\sigma}^2}}|\eta|, \sqrt{\frac{2\gamma_{\text{th}}}{\bar{\sigma}^2}}\right)\right]^B. \tag{8.27}$$

For the lower bound, we simplify the expression by removing the Bessel function terms, resulting in

$$\hat{\mathbb{P}}_m^{\text{out}} = e^{-\frac{B|\eta|^2}{\bar{\sigma}^2\mu^2}} \prod_{b=1}^{B} \frac{1-\mu^2}{1+(L_b-1)\mu^2}\left[1 - e^{-\frac{\gamma_{th}}{\bar{\sigma}^2(1-\mu^2)}}\right]^{L_b}. \tag{8.28}$$

In the high SNR regime, we can further approximate the outage probability as

$$\mathbb{P}_m^{\text{out}} \simeq \prod_{b=1}^{B} \eta_b \left[\frac{\gamma_{\text{th}}}{\bar{\sigma}^2(1-\mu^2)}\right]^{L_b}, \tag{8.29}$$

where $\eta_b = \frac{(1-\mu^2)e^{-\frac{|\eta|^2}{\bar{\sigma}^2\mu^2}}}{1+(L_b-1)\mu^2}$.

Remark 2. The asymptotic outage probability indicates that the diversity order of the system is $\sum_{b=1}^{B} L_b = N$, which means that the system can achieve a diversity gain equal to the number of fluid antenna ports.

RIS phase optimization

The performance of the FAS-RIS system can be further improved by optimizing the phase shifts of the RIS elements. The optimization problem can be formulated as

$$\min_{\boldsymbol{\Phi}} \mathbb{P}^{\text{out}} \text{ such that } |\phi_m| = 1, \ \forall m \in \mathcal{M}. \tag{8.30}$$

Due to the complexity of the outage probability expression, we can simplify the problem by maximizing the deterministic component of the channel η, which is equivalent to

$$\max_{\boldsymbol{\Phi}} \left|\bar{\mathbf{h}}\boldsymbol{\Phi}\mathbf{g}^H\right| \text{ such that } |\phi_m| = 1, \ \forall m \in \mathcal{M}. \tag{8.31}$$

The optimal phase shift for each RIS element can be derived as

$$\phi_m^o = e^{-j(\angle\bar{\mathbf{h}}(m) - \angle\mathbf{g}(m))}, \tag{8.32}$$

where $\angle\bar{\mathbf{h}}(m)$ and $\angle\mathbf{g}(m)$ represent the phases of the mth elements of the LoS components of the RIS-user channel and the BS-RIS channel, respectively.

Remark 3. This optimal solution indicates that the reflection elements of the RIS should be strategically designed to maximize the LoS components of the cascaded BS-RIS-user channel.

To comprehensively evaluate the performance of the FAS-RIS system, we consider two scenarios: one where the channel state information (CSI) is available (CSI-based scheme), and the other where the CSI is difficult to acquire (CSI-free scheme). By presenting the performance of each scheme under various conditions, we aim to showcase the robustness and efficiency of FAS-RIS systems in different environments. Building upon this, we explore throughput optimization to highlight how the system performs under both CSI-based and CSI-free schemes.

8.3.5 Performance analysis

In Fig. 8.2, we examine the impact of the BS transmit power P on the outage probability for both the CSI-based and CSI-free schemes. In the legend, "Simulation" refers to the simulated outage probabilities, while "BCMA" represents the theoretical results based on the block-correlation matrix approximation model [7]. As shown in Fig. 8.2, we observe that the outage probabilities of all schemes decrease as the transmit power P increases. Additionally, we find that for a given P, increasing the number of RIS elements M leads to lower outage probabilities. This improvement occurs because a larger M provides more degrees of freedom through the RIS, enhancing the signal propagation environment. Furthermore, the close agreement between the theoretical and simulated results validates the accuracy of the analytical block-correlation model. This model effectively avoids the heavy computations involved in evaluating the cumulative distribution functions of multidimensional Gaussian distributions, thereby simplifying the performance analysis of FAS-RIS systems [8].

These findings highlight the significant benefits of FAS-RIS in overcoming communication challenges when direct links are obstructed. The RIS enhances the system's ability to control the propagation environment, while the fluid antenna at the user end adapts to changing conditions. Even without perfect CSI, the system maintains robust performance, demonstrating its potential for future wireless networks where adaptability and efficiency are paramount.

To further demonstrate the advantages of the proposed FAS-RIS approach, we compare its performance with state-of-the-art benchmark methods. Compared to traditional fixed-position antenna (FPA) systems without RIS assistance, the FAS-RIS system achieves substantial performance gains: for instance, with $M = 64$ RIS elements and $N = 20$ FAS ports, the outage probability improvement can reach up to 2–3 orders of magnitude at moderate transmit power levels (e.g., $P = 10$ dBm), as the FAS-RIS system exploits both spatial diversity through FAS port selection and environmental control through RIS phase

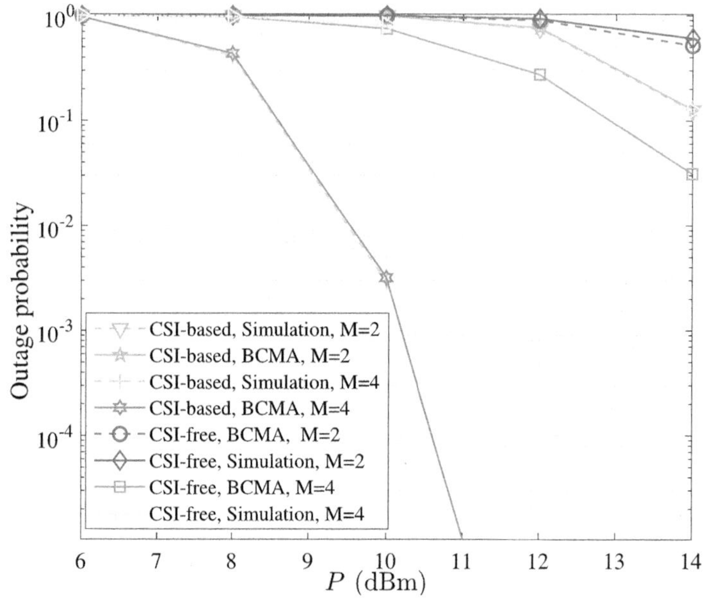

Figure 8.2:
Outage probability versus transmit power P for FAS-RIS systems.

optimization. Compared to conventional RIS-assisted systems with fixed antennas, the FAS-RIS approach provides additional performance enhancement of approximately 5–10 dB in terms of outage probability reduction, particularly in scenarios with high spatial correlation, as the fluid antenna can dynamically select the optimal reception point to maximize the received signal quality. Furthermore, compared to existing FAS-only systems without RIS [7], the integration of RIS provides an additional 3–6 dB improvement in outage performance by intelligently controlling the propagation environment, especially when the direct link is blocked. These performance gains demonstrate the significant advantages of the proposed FAS-RIS integration over existing state-of-the-art methods, making it a promising solution for future 6G wireless networks.

8.3.6 Throughput optimization

In Fig. 8.3, we examine the optimized throughput as a function of the number of FAS ports N, with the transmit power fixed at $P = 10$ dBm and the number of RIS elements set to $M = 2$, for both CSI-based and CSI-free schemes. The curves labeled "GDA" and "PGDA" represent the solutions obtained via gradient ascent and partial gradient ascent algorithms for the CSI-based and CSI-free schemes, respectively, as detailed in [8]. "ES" denotes the optimal solutions from exhaustive search, serving as a benchmark. To provide a fair and practical

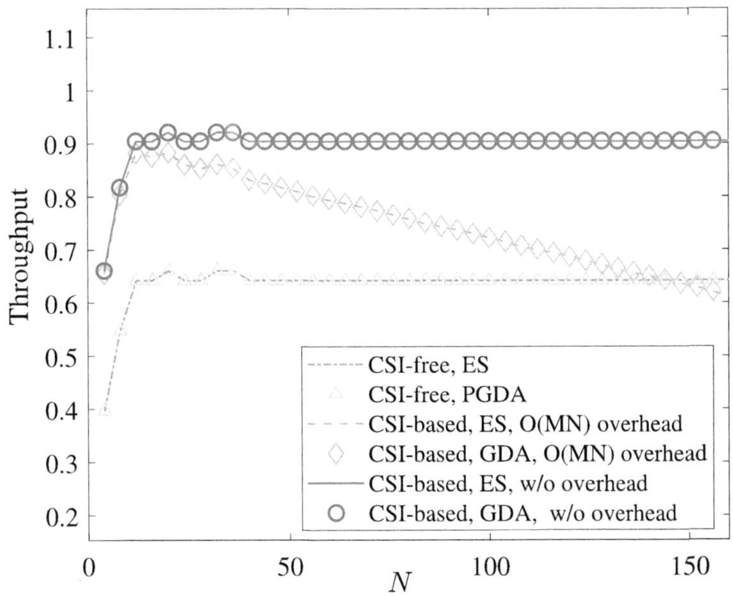

Figure 8.3:
Throughput versus the number of FAS ports N, where $P = 10$ dBm and $M = 2$.

evaluation, we account for the rate loss caused by channel estimation overhead. Recognizing the challenges in estimating channels for both FAS and RIS, we adopt an on–off scheme, resulting in an overall overhead of $O(MN)$, as indicated in the legend.

Observing Fig. 8.3, we find that as the number of FAS ports N increases from 1 to 20, the throughput for both schemes correspondingly improves, highlighting the significant role of FAS in enhancing system performance through increased spatial diversity. However, when channel estimation overhead is considered, especially as N grows beyond 40, the throughput of the CSI-based scheme begins to decrease and can even fall below that of the CSI-free scheme at larger values of N. This decline occurs because a substantial portion of time slots is consumed by pilot transmissions needed for channel estimation, leaving less time for actual data transmission, thereby diminishing the gains from utilizing CSI. These insights emphasize the need to balance the number of FAS ports and RIS elements against practical limitations such as channel estimation overhead. They also showcase the potential advantages of CSI-free schemes in scenarios with short channel coherence times and heavy overhead, as these schemes mitigate the adverse effects of overhead and maintain a reliable throughput.

8.3.7 Computational complexity analysis

For the RIS phase optimization problem formulated in (486), the optimal solution can be obtained in closed form as given in (491), requiring $O(M)$ operations for M RIS elements, making it highly efficient. However, when considering the original outage probability minimization problem (481), the complexity becomes significantly higher: direct optimization involves computing B-dimensional integrals with complexity $O(B \cdot I)$, while employing the block-correlation model reduces the computational complexity from $O(N^3)$ (for full correlation matrix operations) to approximately $O(\sum_{b=1}^{B} L_b^2)$, where L_b is the size of the bth block, representing a significant reduction especially when $B \ll N$.

8.4 Future research directions and potential applications

The rapid development of FAS-RIS technology has opened up numerous opportunities for innovation and practical applications in wireless communications. While significant progress has been made in understanding and implementing these systems, several key challenges and promising research directions remain to be explored. Furthermore, the unique capabilities of FAS-RIS systems make them particularly suitable for a wide range of applications across different domains. This section discusses the critical research directions that warrant further investigation and explores the potential applications that could benefit from this emerging technology.

8.4.1 Future research directions

This section explores several important future research directions for FAS-RIS systems, including material development, channel modeling, system optimization, and technology integration.

Material development and innovation

The development of novel materials is a critical foundation for the advancement of FAS-RIS systems. High-performance liquid metal materials offering superior conductivity and stability are a current research focus. Smart materials capable of dynamically adjusting their electromagnetic properties provide flexible adaptation capabilities for these systems. Environmentally friendly materials serve as alternatives to traditional options, supporting sustainable development goals. Composite materials combining the advantages of different material types show promise in creating new possibilities for system design and implementation.

Channel modeling and estimation

Accurate channel models are essential for evaluating and optimizing FAS-RIS system performance. Three-dimensional channel models considering both spatial and temporal correlations can more precisely describe the unique characteristics of FAS-RIS systems. Machine learning-based approaches offer promising solutions for channel prediction and optimization. Advanced models that account for environmental factors and material properties can enhance the accuracy of system performance prediction and optimization.

System optimization techniques

Advanced optimization techniques are needed to fully exploit the potential of FAS-RIS systems. Joint optimization of FAS positioning and RIS phase shifts presents complex challenges requiring innovative solutions. Energy-efficient control algorithms for large-scale deployments and real-time adaptation strategies for mobile scenarios are particularly important for practical applications. Multiobjective optimization frameworks considering various performance metrics can achieve more balanced and efficient system designs.

Integration with emerging technologies

The integration of FAS-RIS systems with other emerging technologies brings opportunities for wireless communications. In terahertz communication systems, FAS-RIS can help overcome the severe path loss and blockage issues characteristic of high-frequency bands. Applications in quantum communication networks can enhance security and reliability. Integration with holographic MIMO systems could revolutionize spatial multiplexing capabilities. AI-driven network optimization can enable more intelligent and adaptive system operation, particularly in complex and dynamic environments.

8.4.2 Potential applications

The unique capabilities of FAS-RIS systems create numerous opportunities for practical applications in next-generation wireless networks. This section explores key application domains where FAS-RIS technology can make significant contributions to 6G communications and beyond.

Advanced 6G communications

In 6G networks, FAS-RIS systems will be instrumental in achieving the ambitious performance targets of ultra-high data rates exceeding 1 Tbps and massive connectivity supporting over 10 million devices per km^2. The integration of FAS-RIS with ultramassive

MIMO systems can dramatically enhance spatial multiplexing capabilities, while intelligent reflecting surfaces provide unprecedented control over wireless propagation environments.

Dynamic beamforming capabilities are particularly valuable for high-mobility scenarios such as connected vehicles and drones, enabling robust communications in challenging conditions where conventional technologies struggle. Specifically, FAS-RIS systems can facilitate sub-THz and THz communications by intelligently reconfiguring the propagation environment to overcome the severe path loss and molecular absorption issues that characterize these frequency bands. By dynamically adapting antenna positions and reflection phases, FAS-RIS systems can maintain stable connections across variable channel conditions, enabling the extreme-high-throughput applications envisioned for 6G.

Intelligent transportation systems

Smart transportation represents a critical application domain for 6G and beyond. V2X communications can benefit significantly from the adaptive capabilities of FAS-RIS systems, particularly in complex urban environments where buildings and other structures cause signal blockages. The millisecond-level latency requirements of safety-critical applications can be supported through optimized signal paths created by adaptive RIS configurations.

High-speed railway communication systems require robust and reliable connections at speeds exceeding 500 km/h, where conventional technologies struggle with rapid channel variations. The dynamic adaptation capabilities of FAS-RIS systems can maintain stable connections by tracking high-speed trains and continuously optimizing signal paths. Similarly, autonomous vehicle networks can leverage FAS-RIS to enhance perception range and reliability beyond line-of-sight, facilitating cooperative driving and advanced traffic management systems that depend on ultra-reliable low-latency communications.

Smart manufacturing and industry 5.0

In industrial settings aligned with Industry 5.0 principles, FAS-RIS systems can revolutionize manufacturing processes through flexible and reliable wireless connectivity. Factory automation systems require deterministic communication with reliability and submillisecond latency, which can be facilitated by the adaptive nature of FAS-RIS. The technology enables reliable wireless control of precision machinery even in challenging radio environments with high metal content and electromagnetic interference.

Smart manufacturing environments benefit from improved coverage and reliability, allowing thousands of sensors and actuators to communicate without interference. Industrial IoT deployments can achieve better energy efficiency and connectivity through optimized signal paths, extending battery life of wireless sensors. The ability to adapt to changing environments makes FAS-RIS particularly suitable for dynamic industrial settings where traditional fixed antenna systems struggle with changing layouts, moving machinery, and variable propagation conditions.

Healthcare and biomedical applications

Healthcare applications present unique opportunities for FAS-RIS technology in the 6G era. Medical device communications require reliable and secure links with stringent reliability requirements, which can be enhanced through the precise control offered by FAS-RIS systems. Hospital environments, characterized by complex indoor propagation conditions with numerous walls, equipment, and metallic surfaces, benefit from improved coverage and signal quality.

Advanced patient monitoring systems utilizing implantable and wearable devices can achieve more reliable connections through optimized signal paths, enabling continuous health monitoring without communication gaps. Telemedicine networks supporting remote surgery applications, which demand ultra-high definition video streams with guaranteed latency and reliability, can leverage FAS-RIS to maintain stable connections even in challenging environments. The technology's ability to create controlled propagation environments also supports emerging applications like wireless power transfer to implantable devices and precise localization of medical equipment.

Integrated sensing and communications

The integration of sensing and communications represents a key direction for 6G networks, where FAS-RIS systems offer significant advantages. By dynamically configuring both antenna positions and RIS elements, these systems can simultaneously optimize for communication performance and sensing accuracy. Applications include high-precision indoor positioning with centimeter-level accuracy, environmental mapping and reconstruction, and gesture recognition for extended reality applications.

In urban surveillance and monitoring, FAS-RIS can enhance coverage in blind spots while improving detection and tracking capabilities. For disaster response scenarios, rapidly deployable FAS-RIS units can establish emergency communication networks while providing situational awareness through integrated sensing functionality. The joint optimization of sensing and communication resources enabled by FAS-RIS represents a paradigm shift in how wireless systems interact with their environment, supporting the pervasive intelligence vision of 6G networks.

These application domains represent just a fraction of the potential impact that FAS-RIS systems could have on future wireless communications. As the technology matures, new opportunities will emerge at the intersection of various vertical sectors, driving further innovation in this exciting field. The key to realizing this potential lies in addressing fundamental challenges while exploring novel applications that can benefit from the unique adaptive capabilities of FAS-RIS systems.

8.5 Conclusions

This chapter has explored the integration of FAS and RIS for next-generation 6G wireless networks. The FAS-RIS combination offers unprecedented flexibility and adaptability, addressing many limitations of conventional wireless systems. We introduced various FAS-RIS architectures including base station FAS, user equipment FAS, FAS-enabled RIS, and fluid-RIS, each serving specific deployment scenarios. The developed block-correlation approximation model provides an efficient analytical framework for system performance evaluation while maintaining accuracy. Our performance analyses demonstrated how FAS-RIS systems significantly enhance communication reliability, particularly in challenging environments with blockages or high mobility. FAS-RIS systems are positioned to play a pivotal role in realizing 6G performance goals, enabling transformative applications across advanced communications, intelligent transportation, smart manufacturing, and healthcare. While challenges remain in channel estimation, optimization algorithms, and materials development, the seamless integration of dynamic antenna positioning with intelligent environmental control opens new dimensions for wireless optimization that will be essential for future 6G applications.

References

[1] W.K. New, et al., A tutorial on fluid antenna system for 6G networks: encompassing communication theory, optimization methods and hardware designs, IEEE Communications Surveys and Tutorials (2024), https://doi.org/10.1109/COMST.2024.3498855, early access.

[2] T. Wu, et al., Fluid antenna systems enabling 6G: principles, applications, and research directions, IEEE Wireless Communications (2024), in press arXiv:2412.03839.

[3] K.-K. Wong, A. Shojaeifard, K.-F. Tong, Y. Zhang, Performance limits of fluid antenna systems, IEEE Communications Letters 24 (11) (Nov. 2020) 2469–2472.

[4] J. Yao, et al., FAS-driven spectrum sensing for cognitive radio networks, IEEE Internet of Things Journal 12 (5) (Mar. 2025) 6046–6049.

[5] H. Xu, et al., Channel estimation for FAS-assisted multiuser mmWave systems, IEEE Communications Letters 28 (3) (Mar. 2024) 632–636.

[6] X. Lai, et al., On performance of fluid antenna system using maximum ratio combining, IEEE Communications Letters 28 (2) (Feb. 2024) 402–406.

[7] P. Ramirez-Espinosa, D. Morales-Jimenez, K.K. Wong, A new spatial block-correlation model for fluid antenna systems, IEEE Transactions on Wireless Communications 23 (11) (Nov. 2024) 15829–15843.

[8] X. Lai, J. Yao, K. Zhi, T. Wu, D. Morales-Jimenez, K.K. Wong, FAS-RIS: a block-correlation model analysis, IEEE Transactions on Vehicular Technology 74 (2) (Feb. 2025) 3412–3417.

[9] J. Yao, et al., FAS-RIS communication: model, analysis, and optimization, IEEE Transactions on Vehicular Technology (2025), https://doi.org/10.1109/TVT.2025.3537294, early access.
[10] J. Yao, et al., Proactive monitoring via jamming in fluid antenna systems, IEEE Communications Letters 28 (7) (Jul. 2024) 1695–1702.
[11] J. Yao, et al., Exploring fairness for FAS-assisted communication systems: from NOMA to OMA, IEEE Transactions on Wireless Communications 24 (4) (Apr. 2025) 3433–3449.
[12] K.-K. Wong, A. Shojaeifard, K.-F. Tong, Y. Zhang, Fluid antenna system, IEEE Transactions on Wireless Communications 20 (3) (Mar. 2021) 1950–1962.
[13] K.-K. Wong, K.-F. Tong, Fluid antenna multiple access, IEEE Transactions on Wireless Communications 21 (7) (Jul. 2022) 4801–4815.
[14] T. Wu, et al., Fingerprint-based mmWave positioning system aided by reconfigurable intelligent surface, IEEE Wireless Communications Letters 12 (8) (Aug. 2023) 1379–1383.
[15] T. Wu, et al., Exploit high-dimensional RIS information to localization: what is the impact of faulty element?, IEEE Journal on Selected Areas in Communications 42 (10) (Oct. 2024) 2803–2819.
[16] L. Yang, Y. Yang, M.O. Hasna, M.-S. Alouini, Coverage, probability of SNR gain, and DOR analysis of RIS-aided communication systems, IEEE Wireless Communications Letters 9 (8) (Aug. 2020) 1268–1272.
[17] X. Gan, C. Zhong, Y. Zhu, Z. Zhong, User selection in reconfigurable intelligent surface assisted communication systems, IEEE Communications Letters 25 (4) (Apr. 2020) 1353–1357.
[18] X. Yu, et al., Phase shift compression for control signaling reduction in IRS-aided wireless systems: global attention and lightweight design, IEEE Transactions on Wireless Communications 23 (8) (Aug. 2024) 8528–8541.
[19] R. Singh, A. Kaushik, W. Shin, G.C. Alexandropoulos, M. Toka, M. Di Renzo, Indexed multiple access with reconfigurable intelligent surfaces: the reflection tuning potential, IEEE Communications Magazine 62 (4) (Apr. 2024) 120–126.
[20] M. Katwe, A. Kaushik, L. Mohjazi, M. Abualhayja'a, D. Dardari, K. Singh, M.A. Imran, M.M. Butt, O.A. Dobre, An overview of intelligent meta-surfaces for 6G and beyond: opportunities, trends, and challenges, IEEE Communications Standards Magazine 8 (4) (Dec. 2024) 62–69.
[21] S. Mondal, K. Singh, A. Kaushik, C.-P. Li, Performance of RIS-aided fluid antenna-enabled multiuser NOMA non-terrestrial networks, IEEE Journal on Selected Areas in Communications (Sept. 2025), early access.
[22] R. Singh, J. An, A. Kaushik, C. Yuen, Y.-J.A. Zhang, O.A. Dobre, Redefining electromagnetic signal and information theory with programmable surfaces: open critical questions and future horizon, [Online]. Available, https://doi.org/10.36227/techrxiv.175621821.18352632/v1, Aug. 2025.
[23] A. Salem, K.-K. Wong, G. Alexandropoulos, C.-B. Chae, R. Murch, A first look at the performance enhancement potential of fluid reconfigurable intelligent surface, arXiv preprint, arXiv:2502.17116.
[24] H. Xiao, X. Hu, K.-K. Wong, H. Hong, G.C. Alexandropoulos, C.-B. Chae, Fluid reconfigurable intelligent surfaces: joint on-off selection and beamforming with discrete phase shifts, arXiv preprint, arXiv:2503.14601, Mar. 2025.
[25] P. Ramirez-Espinosa, D. Morales-Jimenez, K.-K. Wong, A new spatial block-correlation model for fluid antenna systems, IEEE Transactions on Wireless Communications 23 (11) (Nov. 2024) 15829–15843.

Achievable data rate investigation and power allocation of reconfigurable intelligent surfaces assisted multiuser OFDM communication

Saber Hassouna[a,b], Jalil Kazim[a,b], Muhammad Imran[a,b], and Qammer H. Abbasi[a,b]

[a]James Watt School of Engineering, University of Glasgow, Glasgow, United Kingdom [b]School of Engineering, College of Science and Engineering, University of Glasgow, Glasgow, United Kingdom

With the help of a developing technology called reconfigurable intelligent surface (RIS), it is possible to modify the propagation environment and boost the data rates of wireless communication networks. In this chapter, we optimized the phases of the RIS elements and performed a fair power allocation for each subcarrier over the full bandwidth in a single-input-single-output (SISO) wideband system where the user and the access point (AP) are provided with a single antenna. The used acronyms are listed in Table 9.1 and the notation is summarized in Table 9.2. The data rate or its equivalent channel power is maximized by proposing different low-complex algorithms. The strongest tap maximization (STM) and power methods are compared with the semidefinite relaxations (SDR) method in terms of computational complexity and data rate performance. Runtime and complexity analysis of the suggested methods are computed and compared to reveal the actual time consumption and the required number of operations for each method. Simulation results show that with an optimized RIS, the sum rate is 2.5 times higher than with an unconfigured surface, demonstrating the RIS's tremendous advantages even in complex configurations. The data rate performance of the SDR method is higher than the power method and less than the STM method but with higher computational complexity, more than 6 million complex operations, and 50 minutes of runtime calculations compared with the other STM and power optimization methods.

Electromagnetic Signal and Information Theory
https://doi.org/10.1016/B978-0-44-336354-2.00016-2
Copyright © 2027 Elsevier Inc. All rights are reserved, including those for text and data mining, AI training, and similar technologies.

9.1 Introduction

The use of RIS is a recent approach for enhancing wireless-channel quality and enabling programmable wireless environments [1–3,18]. It enables the receiver to concentrate the signal due to the combined impact of reflections from all RIS elements and the direct line-of-sight (LoS) route between the transmitter and the receiver. This combination produces a focused beam in a specific direction [4]. Unlike wireless repeaters and relays, RISs have unique features such as configurability, no signal amplification requirements, and no need for complex processing, coding, and decoding techniques [5].

The literature has suggested a number of alternative optimization techniques for increasing the data rates in RIS-assisted wireless communication systems, which aims to identify a nearly ideal solution with tolerable computational complexity and run time. Heuristic solutions, such as SDR [6–13], strongest signal path maximization in time domain [18,28] and successive convex approximations [14] are investigated by many authors in the literature. Applying the SDR and the successive convex approximation approaches require efficient optimization tools. Furthermore, the SDR needs Gaussian randomization methods to obtain the rank-one suboptimal solution. Considerable effort has been devoted to developing sophisticated algorithms that can identify high-quality solutions that are close to optimal. However, the computational complexity associated with these algorithms is often quite high, making them prohibitively expensive in practical settings where a large number of RIS elements are present. It is precisely in such situations that RIS is most valuable [15–17].

The preceding discussion has primarily focused on the fundamental motivation for enhancing data rate and reliability through RIS-assisted multiuser OFDM systems. While most current evaluations and prototypes operate at mid-band frequencies (around 3.5–4 GHz), the underlying principles are frequency-agnostic and equally relevant to the spectrum evolution envisioned for 6G and beyond. Given that future networks will rely on both low–mid-band (*cmWave*) and high–frequency (*sub-THz*) ranges, it becomes essential to assess how the proposed RIS-aided architecture and algorithms can be adapted to these complementary spectral regimes. The proposed design and prototypes around 3.5–3.75 GHz (cmWave) align with current 6G roadmaps identifying the 7–15 GHz and adjacent mid-bands as key enabling spectra due to their favorable coverage–bandwidth trade-off. RIS at 3.5–5.5 GHz has been experimentally validated with multi-hundred-element panels, confirming practical control, calibration, and throughput improvements in non-line-of-sight indoor deployments. Toward sub-THz operation, surface losses, phase quantization, synchronization, and beam-squint become more pronounced; thus, refractive or holographic implementations and low-loss metasurface technologies are promising directions to preserve rate and enhance energy efficiency. Overall, the presented RIS-aided MU-OFDM framework serves as a proof-of-concept for current cmWave systems and a scalable pathway toward sub-THz bands.

9.1.1 Contribution and organization of the chapter

The contributions of this chapter are highlighted as follows:

- We maximize the data rate of the SISO wideband system equipped with RIS surface and single antenna for both AP and user equipment. We formulate a joint optimization problem of the subcarriers power allocations and the phase shifts of the RIS elements. Then, we proposed an iterative power method to solve the nonconvex problem resulting from the unit-modular constraint on each RIS element. It is noted that for large surfaces the reflection coefficients can be estimated by the column of the discrete Fourier transform (DFT) matrix [25]. We used the DFT beamforming codebook to select a signal-to-noise ratio (SNR) maximizing configuration to be utilized in the proposed power method.
- To evaluate the performance of the power method in terms of the data rate and complexity computation, it is compared with the SDR method. The SDR approach handles the nonconvex unit-modulus constraint by transforming the passive beamforming vector into a rank-one and positive semidefinite matrix. The nonconvex problem is transformed into a convex standard convex semidefinite program (SDP) problem by utilizing the SDR technique, which may then be addressed by a variety of effective convex optimization tools. The SDR method shows a higher performance data rate than the power method but at the cost of increased computational complexity and runtime. However, it shows lower performance than the STM method. The STM chooses a configuration that is good for channel tap rather than a subcarrier, when there exits a stronger LoS path.
- The goal of this article is to demonstrate that the heuristic approaches like power and STM methods show significantly less computational complexity than the famous SDR method and achieve comparable data rates as a state-of-the-art benchmark scheme. The performance of all the methods is validated by comparing them with the upper bound of the system.

The rest of this chapter is organized as follows. The system and channel models are presented in Section 9.2, while Section 9.3 discusses the data rate optimization using different schemes. Section 9.4 presents simulation results for single and multiuser (MU) and, finally, Section 9.5 gives the summary for the chapter.

Table 9.1: List of acronyms.

Acronym	Definition
AO	Alternating optimization
AP	Access point
DFT	Discrete Fourier transform
LoS	Line-of-sight
MIMO	Multiple-input multiple-output
mm-Wave	Millimeter wave
MU	Multiuser
NLoS	Non-line-of-sight
OFDM	Orthogonal frequency-division multiplexing
PIN	Positive-intrinsic-negative
QCQP	Quadratically constrained quadratic program
RIS	Reconfigurable intelligent surface
SDR	Semidefinite relaxations
SDP	Standard convex semidefinite program
SISO	Single-input-single-output
SNR	Signal-to-noise ratio
STM	Strongest tap maximization
USRP	Universal software radio peripheral
USRPs	Universal software radio peripherals

Table 9.2: List of symbols and notations.

Symbol/Notation	Description
j	Imaginary unit ($\sqrt{-1}$)
$[\mathbf{A}]_{m,n}$	Element in the mth row and nth column of matrix $\mathbf{A}$
$[\mathbf{A}]_{i:j,m:n}$	Returns a matrix comprising rows i through j and columns m through n from $\mathbf{A}$
$(\cdot)^{\mathrm{T}}$	Vector and matrix transpose operator
$(\cdot)^{\mathrm{H}}$	Vector and matrix complex conjugate transpose operator
$\det(\mathbf{A})$	Determinant of a square matrix $\mathbf{A}$
$\mathrm{diag}(\mathbf{A})$	Diagonal of a square matrix $\mathbf{A}$
$\mathrm{Tr}(\mathbf{A})$	Trace of a square matrix $\mathbf{A}$
$\Vert \cdot \Vert$	ℓ_2 norm of a vector
$\Vert \cdot \Vert_{\mathrm{F}}$	Frobenius norm of a matrix
$\mathbf{0}_N$	All-zeros column vector of length N
$\mathbf{0}_{N \times N}$	All-zeros matrix of size $N \times N$
$\mathbf{I}_N$	All-ones column vector of length N

continued on next page

Table 9.2: (*continued*)

Symbol/Notation	Description
$\mathbf{1}_{N\times N}$	All-ones matrix of size $N \times N$
$\mathbf{I}_N$	Identity matrix of size $N \times N$
$\mathbb{E}\{\cdot\}$	Expectation operator
$\Re\{\cdot\}$	Real part
$\Im\{\cdot\}$	Imaginary part
$\odot$	Hadamard product
$\otimes$	Kronecker product
$\dagger$	Pseudo-inverse of a matrix
$\mathbb{R}^N$	N-dimensional real vector space
$\mathbb{R}^{N\times M}$	$N \times M$-dimensional real matrix space
$\mathbb{C}^N$	N-dimensional complex vector space
$\mathbb{C}^{N\times M}$	$N \times M$-dimensional complex matrix space
λ	Wavelength
c	Speed of light
ϑ	Elevation angle measure with respect to positive z-axis
φ	Azimuth angle measure with respect to positive x-axis
$a(\vartheta, \varphi)$	Array response vector in the (ϑ, φ)-direction
N_0	Noise power spectral density
$\mathcal{K}(\vartheta, \varphi)$	Wavenumber vector in the (ϑ, φ)-direction
r_t	Received signal vector at the receiver
x_t	Transmitted signal vector
$J(\Upsilon)$	Fisher information matrix of Υ
$f(d)$	Probability density function of a random variable d
w_θ	RIS reflection coefficients

9.2 System model

A single antenna (transmitter) and a single user (receiver) are used in an SISO wideband system. As depicted in Fig. 9.1, an RIS with N reconfigurable elements is taken into consideration inside the coverage area of the transmitter and receiver. An orthogonal frequency-division multiplexing (OFDM) is used for transmission. In the complex baseband domain, if $x(k)$ is the transmitted discrete-time signal, then the discrete-time received signal at the receiver can be written as [18]

$$r[k] = \sum_{l=0}^{M-1} h_\theta[l]x[k-l] + e[k], \tag{9.1}$$

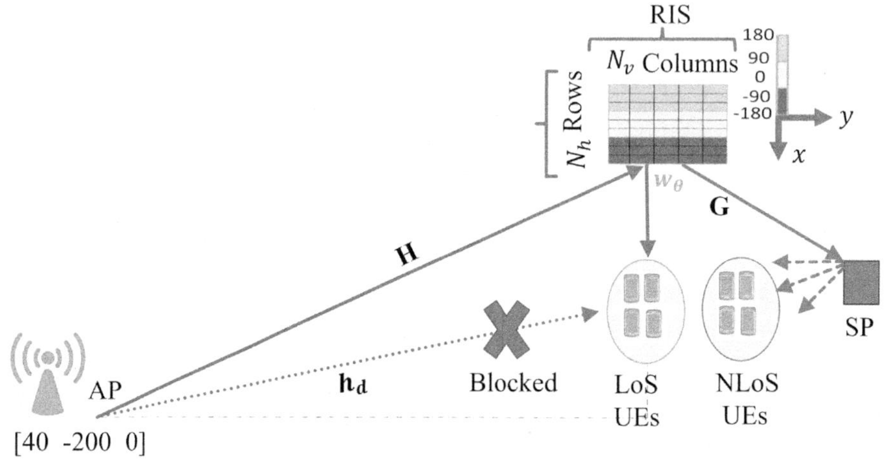

Figure 9.1:
System setup illustration.

where the wideband channel in the time-domain with the RIS configuration $\boldsymbol{\theta}$ is represented as $\{h_\theta[l] : l = 0, \ldots, M-1\}$ and $e[k] \sim \mathcal{N}_\mathbb{C}\left(0, \sigma_e^2\right)$ is the receiver noise. The $h_\theta[l]$ is given by

$$h_\theta[l] = h_d[l] + v_l^\top w_\theta, \tag{9.2}$$

where $h_d[l]$ and $v_l \in \mathbb{C}^N$ are the AP-user uncontrollable direct channel and the AP-RIS-user controllable indirect channels via all RIS N elements, respectively. The reflection coefficients of all elements is described by $w_\theta \in \mathbb{C}^N$. With $K > M$ subcarriers and a length of $M-1$ cyclic prefix, an OFDM transmission is taken into consideration. As a result, a time domain signal with a block of size $K + M - 1$ is transmitted to generate a block of OFDM signal with K of parallel subcarriers by using the DFT:

$$\bar{r}[s] = \bar{h}_\theta[s]\bar{x}[s] + \bar{e}[s], \quad s = 0, \ldots, K-1. \tag{9.3}$$

The above Eq. (9.3) can be illustrated in vector form as follows:

$$\begin{bmatrix} \bar{r}[0] \\ \vdots \\ \bar{r}[K-1] \end{bmatrix} = \begin{bmatrix} \bar{h}_\theta[0] \\ \vdots \\ \bar{h}_\theta[K-1] \end{bmatrix} \odot \begin{bmatrix} \bar{x}[0] \\ \vdots \\ \bar{x}[K-1] \end{bmatrix} + \begin{bmatrix} \bar{e}[0] \\ \vdots \\ \bar{e}[K-1] \end{bmatrix}, \tag{9.4}$$

where $\odot$ describes the Hadamard product. The channel vector $\bar{h}_\theta$ can be realized a function of w_θ as follows:

$$\bar{h}_\theta = F \begin{bmatrix} h_d[0] + v_0^\top w_\theta \\ \vdots \\ h_d[M-1] + v_{M-1}^\top w_\theta \end{bmatrix} = F\left(h_d + V^\top w_\theta\right), \tag{9.5}$$

where $h_d = [h_d[0], \ldots, h_d[M-1]]^\top$ and $V = [v_0, \ldots, v_{M-1}] \in \mathbb{C}^{N\times M}$ are the AP-user uncontrollable direct channel components and the AP-RIS-user controllable indirect propagation channels, respectively. The channel parameters will be defined later, in the channel model. We assume that the channel models are known in the system. The (s, k)th element of the DFT matrix $F \in \mathbb{C}^{K\times M}$ is $e^{-\frac{2\pi ks}{K}}$. The reflection coefficients of the RIS that accounts for the effective phase shifts $(\emptyset_1, \emptyset_2, \ldots, \emptyset_N) \in [-\pi, \pi)$ and amplitude coefficients $(\gamma_1, \gamma_2, \ldots, \gamma_N) \in [0, 1]$ are included in the vector $w_\theta = \left[w_{\theta_1}, \ldots, w_{\theta_N}\right]^\top \in \mathbb{C}^{N\times 1}$. The sum data transmission rate across K subcarriers, given a specific RIS setup and assuming uniform power distribution and known channel at the receiving end, is expressed as

$$\mathcal{R} = \frac{B}{K+M-1}\sum_{s=0}^{K-1}\log_2\left(1 + \frac{P_s\left|f_s^H h_d + f_s^H V^\top w_\theta\right|^2}{BN_o}\right)\frac{\text{bit}}{\text{s}}, \tag{9.6}$$

the channel bandwidth represented by B, the power allocated to the subcarrier s indicated by P_s, the number of channel taps referred to as M, and the f_s^H component, which corresponds to the vth row of DFT matrix F. It is worth noting that the upper bound of (9.6) is provided in [18] as

$$\mathcal{R} \le \frac{B}{K+M-1}\sum_{s=0}^{K-1}\log_2\left(1 + \frac{P_s}{BN_o}\left(\left|f_s^H h_d\right| + \left\|f_s^H V^\top\right\|_1\right)^2\right), \tag{9.7}$$

where $\|\cdot\|_1$ is the L_1-norm.

9.2.1 Channel model

The uncontrollable direct channel is given by [19]

$$h_d = \sum_{\ell=1}^{L_d}\sqrt{\beta_{d,\ell}}\,e^{-j2\pi f_c\tau_{d,\ell}}\begin{bmatrix}\operatorname{sinc}\left(0 + B\left(\varsigma - \tau_{d,\ell}\right)\right) \\ \vdots \\ \operatorname{sinc}\left(M-1 + B\left(\varsigma - \tau_{d,\ell}\right)\right)\end{bmatrix}, \tag{9.8}$$

where $\beta_{d,\ell}$ is the ℓth pathloss from the AP to the user with L_d propagation paths as per Fig. 9.1, $\tau_{d,\ell}$ is the delay of propagation, and the sampling delay over the shortest path is ς. Similarly, the controllable indirect channel is presented as

$$V = \sum_{\ell=1}^{L_a} \sum_{\ell_b=1}^{L_b} \sqrt{\beta_{H,\ell}\beta_{G,\ell_b}} e^{-j2\pi f_c(\tau_{H,\ell}+\tau_{G,\ell_b})} \left(a\left(\varphi_{H,\ell}, \vartheta_{H,\ell}\right) \odot a\left(\varphi_{G,\ell}, \vartheta_{G,\ell_b}\right)\right)$$
$$\times \begin{bmatrix} \operatorname{sinc}\left(0 + B\left(\varsigma - \tau_{H,\ell} - \tau_{G,\ell}\right)\right) \\ \vdots \\ \operatorname{sinc}\left(M - 1 + B\left(\varsigma - \tau_{H,\ell} - \tau_{G,\ell_b}\right)\right) \end{bmatrix}^\top . \tag{9.9}$$

The paths L_a and L_b represent the propagation of signals from the AP to the RIS, and from the RIS to the users, respectively. The attenuation of the combined indirect channel from the AP to the user via the RIS is denoted by $\beta_{H,\ell} \geq 0$, and similarly $\beta_{G,\ell_b} \geq 0$, as depicted in Fig. 9.1. The time delay for propagation along the paths to and from the RIS is $\tau_{H,\ell}$ and τ_{G,ℓ_b}, respectively. The array response vector is denoted by $a(\varphi, \vartheta) = \sqrt{\mathbb{D}(\varphi, \vartheta)} \left[e^{j\mathcal{K}(\varphi,\vartheta)^\top \mathbb{U}_1}, \ldots, e^{j\mathcal{K}(\varphi,\vartheta)^\top \mathbb{U}_N}\right]^\top$, where $\mathbb{D}(\varphi, \vartheta)$ represents the directivity pattern of each element, and ϑ and φ denote the elevation and azimuth angles, respectively. Furthermore, $\mathcal{K}$ is defined as the wavenumber, $\mathcal{K}(\varphi, \vartheta) = \frac{2\pi}{\lambda}[\cos(\vartheta)\cos(\varphi), \cos(\vartheta)\sin(\varphi), \sin(\vartheta)]^\top$. The location of the nth element is represented by $\mathbb{U}_n = [0, i(n)0.25\lambda, j(n)0.25\lambda]^\top$, where $i(n) = \bmod(n-1, N_{hor})$ and $j(n) = \lfloor (n-1)/N_{hor} \rfloor$ correspond to the horizontal and vertical indices of element n. The modulus and truncation operations are represented by $\bmod(\cdot, \cdot)$ and $\lfloor \cdot \rfloor$, respectively. Finally, the vertical and horizontal spacing between the elements is $\lambda/4$.

9.3 Data rate optimization using different methods

The need for faster data rates has grown significantly and exponentially over the past few years. The proliferation of mobile phones and the emergence of services that demand high data rates (such online gaming and video streaming) are the key factors that are continually driving up this demand. Massive multiple-input multiple-output (MIMO) and millimeter wave (mm-Wave) are enabling technologies for future communication systems. These technologies can provide wide bandwidths combined with improved beamforming and spatial-multiplexing benefits from the antenna arrays to satisfy the requirements for increased data rates and accommodate an increasing range of wireless equipments [20]. Despite having potentially large possible rate improvements, these technologies typically have higher hardware and power costs, necessitating a separate evaluation of the overall benefits of their application for each user scenario. In addition, they are unable to directly control the propagation channel, whose randomness nature can occasionally limit the effectiveness of these solutions.

The sum rate in RIS-enhanced multiuser MISO downlink communications was maximized by the authors in [21]. The power allocation and the RIS reflection matrix were optimized alternatively using the majorization–minimization method by using the zero-forcing precoding at the BS. The authors in [22] also looked at the issue of weighted sum rate maximization. Three iterative techniques were developed for optimizing the reflection coefficients in terms of various kinds of RIS reflection components within the alternating optimization (AO) framework in order to produce the transmit beamforming. In [14], the authors first provide a workable transmission protocol for estimating the channels in an MIMO-OFDM system via pilot training, and they then suggest an AO algorithm for optimizing the RIS phase shifts and power allocation matrix. The authors in [19] discuss the sum data increase with the presence of mutual coupling in an SISO OFDM system and provide a simple method for channel estimation and RIS construction that includes realistic reflection amplitudes and binary RIS phase shifts. In order to increase the attainable average sum rate in an RIS-enhanced multiuser system, [23] presented a two-timescale transmission protocol as opposed to optimizing the passive beamforming using the instantaneous CSI. In [24], the authors used the AO approach to increase the achievable data rate of a multistream MIMO communication system supported by RIS. Despite being a straightforward optimization technique to use, it may take several iterations to converge.

A more challenging RIS optimization problem arises in the wideband scenario with K orthogonal subcarriers than in the narrowband case with a single subcarrier. Estimating the ideal vector w_θ and fairly assigning the power to each subcarrier are necessary for maximizing the data rate in (9.6). In a one-bit RIS architecture, each RIS element can alternate between the phase shifts of 90 or -90 degrees. The phase matrices can be created from the DFT columns of the beamforming codebook [25].

Recent community efforts have extended RIS research toward a unified *electromagnetic signal and information theory (ESIT)* framework and identified standardization trajectories for interfaces, calibration, and performance metrics. Notably, TechRxiv's ESIT overview [32] and the IEEE Communications Standards Magazine article [30] summarize ongoing efforts toward active, hybrid, and holographic metasurfaces. For wideband and near-field regimes, refractive and holographic surfaces [31] mitigate beam-squint and improve spatial focusing, while RIS-indexed multiple access [33] expands the multiuser design space by embedding user information directly into the surface configuration states.

9.3.1 Beamforming 2D-DFT codebook

Finding the optimal configuration involves a lengthy analysis and is essentially impossible for large values of N because there are 2^N possible RIS configurations. In existing research, switching between N orthogonal configurations for an RIS with N reflecting elements

Algorithm 1: Proposed power method for optimizing $\left\| \bar{h}_\theta \right\|^2$.

Input: $w_{\theta_{\text{CodeBook}}}$, Ω_0, $h_{\text{equivalent}}$
Output: $w_{\theta_{\text{Optimized}}}$
Select $w_{\theta_{\text{CodeBook}}}$ from the codebook $\left\{W_\theta\right\}$ that gives the maximum SNR in (11) and set
$w_\theta = w_{\theta_{\text{CodeBook}}}$ so $\Omega_0 = \begin{bmatrix} 1 \\ w_\theta \end{bmatrix}$
Set $n = 1$
for $n <$ maximum number of iterations **do**
  Compute $\mathcal{H}_i = h_{\text{equivalent}}\, \Omega_i$ from the initial solution Ω_0 and let the first entry $\mathcal{H}_{i,1} = \text{sign}\left(\mathcal{H}_{i,1}^* \mathcal{H}_{i,1}\right)$ to ensure that the direct path is always assigned to 1 and unconnected with the RIS configurations. Here sign$(\cdot)$ is the signum function
  Set $\Omega_{i+1} = \mathcal{H}_i$ and then quantize the current phases of all candidates $\arg\left([w_\theta]_i\right)$ where $i = 1, \ldots, N$ to be either $e^{j\pi/2}$ or $e^{-j\pi/2}$ if $\arg\left([w_\theta]_i\right) \in [-\pi, 0)$ and $\arg\left([w_\theta]_i\right) \in [0, \pi)$, respectively.
  Iterate until convergence
end for
After convergence $w_{\theta_{\text{Optimized}}}$ is obtained.
Calculate the power allocation for subcarriers P_s as per (9.25).
Evaluate the data rate $\mathcal{R}$ in (9.6) by substituting $w_\theta = w_{\theta_{optimized}}$ and the optimized P_s resulting from (9.25).

depends on accurate configurations. The structure of the optimal RIS phase shifts remains compatible with the 2D-DFT codebook for a given incident and desired reflected angles [25]. Therefore, if an incoming signal is directed towards a particular beam, each column of the codebook beamformer $W_\theta = F(N_v) \otimes F(N_h) \in \mathbb{C}^{N \times N}$ (where $\otimes$ represents the Kronecker product) may correspond to a potential reflection configuration. As a result, for this codebook setup, the likelihood of obtaining a strong SNR configuration is significantly high. The DFT matrices for the columns $F(N_v)$ and $F(N_h)$ can be denoted as

$$F(N_v) = \begin{bmatrix} 1 & 1 & 1 & \cdots & 1 \\ 1 & f_{N_v} & f_{N_v}^2 & \cdots & f_{N_v}^{N_v-1} \\ \vdots & \vdots & \vdots & \ddots & \vdots \\ 1 & f_{N_v}^{N_v-1} & f_{N_v}^{2(N_v-1)} & \cdots & f_{N_v}^{(N_v-1)(N_v-1)} \end{bmatrix}, \tag{9.10}$$

in this context, $f_{N_v} = e^{-j2\pi/N_v} = \cos(2\pi/N_v) - \sin(2\pi/N_v))$. The variables N_v and N_h correspond to the number of vertical and horizontal elements, respectively, within the RIS, as shown in Fig. 9.1. To fit the RIS design, the phase shifts produced by the codebook should be quantized. Therefore, the reflection coefficient w_θ for element i can be either $e^{j\pi/2}$ if $\arg([w_\theta]_i) \in [-\pi, 0)$ or $e^{-j\pi/2}$ if $\arg([w_\theta]_i) \in [0, \pi)$. We looked through the codebook W_θ to find the best configuration that will maximize the SNR in (9.11). The best phase configuration that can be created from the codebook can be represented as $w_{\theta_{\text{CodeBook}}} \in W_\theta$. It is considered the standard or reference point for achieving the maximum SNR,

$$SNR = \frac{P\left|\bar{h}_\theta\right|^2}{BN_o}. \tag{9.11}$$

Algorithm 1 utilizes the optimal configuration with the highest SNR, which is determined by evaluating P across all numbers of subcarriers.

9.3.2 Problem formulation

The objective of this research is to enhance the sum rate received by the user by optimizing the transmit power P and reflection coefficient vector w_θ across all the RIS elements. The optimization problem (OP1) can be formulated mathematically to represent the problem of maximizing the sum rate:

$$\begin{aligned}
&\text{(OP1)} \quad \max_{\boldsymbol{P}, \boldsymbol{w_\theta}} \mathcal{R}(\boldsymbol{P}, \boldsymbol{w_\theta}) = \sum_{s=0}^{K-1} \log_2\left(1 + \frac{P_s\left|f_s^H h_d + f_s^H V^\top w_\theta\right|^2}{BN_o}\right), \\
&\quad \text{such that} \\
&\quad (C1) \quad |[w_\theta]_i| = 1, \ \text{for } i = 1, ..., N, \\
&\quad (C2) \quad \frac{1}{K}\sum_{s=0}^{K-1} P_s \leq P, \\
&\quad (C3) \quad P_s >= 0, \ \text{for } s = 0, ..., K-1.
\end{aligned} \tag{9.12}$$

The first constraint (C1) confirms that there is no pathloss in the reflection of each RIS element. It is necessary to use unit magnitude elements as RIS elements do not amplify or decode signals like relays. Constraint (C2) specifies that power allocations across all subcarriers must not exceed the power budget of the base station and should be ≥ 0, as indicated in constraint (C3). It is important to note that the nonconvexity of the (OP1) over the unit modular constraint (C1) for all RIS elements [14] is addressed by iteratively utilizing the power method until convergence is achieved, leading to the optimal w_θ that can impact all subcarriers. Furthermore, the power in constraint (C2) is allocated to each subcarrier using the well-established water-filling algorithm.

9.3.3 Proposed solutions

Power method

In order to optimize the RIS phase shifts, we employ the power method to identify a low-complex solution. It determines the channel power's dominating eigenvector. The channel power is given as follows:

$$\left\|\bar{h}_{\theta}\right\|^{2}=\underbrace{\begin{bmatrix}1\\ w_{\theta}\end{bmatrix}^{H}}_{\Omega^{H}}\underbrace{\left[h_{d},V^{\top}\right]^{H}\left[h_{d},V^{\top}\right]}_{h_{\text{equivalent}}}\underbrace{\begin{bmatrix}1\\ w_{\theta}\end{bmatrix}}_{\Omega}. \tag{9.13}$$

We consider that the channels h_d and V are known at the receiver. To initiate the power method and optimize the quadratic form of the channel $\left\|\bar{h}_{\theta}\right\|^{2}$, we select $w_{\theta_{\text{CodeBook}}}$ from the codebook $W_{\theta} \in \mathbb{C}^{N\times N}$ as shown in Algorithm 1. The power method determines the dominant eigenvalue by iteratively computing $\Omega_{i+1}=\frac{h_{\text{equivalent}}\,\Omega_i}{\left\|h_{\text{equivalent}}\,\Omega_i\right\|}$ from the initial solution $\Omega_0=\begin{bmatrix}1\\ w_{\theta}\end{bmatrix}$, where we set $w_{\theta}=w_{\theta_{\text{CodeBook}}}$, until convergence, following the steps outlined in Algorithm 1.

Semidefinite relaxation method

The total channel power across the K subcarriers is given by

$$\left\|\bar{h}_{\theta}\right\|^{2}=\sum_{v=0}^{K-1}\left|f_v^{H}h_d+f_v^{H}V^{\top}w_{\theta}\right|^{2}. \tag{9.14}$$

The channel gain fluctuates as a function of the passive beamforming vector w_{θ}. In order to increase the data rate in (9.6), we develop an optimization problem, relying on (9.14), using beamforming reflection coefficient vector w_{θ} and fairly allocating the power over all subcarriers. The optimization problem below is specifically reformulated as follows:

$$\begin{aligned}
&(\text{OP2}) \quad \max_{P,w_{\theta}} \sum_{v=0}^{K-1}\left|f_v^{H}h_d+f_v^{H}V^{\top}w_{\theta}\right|^{2},\\
&\text{such that}\\
&(\text{C1}) \quad \left|[w_{\theta}]_i\right|=1, \quad \forall i\le N,\\
&(\text{C2}) \quad \frac{1}{K}\sum_{v=0}^{K-1}P_v\le P,\\
&(\text{C3}) \quad P_v\ge 0, \quad \forall v<K,
\end{aligned} \tag{9.15}$$

where unnecessary terms have been removed for the sake of clarity. Keep in mind that the sum channel gain maximization at the user is the optimal value of (OP2). Through the use of the semidefinite relaxation approach, we are able to solve the problem (OP2) suboptimally. Thus, the (OP2) objective function of (9.15) can be rewritten denoting:

$$\widetilde{U} = \sum_{v=0}^{K-1} f_v^H V V^H, \tag{9.16}$$

$$\tilde{u} = \sum_{v=0}^{K-1} f_v^H V h_d, \tag{9.17}$$

$$\widetilde{D} = \sum_{v=0}^{K-1} \left| f_v^H h_d \right|^2, \tag{9.18}$$

as

$$\begin{aligned} \text{(OP3)} \quad & \max_{w_\theta} \; w_\theta^H \widetilde{U} w_\theta + w_\theta^H \tilde{u} + \tilde{u}^H w_\theta + \widetilde{D} \\ & \text{such that} \\ & |[w_\theta]_i| = 1, \forall i = 1, \ldots, N, \\ & \frac{1}{K} \sum_{v=0}^{K-1} P_v \le P, \\ & P_v \ge 0 \quad \forall v < K \end{aligned} \tag{9.19}$$

Problem (OP3) is easily distinguishable as a nonconvex quadratically constrained quadratic program (QCQP) problem, which can be rewritten as a homogeneous QCQP problem, in particular, by setting

$$Q = \begin{bmatrix} \widetilde{U} & \tilde{u} \\ \tilde{u}^H & 0 \end{bmatrix}, \widetilde{w}_\theta = \begin{bmatrix} w_\theta \\ t \end{bmatrix}, \tag{9.20}$$

where t is an auxiliary variable. Let us define $\phi = \widetilde{w}_\theta \widetilde{w}_\theta^\top$ with $\text{rank}(\phi) = 1$. Optimization problem (OP3) is changed into the following problem:

$$\begin{aligned} \text{(OP4)} \quad & \max_{\phi} \; \text{tr}(Q\phi) + \widetilde{D} \\ & \text{such that} \\ & [\phi]_{i,i} = 1, \quad \forall i = 1, 2, \ldots, N+1, \\ & \phi \ge 0. \end{aligned} \tag{9.21}$$

It is noted that due to the rank-one constraint, (OP4) is still not convex. As a result, we apply the SDR method to overcome the rank-one constraint, transforming (OP4) into a convex SDP that can be solved effectively with the help of current convex optimization solvers like CVX [26]. It is important to note that the ideal objective value of (OP4) acts as an upper bound on that of (OP2). However, the best solution ϕ^* to (OP4) might not be a rank-one answer. As a consequence, using the Gaussian randomization method, we produce the following rank-one solution:

$$\phi^* = \Lambda \Sigma \Lambda^H \tag{9.22}$$

where $\Lambda = [\rho_1, \ldots, \rho_{N+1}]$ and $\Sigma = \text{diag}(\mathbb{A}_1, \ldots, \mathbb{A}_{N+1})$ are unitary and diagonal matrices, respectively, both of size $(N+1) \times (N+1)$. Then we obtain a suboptimal solution to (OP4) as $\widetilde{w}_\theta^* = \Lambda \Sigma \mathbb{M}$, where $\mathbb{M} \in \mathbb{C}^{(N+1)\times 1}$ is a random vector generated according to $\mathcal{N}_{\mathbb{C}}(0, I_{N+1})$, which denotes the circularly symmetric complex Gaussian (CSCG) distribution with zero mean and covariance matrix I_{N+1}. Finally, we achieve a suboptimal solution to (OP2) as $w_\theta{}^* = e^{j\angle\left([\widetilde{w}_\theta{}^*]_{1:N}/[\tilde{w}_\theta]_{N+1}\right)}$.

Strongest tap maximization method

According to STM, the received signal power should be concentrated in the time domain rather than distributed over the K subcarriers. Therefore, choosing a configuration that is suitable for a single channel tap is preferable to that which is suitable for a single subcarrier. This is especially valid if one of the LoS propagation paths is significantly more powerful than all other paths. It should be noted that the total energy of the incoming signals, computed in the frequency domain, is the objective function in (9.15). The optimization problem (OP2) is thus equivalent to the following problem:

$$\begin{aligned} \text{(OP5)} \quad & \max_{w_\theta} \sum_{l=0}^{M-1} h_d[l] + v_l^{\mathrm{T}} w_\theta \\ & \text{such that} \\ & |[w_\theta]_i| = 1, \quad \forall i = 1, \ldots, N. \end{aligned} \tag{9.23}$$

Because practical wireless communication systems have $M \ll K$, it should be noted that computing the objective function in (9.23) is more effective than calculating it in (9.15). To start, we determine the vector w_θ that increases the magnitude of every channel tap. We then select the solution resulting in the largest magnitude with respect to the channel tap index l,

$$\ell^{\text{opt}} = \underset{\ell \in \{0,\ldots,M-1\}}{\arg\max} \left| h_d[\ell] + v_\ell^{\mathrm{T}} \omega_\ell \right|^2. \tag{9.24}$$

Consequently, we obtain the following suboptimal solution to the problem (OP2) as $w_{\theta_l} = \left[e^{\mathrm{j}(\arg(h_d[l]) - \arg([v_l]_1))}, \ldots, e^{\mathrm{j}(\arg(h_d[l]) - \arg([v_l]_N))}\right]^{\mathrm{T}}$, where the argument of a complex number is

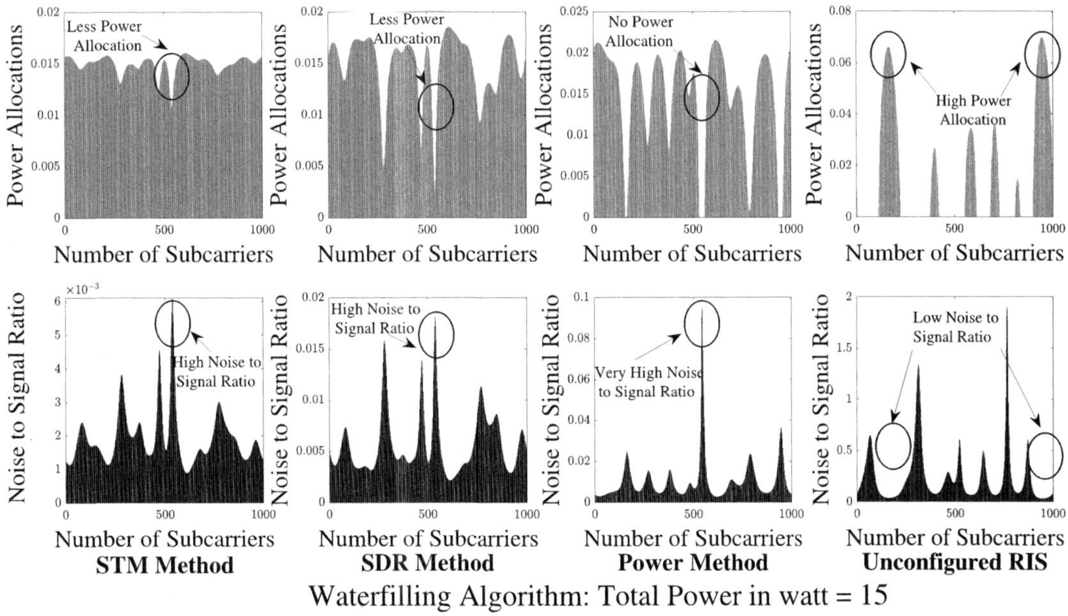

Figure 9.2:
Fair power allocations for all subcarriers in STM method, SDR method, power method, and the case when the RIS surface is unconfigured.

provided by arg(·) and the nth entry of v_l is signified by $[v_\ell]_n$. The phase of every component in the inner product $v_l^{\mathrm{T}} \omega_l$ is rotated in this solution to line up with the phase of $h_d[\ell]$.

Water-filling algorithm

All subcarriers $P_0, \ldots, P_{K-1}$ have a power distribution that satisfies $P = \frac{1}{K} \sum_{s=0}^{K-1} P_s$, where $P_s = E\left\{|\bar{x}[v]|^2\right\}$ is the power assigned to subcarrier s. The symbol $E\{\cdot\}$ represents the expectation operator. To determine the optimal amount of power assigned to each subcarrier, the water-filling algorithm [14] can be employed, giving

$$P_s = \max\left(\eta - \frac{BN_o}{\left|f_s^H h_d + f_s^H V^\top w_\theta\right|^2}, 0\right), \tag{9.25}$$

where the parameter $\eta \geq 0$ is chosen to satisfy $\frac{1}{K} \sum_{s=0}^{K-1} P_s = P$. Consequently, the power allocated for the sth subcarrier P_s is dependent on the water level parameter η and can be rewritten as follows:

$$P_s = \begin{cases} \eta - \frac{BN_o}{|f_s^H h_d + f_s^H V^\top w_\theta|^2} & \text{if } \eta \geq \frac{BN_o}{|f_s^H h_d + f_s^H V^\top w_\theta|^2}, \\ 0 & \text{if } \eta < \frac{BN_o}{|f_s^H h_d + f_s^H V^\top w_\theta|^2}. \end{cases} \tag{9.26}$$

The water-filling algorithm is used to allocate the transmitted power fairly to all subcarriers in the different optimization methods, STM, SDR, power method, and the case when the surface is unconfigured. In Fig. 9.2, we show how the water-filling algorithm allocates 15 W of power for the entire number of subcarriers fairly, taking into consideration the channel gain status. The black circles in the figure refer to the amount of power allocated by the algorithm to some subcarriers taking into consideration their corresponding noise-to-signal ratio status.

9.3.4 Computational complexity

This section discusses the computational complexity of the proposed optimization methods for allocating fair power to all subcarriers and tuning phases of the RIS. To estimate the Big-O ($\mathcal{O}$) complexity of these methods, we followed the same procedure as in [20]. The complexity of the power method can be calculated by considering the steps in Algorithm 1. The algorithm requires $(N+1)^2$ complex multiplications for the term $h_{\text{equivalent}}$, while the term $h_{\text{equivalent}}\,\Omega_i$ necessitates $(N+1)$ complex multiplications and phase quantization demands N complex computations. Additionally, allocating power to all subcarriers requires $2(KM)$ complex multiplications. Thus, the estimated overall complexity per iteration, denoted as $\mathcal{O}_{\text{Power}}$, is

$$\mathcal{O}_{\text{Power}} = \mathcal{O}\left(\left((N+1)^2 + (N+1) + N\right) I_R + 2(KM)\right), \tag{9.27}$$

where I_R represents the maximum number of iterations required to achieve the optimal data rate. The complexity of the SDR method can be determined by calculating the complex operations of the following terms. The terms Q and ϕ in (9.20) require $(N+1)^2$ complex multiplications for each of them. The Gaussian randomization method needs $(N+1)^{4.5}$ complex multiplications to compute the best solution ϕ^*. Moreover, power allocation for all subcarriers, as per SDR method in Fig. 9.2, needs $2(KM)$ complex multiplications. Consequently, the overall estimated complexity is given as $\mathcal{O}_{SDR}$, which is

$$\mathcal{O}_{SDR} = \left(2(N+1)^2 + (N+1)^{4.5} + 2(KM)\right). \tag{9.28}$$

The complexity in (9.28) meets with the computational complexities estimated in [14] and [6]. The complexity in [14] is $\mathcal{O}\left(NKI_{SA} + N^{4.5}K^{3.5}I_T\right)$, where I_{SA} denotes the number of iterations that the successive approximation requires and I_T is the total time that is needed to

Table 9.3: Proposed methods complexities and qualities.

Method	Complexity	Objective Function	Quality
Power	$\mathcal{O}((N+1)^2+(N+1)+N)I_R+2(KM))$	Equivalent power of the channel	Low complex and suboptimal solution
SDR	$\mathcal{O}(2(N+1)^2+(N+1)^{4.5}+2(KM))$	Equivalent power of the channel	High complex and suboptimal solution
STM	$\mathcal{O}((N)+2(KM))$	Equivalent power of the channel	Low complex and suboptimal

solve the problem during the successive convex approximation algorithm. Furthermore, the complexity is calculated as $\mathcal{O}((N+1)^{4.5})$ in [6]. Consequently, the complexity in (9.28) coincides with the results in [14] and [6] in terms of the complexity degree of $\mathcal{O}(N^{4.5})$. Furthermore, the complexity for STM method is computed using the same procedure,

$$\mathcal{O}_{STM} = \mathcal{O}(N+2(KM)). \tag{9.29}$$

The methods complexities are compared in Table 9.3. It is obvious from Table 9.3 that the SDR complexity is the highest in terms of $\mathcal{O}(N^{4.5})$ while the STM is the lowest in terms of $\mathcal{O}(N)$ and the power method is lying in between SDR and STM with Complexity $\mathcal{O}(N^2)$.

9.3.5 Energy–rate trade-off and emerging RIS variants

Recent developments highlight that maximizing data rate with passive RIS must be balanced against the overall energy consumption introduced by control circuits, biasing networks, and training overhead. Active RIS (ARIS) and holographic or radiating metasurfaces (HMS/Holographic MIMO) offer new operating points on the rate–energy frontier. Specifically, ARIS can inject or amplify signals to partially overcome double-fading, yielding higher sum-rate at the expense of additional power draw, whereas subconnected ARIS architectures can improve energy efficiency under practical power budgets by reducing per-element RF chains. HMS further pushes surface density toward near-continuous apertures, enabling fine-grained wavefront control with promising spectral- and energy-efficiency gains when jointly optimized with resource allocation. These alternatives complement the passive RIS assumed in this chapter and indicate that system design should target *rate under an energy constraint* (or vice versa), depending on deployment. For further insight, readers can refer to [29], which discuss the rate–energy balance and emerging RIS paradigms.

Table 9.4: Simulation parameters.

Parameter	**Value**
Carrier frequency f_c	4e9 Hz
Speed of light c	3e8 m/s
Wavelength λ	0.075 m
Total number of elements N	400
Number of horizontal elements N_{hor}	20
Number of vertical elements N_{ver}	20
Vertical and horizontal element spacing	$\lambda/4$
AP location in meters, coordinates $[x, y, z]$	$\begin{bmatrix} 40 & -200 & 0 \end{bmatrix}^T$
User location in meters, coordinates $[x, y, z]$	$\begin{bmatrix} 20 & 0 & 0 \end{bmatrix}^T$
RIS location in meters, coordinates $[x, y, z]$	$\begin{bmatrix} 0 & 0 & 0 \end{bmatrix}^T$
Pathloss NLoS	$34.53 + 38\log_{10}(d)$
Pathloss LoS	$30.18 + 26\log_{10}(d)$
Channel taps M	23
Total number of subcarriers K	1000
Transmitted power P	15 W
Bandwidth B	15e6 Hz

9.4 Simulation results

9.4.1 Simulation results for single user

We analyze the performance of our suggested approaches for increasing the RIS-aided SISO-OFDM system achievable data rate by providing numerical results from Monte Carlo simulations. The simulation parameters are considered by using the 3GPP channel model [27] and Table 9.4. In this study, we take into account an OFDM system with number of subcarriers $K = 1000$ and a more realistic multipath channel model with channel taps $M = 23$ at random delays and an exponential power decay profile. The AP, user, RIS location coordinates in meters are shown in Fig. 9.1. We considered LoS channels between the AP-RIS-user indirect link as per Fig. 9.1, to guarantee effective transmissions and large channel gains. However, the NLoS propagation is assumed between the AP and the user direct link. We contrast the achieved data rate of the suggested methods with the system's upper bound. For comparison with the system upper bound, we consider the power method, SDR, STM benchmark schemes, the random phase shifts, and the case when the surface is unconfigured, where we replaced the RIS with a passive metal sheet causing zero phase-shifts for the entire elements. We search in the codebook W_θ for the best configuration that maximize the SNR in (9.11). Fig. 9.3 depicts the related strongest and weakest SNR with a difference of around 20 dB. The

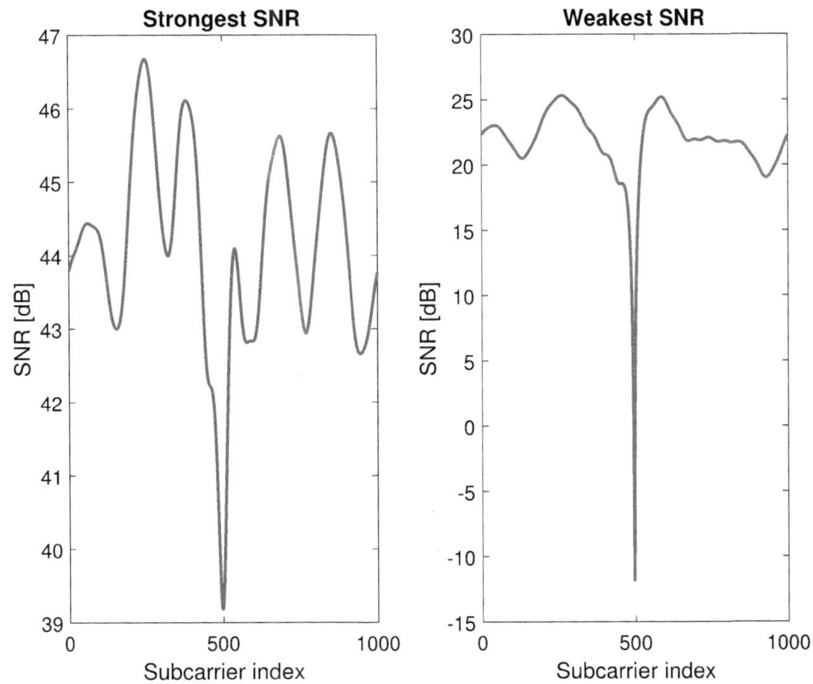

Figure 9.3:
The best confguration in the codebook that gives the maximum SNR and the worst confguration that gives the minimum SNR.

strongest SNR configuration is further utilized in Algorithm 1. The strong SNR which is achieved by the beamforming codebook increases the data rate as per Figs. 9.5 and 9.6.

In order to validate the performance of the proposed method, Fig. 9.4 shows the performance of the iterative power method and the benchmark schemes at different SNR values, with $N = 400$. It is noticed that all the schemes with RIS (SDR method, power method and random phase) outperform the scheme without RIS (unconfigured surface), owing to the RIS-enhanced average channel power between the AP and the user. Additionally, the proposed iterative power method and the SDR scheme both significantly outperform the random phase scheme in terms of achievable rate because of the tuned RIS coefficients that help the direct and the reflecting channel to be superimposed more constructively at the receiver. The SDR method shows slightly superior performance to the power method at low and high SNR regimes. Nevertheless, the performance gap between them is very small and can be sacrificed for the sake of practical implementations and lower complexity.

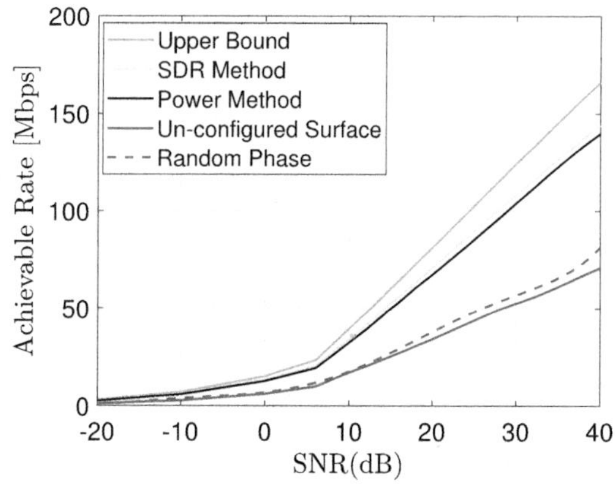

Figure 9.4:
Achievable data rate versus SNR.

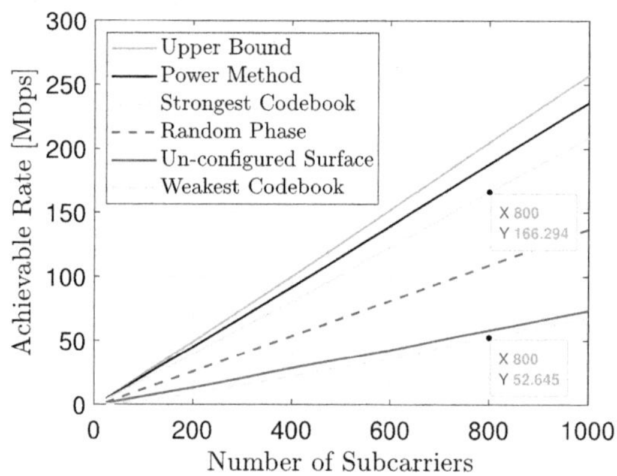

Figure 9.5:
Comparison of the achievable data rate versus number of subcarriers for different schemes when $N = 4096$.

The data rate of the proposed methods is compared in Fig. 9.6 considering LoS channels between the AP-RIS-user indirect link as per Fig. 9.1. The heuristic STM method outperforms both the SDR and the power methods while the data rate is the lowest in case of the unconfigured RIS (metal sheet). The achievable data rate is dramatically boosted up as the range of subcarriers increases. In comparison to unconfigured RIS (metal sheet) configuration, the proposed approaches improve the sum-data rate dramatically with an optimized RIS because of the adjusted RIS coefficients that allow the direct and reflecting channels to be integrated at the receiver more effectively and constructively. The power method shows the lowest data rate among the other methods; however, it is a practical way to configure the RIS surface by implementing the strongest SNR configuration selected from the W_θ codebook.

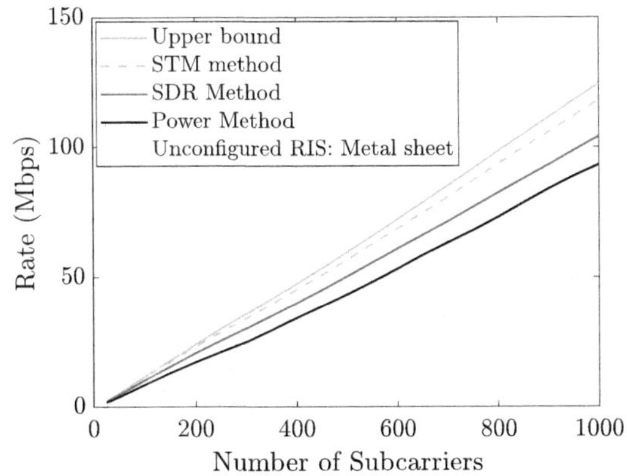

Figure 9.6:
Comparison of data rate versus the number of subcarriers when considering LoS channel between the AP-RIS-user composite indirect link when $N = 400$.

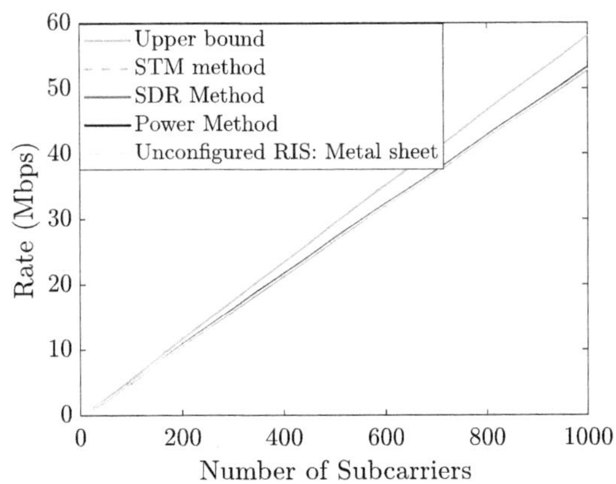

Figure 9.7:
Comparison of data rate versus the number of subcarriers when considering NLoS channel between the AP-RIS-user composite indirect link.

The SDR method outperforms the power method by optimizing the quadratic form of the channel power; however, the STM method shows the highest achievable data rate since it implements the dominant LoS in the composite AP-RIS-user channel.

Unlike the scenario in Fig. 9.6, Fig. 9.7 incorporates the example when the path to the RIS encounters an NLoS, which sheds light on two influences. There is no dominant path, and the path via the RIS is weaker. The upper bound and the unconfigured RIS surface (metal sheet) are much closer together due to the weaker path through the RIS, whereas the absence of the dominating path leads to the impossibility to identify a single RIS configuration that

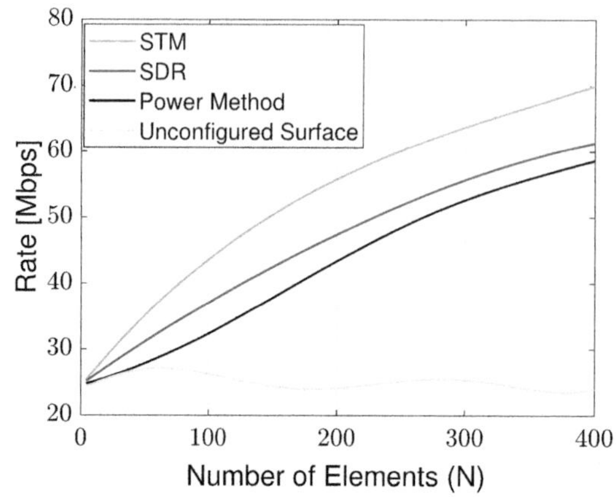

Figure 9.8:
Comparison of data rates for different values of N.

works for the total number of subcarriers. Fig. 9.7 results agree with [18] where the authors demonstrated the same scenario for the STM method and the upper bound cases.

An important parameter that should be taken into account when investigating the data rate is the dimension of the RIS surface. The RIS enhances the data rate and it increases further when the surface is large as per Fig. 9.8. We compare the achievable rate of different schemes versus the number of RIS elements at SNR $= 20$ dB. It is clear that the rate performance of the unconfigured surface (zero phase shifts) scheme at the RIS is independent of the number of elements. However, the tuned RIS phase profiles in the other schemes (STM, SDR and power method) enhanced the data rate with the increase number of RIS elements. The highest average data rate over the entire bandwidth is shown when the number of elements on the surface is $N = 400$, while the lower data rate is obtained at $N = 100$ or less.

Another crucial factor that should be considered is the complexity and runtime. Table 9.3 reveals the complexity level for each method. The complexity in SDR method is $\mathcal{O}\left(N^{4.5}\right)$ while it is much less in both power and STM methods, namely $\mathcal{O}\left(N^{2}\right)$ and $\mathcal{O}(N)$, respectively. We illustrated the complexity level and the runtime consumption for each method in Figs. 9.9 and 9.10, respectively. We noticed that the SDR method complexity is the highest (more than 6 millions of complex multiplications) with a runtime larger than 50 minutes; however, the complexity in the other two methods, STM and the power method, is less than that of the SDR with few minutes runtime consumption.

The above simulation results considered low finite resolution phase shift configurations. Two phase configurations (1-bit RIS) $\theta_n \in \{0, \pi\}$ or $\left\{\frac{\pi}{2}, \frac{-\pi}{2}\right\}$ are used to maximize the SNR at the user. It is preferable, for more realistic and practical scenarios in the design, to select the

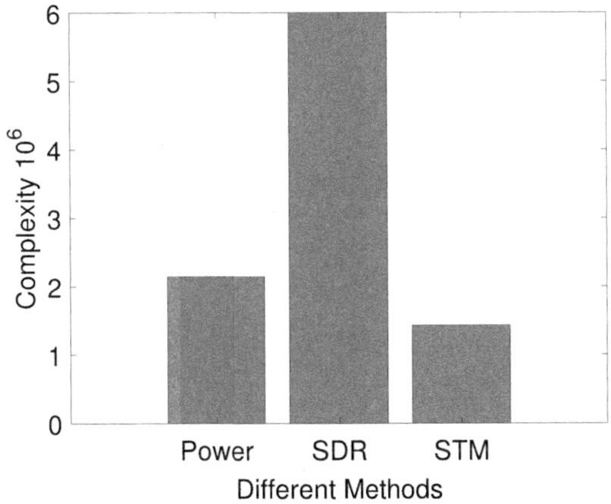

Figure 9.9:
Complexity level for each optimization method when $N = 400$.

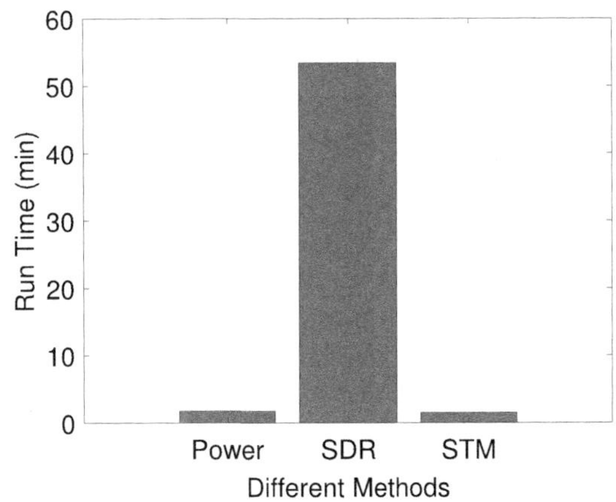

Figure 9.10:
Runtime consumption for each optimization method when $N = 400$.

phase shifts which are limited to a certain number of states. The higher resolutions necessitate complex RIS design with different bias voltages dedicated for each state. For example, in 2-bit RIS scenario, four phase configurations are required for each RIS cell or element, while the 3-bit RIS design switches over to eight-phase configurations. The channel gains $\beta_{H,\ell}$ and β_{G,ℓ_b} move along the unit circle with these phase shifts configurations, while the path $\beta_{d,\ell}$ cannot be moved since it is the direct channel path as per Fig. 9.1. Consequently, if 1-bit RIS design is considered then one should select the 0 or π phase shift. If the 0 phase shift is selected, then the gains $\beta_{H,\ell}$ and β_{G,ℓ_b} will remain at the location where they are on the unit circle; however, if π is chosen then the gains will move to the opposite side of the

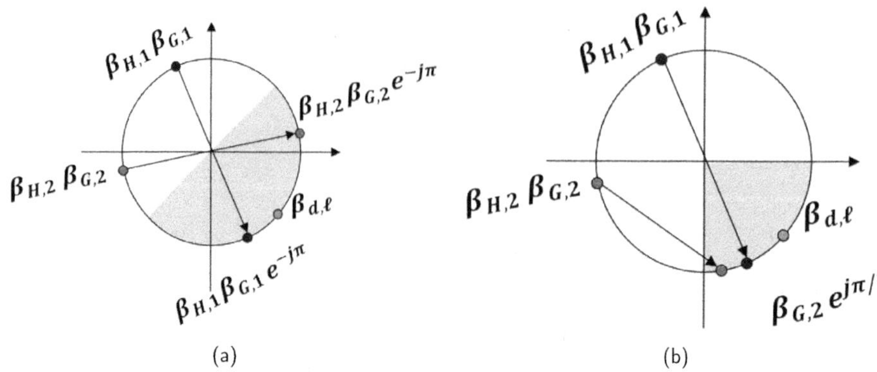

Figure 9.11:
(a) Two configurations, $\theta_n \in \{0, \pi\}$; (b) Four configurations, $\theta_n \in \{0, \pi, \pi/2, -\pi/2\}$.

circle. The key point is to shift theses signal gains $\beta_{H,\ell}$ and β_{G,ℓ_b} in the same half-space of the unit circle as the direct channel gain so, one can add all the terms $(\beta_{d,\ell}, \beta_{H,\ell}, \text{and } \beta_{G,\ell})$ constructively. The same procedure can be repeated for higher resolutions of phases where all the signals are added up in the same quadrant of the unit circle as the direct path. Therefore, higher phase resolutions are not required as long as the signals can be added constructively without cancelling each other. An illustration is shown on the unit circle for two and four finite phase resolutions when the RIS has two elements only for simplicity [18]. All the signals are gathered in the same half-space in the two configurations case while they are in the same quadrant in the four configurations case as per Fig. 9.11.

The finite phase resolutions for different bit-RIS designs are studied in the case of the STM scheme. The obtained results of the sum data rate in case of higher-bit RIS are very close to the 1- and 2-bit RIS, and this is expected since the path phases of the indirect channels are aligned with the dominant LoS path of the direct channel gain in the STM method. Consequently, they are added up constructively at the receiver and hence increase the SNR and the sum rate. However, choosing higher-bit RIS design increase the complexity from $\mathcal{O}(N + 2(KM))$ to $\mathcal{O}(2N + 2(KM))$ due to assigning more phases for each RIS individual element and the switching between of them to aim the direction of the beamforming towards the receiver. Furthermore, the increase in the computational complexity $\mathcal{O}(2N)$ instead of $\mathcal{O}(N)$ is negligible and will increase the run time few seconds only. The challenge is clearly evident in the RIS hardware when higher-bit RIS is required. Generally speaking, it is required more positive-intrinsic-negative (PIN) or varactor diodes for each RIS cell or element to achieve more states of phases so, it not recommended to complicate the design and increase the cost by selecting high-bit RIS while the 1-bit RIS phase profile can give comparable sum data rate performance. Fig. 9.12 shows the data rate for different bit RIS designs with slight difference in the performance for the STM scheme when $N = 900$ elements.

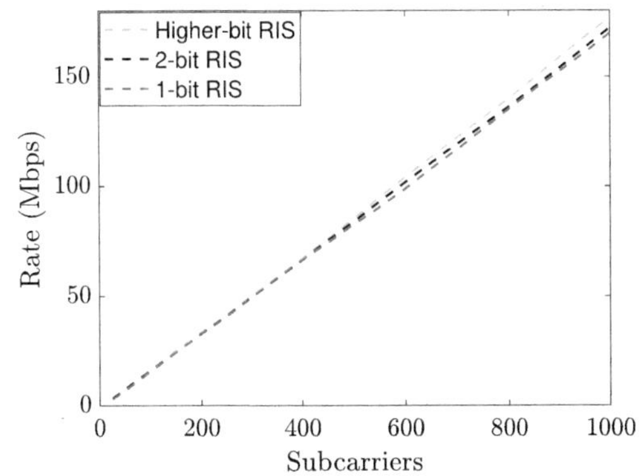

Figure 9.12:
STM method for multibit RIS design.

9.4.2 Simulation results for MU

Considering the MU scenario, in Fig. 9.13, the heuristic STM technique surpasses the SDR, the iterative power, the random, and the RIS when is it is replaced by an absorbed material and passive metal sheet in terms of data rate performance. However, the data rate is at its minimum when dealing with an unconfigured RIS, represented by a absorbed material and metal sheet. Nevertheless, the STM outperforms all other methods for the users who enjoys LoS path from the RIS and the achievable rate is degraded for the NLoS users from the RIS. This is due to the fact that the STM approach of maximizing the strongest tap, aiming to enhance the magnitude of the largest element in the channel vector $\boldsymbol{h}_d + \boldsymbol{V}^{\mathrm{T}} w_\theta$. While effective for LoS channels, this method does not perform well for NLoS channels. Compared

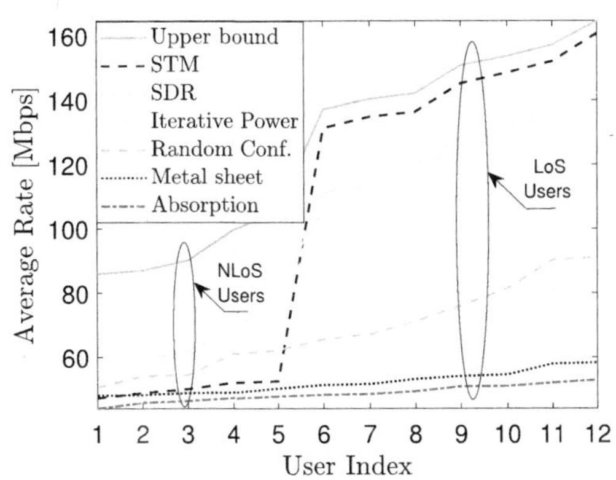

Figure 9.13:
The data rates achieved for 12 users using different schemes in LoS and NLoS propagations channels.

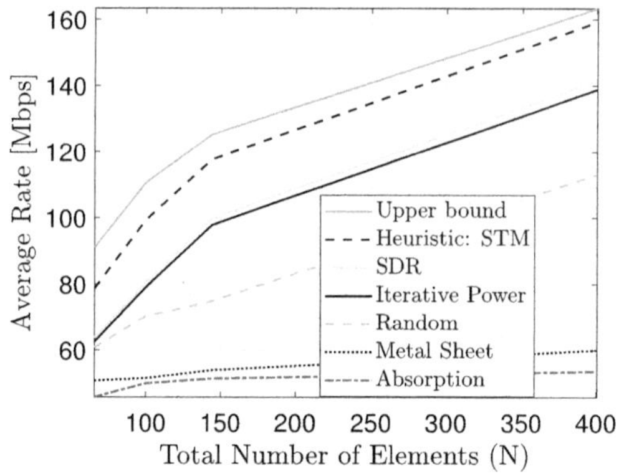

Figure 9.14:
Achievable rates versus number of reflecting elements N using different schemes.

to an unconfigured RIS, the proposed approaches (STM, SDR, and iterative power) yield a significant enhancement in the achieved data rate through the optimized RIS configurations with a 2.5× increase. This improvement is due to the adjusted RIS coefficients, which facilitate a more efficient and constructive integration of both direct and reflected channels at the user. The iterative power-based algorithm, although providing a lower data rate than the SDR algorithm, offers a practical and realistic means to tune and program the RIS surface. This is achieved by applying the configuration with the strongest SNR, chosen from the w_θ codebook. The SDR algorithm surpasses both the iterative power method and the random phase case. However, it comes with the trade-off of higher computational complexity.

In Fig. 9.14, we juxtapose the achievable data rates of various strategies against the number of RIS elements, with SNR of 25 dB. Initial observations reveal two significant patterns. Firstly, it becomes evident that the rate achieved with the random phase shift approach on the RIS remains unaffected by the number of elements. This outcome is anticipated, given that this method solely relies on aperture gain, lacking any passive beamforming gain. However, the random configuration is still higher in the achieved data rates than the unconfigured metal and absorbing sheets cases. Furthermore, a distinct trend emerges: STM, SDR, and iterative power-based methods exhibit better performance compared to the random phase involving the RIS, as well as the unconfigured surfaces.

Scalability for very-large surfaces ($N \gg 400$)

Fig. 9.14 evidences monotonic rate gains with the number of RIS elements N for optimized schemes (STM/SDR/Power), whereas unconfigured surfaces saturate due to aperture limitations. For $N > 400-1000$, both the channel training overhead and the wideband beam-squint or near-field effects become dominant. The proposed codebook-initialized power method maintains a per-iteration cost of $\mathcal{O}(N^2)$; see Table 9.3, which remains tractable compared with $\mathcal{O}(N^{4.5})$ for SDR-based optimization and avoids repeated alternating optimization across subcarriers. In the near-field wideband regime expected with meter-scale apertures or user proximity, refractive or holographic surfaces and near-field beamforming designs [31] have been shown to mitigate beam-squint and enable multiuser focusing, forming a natural extension of the presented MU-OFDM framework.

9.4.3 Indoor field measurements

We present a novel RIS prototype operating in the 3.75 GHz range with 4096 configurable elements. We propose a realistic receiver-RIS Wi-Fi feedback link and an effective method for configuring the RIS over the air by taking advantage of the geometrical array features. Our 1-bit RIS prototype offers an 18 dB power gain over the case when it is replaced with a metal plate. The indoor measurements have been carried out where the transmitter and receiver are separated by a concrete wall.

- We have developed and tested a unique feedback-based adaptive reflection coefficient optimization algorithm that enables smart reflection without changing the current communication standards.
- We have performed indoor NLoS trials and demonstrated an 18 dB channel gain improvement compared to a metal plate reference case.

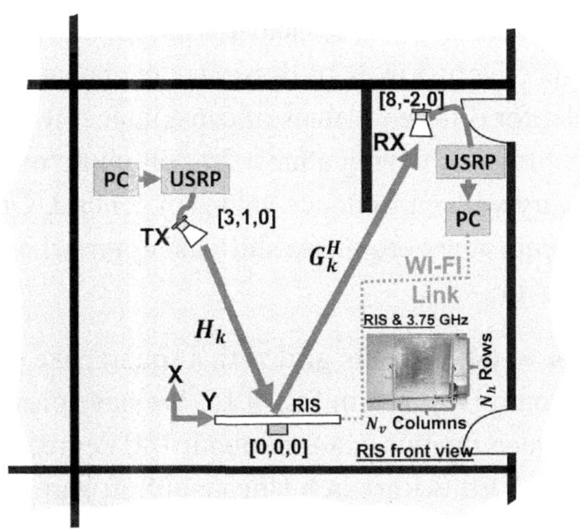

Figure 9.15:
Indoor measurement setup at MAST lab.

The prototype of the RIS-aided wireless communication system seen in Fig. 9.15, which is made up of personal computers (PCs), universal software radio peripherals (USRPs), an

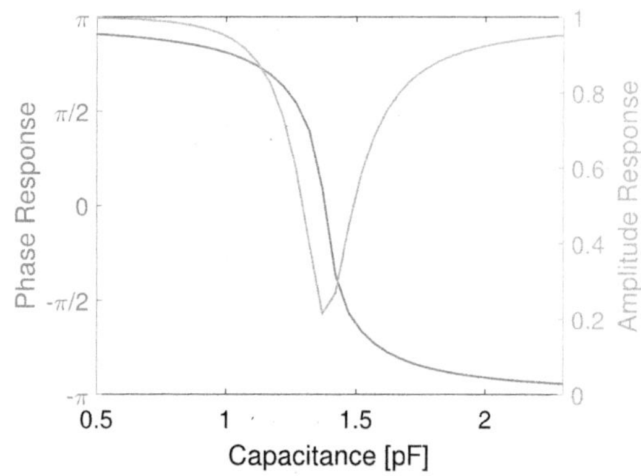

Figure 9.16:
RIS element response versus different capacitance values of the PIN diode at $f = 3.75$ GHz.

RIS, and the real-time RIS-UE feedback module. The key parts of the transmitter are the host computer, a universal software radio peripheral (USRP), and an antenna. The receiver has the same essential parts. These transceivers represent a UE and an access AP or BS, respectively. The locations of the transmitter, receiver, and RIS are measured in meters. A single passive antenna on the transmitter is used to transmit 64-subcarrier OFDM signal. The USRP is used to transport the data. The receiver decodes the data stream, and the host computers recover the transmitted OFDM signal. The RIS element structure can create a 180-degree phase difference in the entire bandwidth around the center frequency, for instance, employing capacitance values of 1.3 and 1.4 pF, for which the phase shifts are around $\pi/2$ and $-\pi/2$, respectively. Fig. 9.16 illustrates the behavior of the amplitude and the phase shift, respectively, for different values of capacitance. It is observed that although a reflecting element is capable of achieving almost 2π full phase tuning, its phase shift and reflection amplitude both vary with capacitance values in general. One can also observe that the minimum amplitude occurs near zero phase shift and approaches unity (the maximum value) at the phase shift of π or $-\pi$.

The fabricated RIS has a 64×64 elements grid with a square size of $192\,\text{cm} \times 192\,\text{cm}$. The front view of the RIS board is shown in Fig. 9.15. We have created a 1-bit prototype with a phase difference between the two states of about 180 degrees. A continuous OFDM transmission is tested with two Ettus Research Universal Software Radio Peripheral USRP X300's at the transmitter and receiver connected with standard horn antenna for each of them.

The RIS will either scatter or reflect the incident wave depending on its configurations. The SNR measures the communication performance where both transmitter and receiver are equipped with a horn single-antenna as per Fig. 9.15. Consequently, the RIS can help the communication system by maximizing the power of the received signal by optimizing its

Algorithm 2: DFT Codebook Algorithm.

Initialize the configuration vector $w_{\theta_0} \in \mathbb{C}^{N\times 1}$ from the codebook $\left\{W_{\theta_N} \in \mathbb{C}^{N\times N}\right\}$
Receive initial feedback of the RX signal power y_0 to calculate the initial data rate
Search in the codebook W_{θ_N}
for $w_{\theta_n}, n \in [1, N]$ **do**
  $w_{\theta_N} \leftarrow \left[w_{\theta_1}, w_{\theta_2}, \ldots, w_{\theta_N}\right]$ where w_{θ_n} is the nth column of the codebook matrix W_{θ_N}.
  Receive the signal power y_n using configuration w_{θ_n}
  if $y_{n-1} \geq y_n$ **then**
    $w_{\theta_n} \leftarrow w_{\theta_{n-1}}$
  end if
end for
Evaluate the data rate with the new configuration.

reflection coefficients. The optimal phase vector $w_{\theta_c} = \underset{|[w_\theta]_i|=1}{\operatorname{argmax}} \left|(\mathrm{G}_k \odot \mathrm{H}_k)^{\mathrm{T}} w_\theta\right|^2$ can be approximated by the columns of the DFT matrix when the RIS surface is large. Consequently, each column of the codebook beamform $W_{\theta_N} = F(N_v) \otimes F(N_h) \in \mathbb{C}^{N\times N}$ can be a possible reflection configuration for an incident signal in a certain beam direction. Let $\otimes$ denote the Kronecker product. We searched in the codebook for the best configuration that maximizes the achievable data rate in (9.6) using Algorithm 2.

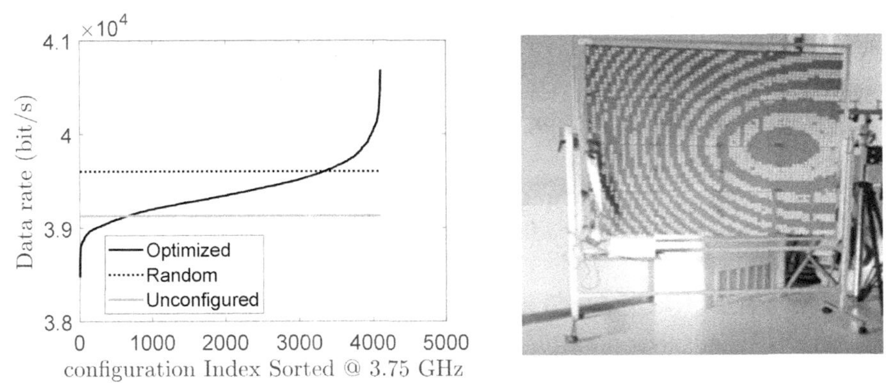

Figure 9.17:
Achievable data rate measurements at $f = 3.75$ GHz.

Fig. 9.17 shows the measurements of the data rates against different configurations at 3.75 GHz. It is compared with the rate when the RIS surface is unconfigured. Algorithm 2, using the feedback iterations, searches in the codebook $\left\{W_{\theta_N}\right\}$ for the configuration that increases the received signal power and then the data rate.

9.5 Summary

In this chapter, we showed that the RIS can be configured to provide significantly higher data rates, in comparison with a uniform surface, e.g., metal sheets. The codebook approach was implemented to generate varieties of phase configurations for the RIS surface in which we searched for the best configuration to be used in the optimization process. A problem formulation was described, and a solution was presented using different optimization methods and the water-filling algorithm. The data rate which is produced by different methods was compared with the upper bound. We noticed that all methods give performance near the upper bound, however, the data rate declines with the decrease in the dimensional area of the surface (number of elements per dimension). In addition, we have proved that the STM and the power methods are simple ways to configure the RIS with lower complexity and less runtime consumption. Selecting higher-bit RIS phase resolutions slightly enhances the sum data rate but at the cost of the hardware and computational complexities. The 1-bit RIS gives performance very close to the higher-bit phases which is enough to align the direct and the indirect paths constructively at the receiver. The RIS is a promising technology and is predicted to have a significant contribution in the 6G networks, so the search for low complex algorithms with reasonable data rate outcomes is crucial and necessary for realistic communications models.

The practicality of the proposed RIS-aided system and its potential use-cases can be summarized as follows:

- **Deep-indoor coverage and NLoS restoration.** RIS steering enables constructive reflection through favorable paths, improving rate and reliability compared to unconfigured sheets, as validated in our cmWave trials.
- **Cell-edge performance in MU-OFDM.** The codebook-guided STM and power schemes provide notable multiuser rate gains with manageable runtime, suitable for RIC/xApp closed-loop implementations.
- **Industrial IoT and smart environments.** Deterministic layouts allow stable beam configurations; energy-aware scheduling can switch between passive and active tiles per application demand [30].
- **Overloaded access (RIS-indexed multiple access).** Embedding user indices within surface states expands user multiplexing beyond conventional OFDM limits [32].
- **UAV/backhaul and NTN edge links.** Active or holographic surfaces can trade modest power consumption for higher rates in challenging aerial and backhaul scenarios [31].

Acknowledgments

This work has been supported by the CHEDDAR: Communications Hub for Empowering Distributed ClouD Computing Applications and Research funded by the UK EPSRC under grant numbers EP/Y037421/1 and EP/X040518/1.

References

[1] M. Di Renzo, A. Zappone, M. Debbah, M.S. Alouini, C. Yuen, J. De Rosny, S. Tretyakov, Smart radio environments empowered by reconfigurable intelligent surfaces: how it works, state of research, and the road ahead, IEEE Journal on Selected Areas in Communications 38 (11) (2020) 2450–2525.

[2] C. Liaskos, A. Tsioliaridou, S. Nie, A. Pitsillides, S. Ioannidis, I.F. Akyildiz, On the network-layer modeling and configuration of programmable wireless environments, IEEE/ACM Transactions on Networking 27 (4) (2019) 1696–1713.

[3] S. Hassouna, M.A. Jamshed, J. Rains, J.U.R. Kazim, M.U. Rehman, M. Abualhayja, L. Mohjazi, T.J. Cui, M.A. Imran, Q.H. Abbasi, A survey on reconfigurable intelligent surfaces: wireless communication perspective, IET Communications 17 (5) (2023) 497–537.

[4] E. Björnson, Ö. Özdogan, E.G. Larsson, Intelligent reflecting surface versus decode-and-forward: how large surfaces are needed to beat relaying?, IEEE Wireless Communications Letters 9 (2) (2019) 244–248.

[5] M. Di Renzo, K. Ntontin, J. Song, F.H. Danufane, X. Qian, F. Lazarakis, J. De Rosny, D.T. Phan-Huy, O. Simeone, R. Zhang, M. Debbah, Reconfigurable intelligent surfaces vs. relaying: differences, similarities, and performance comparison, IEEE Open Journal of the Communications Society 1 (2020) 798–807.

[6] S. Lin, B. Zheng, G.C. Alexandropoulos, M. Wen, F. Chen, Adaptive transmission for reconfigurable intelligent surface-assisted OFDM wireless communications, IEEE Journal on Selected Areas in Communications 38 (11) (2020) 2653–2665.

[7] Q. Wu, R. Zhang, Intelligent reflecting surface enhanced wireless network via joint active and passive beamforming, IEEE Transactions on Wireless Communications 18 (11) (2019) 5394–5409.

[8] M. Cui, G. Zhang, R. Zhang, Secure wireless communication via intelligent reflecting surface, IEEE Wireless Communications Letters 8 (5) (2019) 1410–1414.

[9] Z. Chu, W. Hao, P. Xiao, J. Shi, Intelligent reflecting surface aided multi-antenna secure transmission, IEEE Wireless Communications Letters 9 (1) (2019) 108–112.

[10] P. Wang, J. Fang, X. Yuan, Z. Chen, H. Li, Intelligent reflecting surface-assisted millimeter wave communications: joint active and passive precoding design, IEEE Transactions on Vehicular Technology 69 (12) (2020) 14960–14973.

[11] G. Yang, X. Xu, Y.C. Liang, M. Di Renzo, Reconfigurable intelligent surface-assisted non-orthogonal multiple access, IEEE Transactions on Wireless Communications 20 (5) (2021) 3137–3151.

[12] G. Zhou, C. Pan, H. Ren, K. Wang, Z. Peng, Secure wireless communication in RIS-aided MISO system with hardware impairments, IEEE Wireless Communications Letters 10 (6) (2021) 1309–1313.

[13] M. Zeng, X. Li, G. Li, W. Hao, O.A. Dobre, Sum rate maximization for IRS-assisted uplink NOMA, IEEE Communications Letters 25 (1) (2020) 234–238.

[14] Y. Yang, B. Zheng, S. Zhang, R. Zhang, Intelligent reflecting surface meets OFDM: protocol design and rate maximization, IEEE Transactions on Communications 68 (7) (2020) 4522–4535.

[15] P. Nuti, B.L. Evans, Spectral efficiency vs complexity in downlink algorithms for reconfigurable intelligent surfaces, in: 2021 IEEE International Conference on Communications Workshops (ICC Workshops), IEEE, 2021, June, pp. 1–7.

[16] P. Nuti, E. Balti, B.L. Evans, Spectral efficiency optimization for mmWave wideband MIMO RIS-assisted communication, in: 2022 IEEE 95th Vehicular Technology Conference (VTC2022-Spring), IEEE, 2022, June, pp. 1–6.

[17] S. Hassouna, J. Rains, J.R. Kazim, M.U. Rehman, M. Imran, Q.H. Abbasi, Discrete phase shifts for intelligent reflecting surfaces in OFDM communications, in: 2022 International Workshop on Antenna Technology (iWAT), IEEE, 2022, May, pp. 128–131.

[18] E. Björnson, H. Wymeersch, B. Matthiesen, P. Popovski, L. Sanguinetti, E. de Carvalho, Reconfigurable intelligent surfaces: a signal processing perspective with wireless applications, IEEE Signal Processing Magazine 39 (2) (2022) 135–158.

[19] E. Björnson, Optimizing a binary intelligent reflecting surface for OFDM communications under mutual coupling, in: WSA 2021, 25th International ITG Workshop on Smart Antennas, VDE, 2021, November, pp. 1–6.
[20] N.S. Perović, L.N. Tran, M. Di Renzo, M.F. Flanagan, Achievable rate optimization for MIMO systems with reconfigurable intelligent surfaces, IEEE Transactions on Wireless Communications 20 (6) (2021) 3865–3882.
[21] C. Huang, A. Zappone, M. Debbah, C. Yuen, Achievable rate maximization by passive intelligent mirrors, in: 2018 IEEE International Conference on Acoustics, Speech and Signal Processing (ICASSP), IEEE, 2018, April, pp. 3714–3718.
[22] H. Guo, Y.C. Liang, J. Chen, E.G. Larsson, Weighted sum-rate maximization for intelligent reflecting surface enhanced wireless networks, in: 2019 IEEE Global Communications Conference (GLOBECOM), IEEE, 2019, December, pp. 1–6.
[23] M.M. Zhao, Q. Wu, M.J. Zhao, R. Zhang, Intelligent reflecting surface enhanced wireless networks: two-timescale beamforming optimization, IEEE Transactions on Wireless Communications 20 (1) (2020) 2–17.
[24] S. Zhang, R. Zhang, Capacity characterization for intelligent reflecting surface aided MIMO communication, IEEE Journal on Selected Areas in Communications 38 (8) (2020) 1823–1838.
[25] X. Pei, H. Yin, L. Tan, L. Cao, Z. Li, K. Wang, K. Zhang, E. Björnson, RIS-aided wireless communications: prototyping, adaptive beamforming, and indoor/outdoor field trials, IEEE Transactions on Communications 69 (12) (2021) 8627–8640.
[26] M. Grant, S. Boyd, Y. Ye, 2009, CVX: Matlab software for disciplined convex programming.
[27] Spatial channel model for Multiple Input Multiple Output (MIMO) simulations (Release 16), available at: https://portal.3gpp.org/desktopmodules/Specifications/SpecificationDetails.aspx?specificationId=1382.
[28] B. Zheng, R. Zhang, Intelligent reflecting surface-enhanced OFDM: channel estimation and reflection optimization, IEEE Wireless Communications Letters 9 (4) (2019) 518–522.
[29] R. Singh, J. An, A. Kaushik, C. Yuen, A. Zhang, O.A. Dobre, Redefining electromagnetic signal and information theory with programmable surfaces: open critical questions and future horizon, Authorea Preprints [Online]. Available at: https://doi.org/10.36227/techrxiv.175621821.18352632, Aug. 2025.
[30] M. Katwe, A. Kaushik, L. Mohjazi, M. Abualhayja'a, D. Dardari, K. Singh, M.A. Imran, M.M. Butt, O.A. Dobre, An overview of intelligent meta-surfaces for 6G and beyond: opportunities, trends, and challenges, IEEE Communications Standards Magazine 8 (4) (Dec. 2024) 62–69.
[31] Z. Lin, S. Zeng, A. Kaushik, H. Zhang, Beamforming design for wideband near-field communications with reconfigurable refractive surfaces, IEEE Transactions on Communications (Aug. 2025), Early Access.
[32] R. Singh, A. Kaushik, W. Shin, G.C. Alexandropoulos, M. Toka, M. Di Renzo, Indexed multiple access with reconfigurable intelligent surfaces: the reflection tuning potential, IEEE Communications Magazine 62 (4) (Apr. 2024) 120–126.
[33] A. Kaushik, M. Di Renzo, Electromagnetic Signal and Information Theory: From Synergizing Three Realms to Evolution of Next Generation Communications, Elsevier Academic Press, 2025, Book Editors.

CHAPTER 10

Performance of RIS-assisted IoT network with hardware imperfections

Deepak Kumar[a] and Vimal Bhatia[b]

[a]Department of Electronics & Communication Engineering, Motilal Nehru National Institute of Technology Allahabad, Prayagraj, India [b]Department of Electronics & Communication Engineering, Indian Institute of Technology Indore, Prayagraj, India

10.1 Introduction

Reconfigurable intelligent surface (RIS) is an innovative technology for beyond fifth-generation (B5G) wireless communication systems that can dynamically reconfigure the reflected electromagnetic signal. It comprises a planar meta-surface consisting of reflecting elements that can manipulate signals through reflection, refraction, or scattering to enhance wireless network performance [1–4]. In addition, the RIS utilizes low-power electronic components, such as varactor and PIN diodes, to achieve phase shifting while maintaining minimal power consumption. Moreover, it operates without direct-current (DC) power and generates negligible thermal noise [5,6]. A key advantage of passive RIS is its negligible noise contribution, which allows for a significant array gain. In general, each passive RIS element comprises a reflective patch connected to an impedance-tunable circuit that enables phase shifting [7,8]. Specifically, in an RIS-assisted single-user single-input single-output system, the achievable signal-to-noise ratio (SNR) gain provided by an N-element RIS scales proportionally to N^2 [9].

In addition, B5G wireless communication systems are anticipated to have stringent requirements for massive connectivity and ultra-high data rates [10,11]. The Internet of Things (IoT) network plays a crucial role in the deployment of B5G systems, offering the capability to meet both of these demands effectively [12–15]. The need for massive connectivity is primarily driven by the expansion of the IoT and various real-world applications, such as health monitoring, emergency detection, smart cities, and smart manufacturing. IoT devices often face energy constraints due to limited battery life and the unavailability of direct power connections. A promising solution to this challenge is wireless-powered communication, which harnesses energy from a dedicated source or power station through received radio-frequency (RF) signals using the wireless energy transfer (WET) protocol [16]. RF energy harvesting

Electromagnetic Signal and Information Theory
https://doi.org/10.1016/B978-0-44-336354-2.00017-4
Copyright © 2027 Elsevier Inc. All rights are reserved, including those for text and data mining, AI training, and similar technologies.

is an emerging technology in wireless communication systems that allows devices to capture and convert ambient radio frequency signals into usable electrical power. This technique is particularly advantageous for energy-constrained applications such as IoT networks, wireless sensor nodes, and smart wearables, where frequent battery replacements are not feasible. By harnessing RF signals from sources like cellular networks, Wi-Fi, and dedicated power transmitters, RF energy harvesting minimizes reliance on conventional power supplies, improving device sustainability and extending network lifespan. The process involves an antenna to receive RF signals, a rectifier circuit to convert them into direct current power, and an energy storage unit for effective power management. While challenges such as low energy conversion efficiency and distance limitations remain, advancements in RIS, beamforming, and circuit design are accelerating its adoption in B5G networks. RF energy harvesting is instrumental in fostering self-sustaining, energy-efficient communication systems, contributing to the realization of green and sustainable wireless networks. Furthermore, integrating RIS into the IoT network presents a viable solution for enhancing electromagnetic signal strength, expanding coverage, and massive connectivity. This approach can meet the stringent demands of B5G communication systems. Therefore, this chapter shows the performance of an RIS-assisted IoT network with energy harvesting capability.

10.1.1 Background

In [17–28], different aspects of a passive RIS-assisted IoT network are examined in detail. In [17], an RIS-assisted IoT network with energy harvesting capabilities is introduced. An alternating optimization approach is utilized to enhance sum throughput and fine-tune RIS phase shifts. In [18], a new transmission policy is introduced for an RIS-assisted IoT network utilizing a linear energy harvesting model, where IoT devices communicate via time division multiple access. A joint optimization of wireless information transfer (WIT) and wireless energy transfer (WET) is proposed, implementing a harvest-then-transmit strategy to maximize the system's sum throughput. In [29], a novel federated learning-based framework for optimizing RIS-assisted unmanned aerial vehicle networks is proposed. In [19], an unmanned aerial vehicle-assisted RIS-IoT communication system is proposed, incorporating a nonlinear energy harvesting model with defined minimum and maximum energy bounds. The network's performance is evaluated under finite block length codes by deriving the average outage probability (OP) and the average block error rate. Chu et al. [20] analyzed practical factors, including transceiver hardware impairments and RIS phase shift errors, and their impact on the performance of an RIS-assisted IoT network. A system sum throughput maximization problem is formulated to assess overall performance in both scenarios. Wu et al. [21] introduce an optimization technique for RIS phase shifts and base station beamforming, aimed at minimizing the uplink transmit power in an RIS-assisted IoT network. The core concept of the proposed scheme, termed Riemannian conjugate gradient-based joint optimization, is to jointly optimize RIS phase shifts and BS beamforming vectors using the Riemannian conjugate gradient method. In [22], the authors presented an RIS-assisted wireless-powered heterogeneous network comprising two distinct groups of devices. Specifically, one group, known as energy-harvesting devices, is powered by external energy sources, while the other group, referred to as non-energy-harvesting devices, relies on internal energy supplies. The RIS plays a crucial role in facilitating WET in the downlink and WIT in the uplink. Zhu et al. [23] proposed deploying an RIS to enhance energy harvesting and data transmission efficiency. To improve the RIS's self-sustainability, it harvests energy from the energy source using the harvest-then-transmit protocol. In [24], an alternating optimization algorithm with semidefinite programming relaxation is proposed to optimize the RIS phase shifts, while a closed-form solution is derived using the Lagrange dual method. Chu et al. [25] explored a wireless-powered intelligent radio environment, proposing a fractional nonlinear energy harvesting scheme to facilitate an RIS-assisted wireless-powered IoT network. In [26], the use of an unmanned aerial vehicle-mounted RIS is proposed for data collection, with an analysis of coverage probability in remote areas lacking static telecommunication infrastructure. In [27], the authors designed a transmitter utilizing a single RF chain for low-complexity RIS-assisted multiuser downlink communication. Specifically, in the proposed architecture, multiuser information is precoded and modulated at the RIS by appropriately adjusting the phase shifts and amplitudes of individual RIS elements. In [28], the performance of an RIS-assisted IoT network with a nonlinear energy harvesting model is analyzed for both passive and active RIS configurations. The lower bound of the outage performance is derived using the Meijer G-function. In [30], the authors introduce a novel approach that seamlessly integrates the capabilities of indexed modulation and RIS to enhance multifunctional IoT applications. In particular, the authors have proposed an RIS-assisted indexed modulation framework, where the RIS elements themselves serve as indexed modulation entities. In [31], the performance of a simultaneous wireless information and power transfer enabled full-duplex IoT network with hardware and self-interference imperfection is explored in terms of outage probability, system throughput, and energy efficiency. In [32], the performance of RIS-aided nonorthogonal multiple access network is studied. The closed-form expression of outage probability and asymptotic outage probability are derived under Nakagami-m fading channels.

10.1.2 Motivation

An RIS-assisted IoT network enables adaptive control of the wireless environment by dynamically managing signal reflection and propagation, thereby improving coverage, reliability, and energy efficiency. Within 6G networks, this technology can support diverse use cases such as smart cities, ensuring seamless connectivity among sensors and infrastructure, industrial IoT providing low-latency and reliable links for automation, and smart healthcare facilitating stable communication for remote monitoring. Additionally, it enhances precision agriculture through better rural connectivity, enables high data rate augmented reality/virtual reality experiences, and promotes sustainable, energy-efficient wireless communication, making it a key enabler for next-generation IoT ecosystems. The integration of energy harvesting with RIS technologies in IoT systems is motivated by the growing demand for energy-efficient, sustainable, and intelligent wireless communication. As IoT networks continue to expand, reliance on conventional battery-powered devices becomes less viable due to maintenance complexities and limited energy resources [33]. Energy harvesting allows devices to capture energy from RF sources, minimizing dependence on external power and extending operational lifespan. Simultaneously, RIS enhances signal propagation, improving energy efficiency and expanding network coverage. This combination is especially vital for B5G networks, which require ultrareliable, low-latency communication and large-scale connectivity. By integrating energy harvesting with RIS, IoT systems can achieve self-sustaining functionality, lower power consumption, and promote environmentally friendly communication solutions, making them highly suitable for smart cities, industrial automation, and remote monitoring. Considering this, the main contributions of this chapter are as follows:

- The analytical expression of outage probability for the considered network is derived using the Gaussian–Chebyshev quadrature approximation method. Additionally, to provide valuable insights into the impact of system parameters, the asymptotic outage probability at high signal-to-noise ratio (SNR) is formulated. From this, the diversity order of the proposed network is determined. These expressions reflect the influence of the number of reflecting elements, hardware imperfections, severity, and various other system parameters.
- The analytical expression for the ergodic capacity of the considered network is derived using the Gaussian–Chebyshev quadrature approximation method and is expressed in terms of the Meijer-G function. Additionally, the asymptotic ergodic capacity is also formulated using the same approximation method.
- The analytical expression for the symbol error rate under various digital modulation schemes is derived for the considered network using the Gaussian–Chebyshev quadrature approximation method. The effects of constellation size, the number of reflecting elements, hardware impairments, energy conversion efficiency, and other system parameters are analyzed.
- Finally, Monte Carlo simulations are conducted to validate the accuracy of the derived analytical expressions, providing valuable insights.

10.2 System model

An RIS-aided IoT network is illustrated in Fig. 10.1 that comprises a power station, N IoT devices, labeled M_n $(n = 1, \ldots, N)$, an RIS, and an access point. The IoT devices are wirelessly powered using the WET protocol, receiving RF signals from the dedicated power station, while the M_n IoT device transmits its information signal to the access point through the RIS. The RIS is mounted on a tall building and consists of L reflecting elements, while the power station, access point, and IoT devices are each equipped with a single antenna [17,23]. The availability of a direct link between the IoT devices and access point is not present due to deep fading or shadowing effect. The IoT devices and access point experience transceiver hardware imperfections, leading to the introduction of distortion noise in the system. The channel fading coefficients of the link power station to M_n IoT device, M_n IoT device to RIS, and RIS to access point are denoted as h_{sn}, h_{nq}, and h_{qb}, respectively, where $s, q, b \in$ {power station, RIS, access point} and $q = 1, \ldots, L$. The fading coefficients follow independent and nonidentically distributed Rayleigh fading. Furthermore, the fading coefficients are assumed quasistatic, i.e., for a particular frame duration T, the channel remains the same.

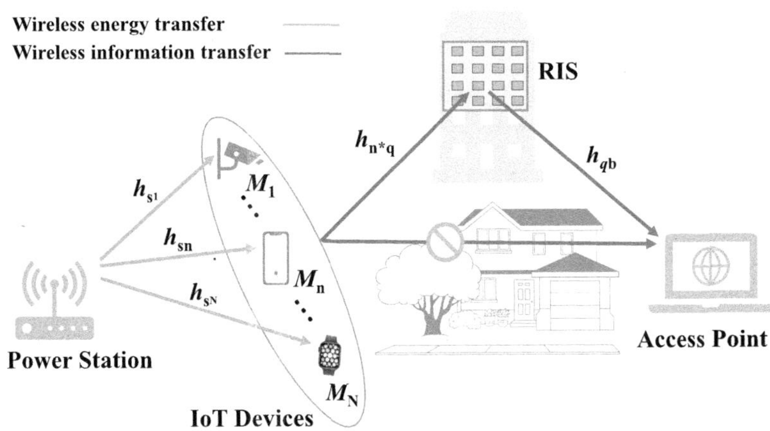

Figure 10.1:
Reconfigurable intelligent surface assisted IoT network.

A harvest-then-transmit strategy is implemented at the energy-constrained IoT nodes, where each node first harvests energy from the received RF signal transmitted by the power station using the WET protocol. However, the access point has a dedicated power source. The IoT node employs the harvested energy for wireless information transmission. The amount of harvested energy at the nth IoT node is given as

$$E_n = \eta T P_s |h_{sn}|^2, \tag{10.1}$$

where η represents energy conversion efficiency, T represents the time period, and P_s denote the transmit power at the power station, respectively. Furthermore, an IoT node selection strategy is implemented to maximize the harvested energy E_n, enhancing system performance

[34]. In particular, the best IoT device represented as $n^\star$ is selected such that

$$n^\star = \arg \max_{n \in \{1,\ldots,N\}} \{|h_{sn}|^2\}. \tag{10.2}$$

With this IoT device selection strategy, the harvested energy at $M_{n^\star}$ can be expressed as

$$E_{n^\star} = \eta T P_s |h_{sn^\star}|^2. \tag{10.3}$$

A generic harvest-then-transmit protocol is employed, and we consider that the whole time period $T = 1$ [17]. The transmit power at the $M_{n^\star}$ IoT device can be expressed as

$$P_{n^\star} = \eta P_s |h_{sn^\star}|^2. \tag{10.4}$$

The IoT device $M_{n^\star}$ transmits its information signal to the access point by using the harvested power. The signal received at the access point can be given as

$$y_b = \sqrt{P_{n^\star}} \sum_{q=1}^{L} h_{n^\star q} e^{j\theta_q} h_{qb} (x_{n^\star} + \Phi_{n^\star b}) + w_b, \tag{10.5}$$

where $\theta_q \in [0, 2\pi]$ represents the phase-shift at the qth reflecting element, $x_{n^\star}$ is the unit energy symbol transmitted by the IoT device $M_{n^\star}$, and $w_b \sim \mathcal{CN}(0, \sigma^2)$ is the additive white Gaussian noise (AWGN) at the access point. The term $\Phi_{n^\star b} \sim \mathcal{CN}(0, \phi^2)$ in (10.5) represents the distortion noise due to the hardware impairments at both the IoT and access point, such that $\phi = \sqrt{\phi_{n^\star}^2 + \phi_b^2}$ [35]. The term ϕ characterizes the aggregate level of hardware impairments, and $\phi_{n^\star}^2$ and ϕ_b^2 represent the level of hardware impairments in the transmitter and the receiver at the IoT and access point, respectively. The received instantaneous signal-to-noise-plus distortion ratio (SNDR) at the access point can be expressed as

$$\Lambda_b = \frac{\eta \zeta |h_{sn^\star}|^2 |\sum_{q=1}^{L} h_{n^\star q} e^{j\theta_q} h_{qb}|^2}{\eta \zeta |h_{sn^\star}|^2 |\sum_{q=1}^{L} h_{n^\star q} e^{j\theta_q} h_{qb}|^2 \phi^2 + 1}, \tag{10.6}$$

where $\zeta = \frac{P_s}{\sigma^2}$ denotes the transmit signal-to-noise ratio (SNR).

10.3 Performance analysis

This section analyzes the performance of the passive RIS-assisted IoT network by deriving the analytical expressions for outage probability, asymptotic outage probability, ergodic capacity, asymptotic ergodic capacity, and symbol error rate under various digital modulation schemes.

10.3.1 Outage probability

In a delay-limited system, communication must be completed within a strict time constraint, making reliability a key concern [36]. The outage probability in such a system quantifies the likelihood that the instantaneous SNR or SINR falls below a required threshold, resulting in transmission failure within the specified time frame. This is particularly critical for applications such as real-time IoT communications, where retransmissions due to outages are not an option. Reducing outage probability in delay-limited systems is essential for ensuring low-latency and high-reliability communication, making it a vital performance metric for next-generation wireless networks, including B5G. For the considered network, the outage event is declared when the received instantaneous SNDR at the access point is lower than a predefined threshold Λ_{th} which is given as

$$\mathbb{P}_{\text{out}}\left(\Lambda_{\text{th}}\right) = \Pr\left[\Lambda_b < \Lambda_{\text{th}}\right], \tag{10.7}$$

where $\Lambda_{\text{th}} = 2^{r_{\text{th}}} - 1$ and r_{th} denotes the threshold data-rate. On invoking (10.6) in (10.7), outage probability can be given as

$$\mathbb{P}_{\text{out}}\left(\Lambda_{\text{th}}\right) = \Pr\left[\frac{\eta\zeta\left|h_{sn^\star}\right|^2\left|\sum_{q=1}^{L} h_{n^\star q} e^{j\theta_q} h_{qb}\right|^2}{\eta\zeta\left|h_{sn^\star}\right|^2\left|\sum_{q=1}^{L} h_{n^\star q} e^{j\theta_q} h_{qb}\right|^2\phi^2 + 1} < \Lambda_{\text{th}}\right]. \tag{10.8}$$

Letting $X \triangleq \left|h_{sn^\star}\right|^2$ and $Y \triangleq \sum_{q=1}^{L}\left|h_{n^\star q} h_{qb}\right|^2$ for an coherent phase shift at the RIS, (10.8) can be expressed as

$$\mathbb{P}_{\text{out}}\left(\Lambda_{\text{th}}\right) = \Pr\left[X < \frac{\Lambda_{\text{th}}}{\eta\zeta Y\left(1 - \phi^2\Lambda_{\text{th}}\right)}\right]. \tag{10.9}$$

Further, the outage probability $\mathbb{P}_{\text{out}}\left(\Lambda_{\text{th}}\right) \triangleq F_{\Lambda_b}\left(\Lambda_{\text{th}}\right)$ can be determined as

$$F_{\Lambda_b}\left(\Lambda_{\text{th}}\right) = \begin{cases} \psi_{\text{out}}\left(\Lambda_{\text{th}}\right), & \text{for } \Lambda_{\text{th}} < \frac{1}{\phi^2}, \\ 1, & \text{for } \Lambda_{\text{th}} \geq \frac{1}{\phi^2}. \end{cases} \tag{10.10}$$

The term $\psi_{\text{out}}\left(\Lambda_{\text{th}}\right)$ can be evaluated as

$$\psi_{\text{out}}\left(\Lambda_{\text{th}}\right) = \int_0^\infty F_X\left(\frac{\Lambda_{\text{th}}}{\eta\zeta\left(1 - \phi^2\Lambda_{\text{th}}\right) y}\right) f_Y(y)\text{d}y. \tag{10.11}$$

The probability density function of the random variable Y and the cumulative distribution function of X are expressed as

$$f_Y\left(y\right) = \frac{1}{2\Gamma\left(\alpha\right)\lambda^\alpha}\left(\sqrt{y}\right)^{\alpha-2} e^{-\sqrt{y}/\alpha} \tag{10.12}$$

and

$$F_X(x) = \left[1 - e^{-\frac{x}{\Omega_{sn^\star}}}\right]^N, \tag{10.13}$$

respectively, where $\alpha = L\pi^2/\left(16 - \pi^2\right)$ and $\lambda = (4/\pi - \pi/4)\sqrt{\Omega_{n^\star q}\Omega_{qb}}$ [37]. Invoking the corresponding cumulative distribution function $F_X(\cdot)$ and the probability density function $f_Y(\cdot)$ in (10.11), $\psi_{\text{out}}(\Lambda_{\text{th}})$ is express as

$$\psi_{\text{out}}(\Lambda_{\text{th}}) = \sum_{p=0}^{N}\binom{N}{p}\frac{(-1)^p}{2\lambda^\alpha\Gamma(\alpha)}\int_0^{\frac{1}{\phi^2}}\left(\sqrt{y}\right)^{\alpha-2}e^{-\frac{\sqrt{y}}{\lambda}-\frac{p\Lambda_{\text{th}}}{\eta\zeta\Omega_{sn^\star}\left(1-\phi^2\Lambda_{\text{th}}\right)y}}\text{d}y. \tag{10.14}$$

The assessment of (10.14) is mathematically intractable. Therefore we resort to the Gaussian–Chebyshev quadrature method which is given as [38]

$$\int_a^b f(y)\text{d}y \approx \frac{(b-a)\pi}{2V}\sum_{m=1}^{V}\sqrt{1-x_m^2}\,f(y_m), \tag{10.15}$$

where $y_m = \frac{(b-a)x_m}{2} + \frac{(b+a)}{2}$, $x_m = \cos\left(\frac{(2m-1)\pi}{2V}\right)$, and V is a complexity accuracy trade-off parameter. Using the Gaussian–Chebyshev quadrature method in (10.14), we obtain the analytical expression of outage probability as

$$\Psi_{\text{out}}(\Lambda_{\text{th}}) \approx \sum_{m=1}^{V}\sum_{p=0}^{N}\binom{N}{p}\frac{(-1)^p\pi\sqrt{1-x_m^2}}{4\lambda^\alpha\phi^2 V\Gamma(\alpha)}\left(\sqrt{y_m}\right)^{\alpha-2}e^{-\frac{\sqrt{y_m}}{\lambda}-\frac{p\Lambda_{\text{th}}}{\eta\zeta\Omega_{sn^\star}\left(1-\phi^2\Lambda_{\text{th}}\right)y_m}}, \tag{10.16}$$

where $y_m = \frac{(1+x_m)}{2\phi^2}$ and $x_m = \cos\left(\frac{(2m-1)\pi}{2V}\right)$. From (10.16), it can be clearly observed that the outage probability of the considered network depends on the system parameters, namely the number of reflecting elements (L), hardware imperfections (ϕ), predefined threshold (Λ_{th}), and energy conversion efficiency (η).

10.3.2 Asymptotic outage probability

The asymptotic outage probability is analyzed to determine the diversity order of the considered network, which is evaluated by approximating the outage probability at high SNR. By employing the high SNR approximation of the lower incomplete Gamma function $\Upsilon(c, x) \underset{x\to 0}{\approx} (\frac{x^c}{c})$ [39], the asymptotic outage probability is obtained as

$$\mathbb{P}_{\text{out}}^{\infty}(\Lambda_{\text{th}}) \approx \sum_{q=0}^{N} \binom{N}{q} \frac{(-1)^{q+1}}{\alpha\left(\frac{4}{\pi}-\frac{\pi}{4}\right)^{\alpha}\Gamma(\alpha)} \left(\frac{1}{\Omega_{n^\star q}\Omega_{qb}}\right)^{\frac{\alpha}{2}} \left(\frac{q}{\Omega_{sn^\star}}\right)^{\frac{\alpha}{2}} \Gamma\left(-\frac{\alpha}{2}+1\right) \times \left(\frac{\Lambda_{\text{th}}}{\eta\zeta\left(1-\phi^2\Lambda_{\text{th}}\right)}\right)^{\frac{\alpha}{2}}. \tag{10.17}$$

From (10.17), it can be clearly observed that the diversity order of the considered network is 1.5α, i.e., $1.5\left(\frac{L\pi^2}{16-\pi^2}\right)$, which depends on the number of reflecting elements.

10.3.3 Ergodic capacity

Ergodic capacity is a key performance metric in wireless communication that quantifies the average achievable data rate of a channel over time, accounting for random variations caused by fading. It serves as an indicator of a system's long-term spectral efficiency by averaging the instantaneous channel capacity over the statistical distribution of the channel fading. Mathematically, the ergodic capacity is defined as [39]

$$\begin{aligned} C_{\text{e}} &= \mathbb{E}\{\log_2(1+\Lambda_b)\} \\ &= \int_0^{\infty} \log_2(1+z)\, f_{\Lambda_b}(z)\, \mathrm{d}z. \end{aligned} \tag{10.18}$$

The term $\log_2(1+z)$ can be expressed in terms of Meijer-G function as $\log_2(1+z) = \mathrm{G}_{2,2}^{1,2}\left[z\middle|_{1,0}^{1,1}\right]$. Differentiating $F_{\Lambda_b}(z)$, the probability density function of Λ_b is expressed as

$$f_{\Lambda_b}(z) = \sum_{m=1}^{V}\sum_{p=0}^{N}\binom{N}{p}\frac{(-1)^{p+1}\pi\sqrt{1-x_m^2}\left(\sqrt{y_m}\right)^{\alpha-4} p}{4\lambda^{\alpha}\phi^2 V\Gamma(\alpha)\eta\zeta\Omega_{sn^\star}} e^{-\frac{\sqrt{y_m}}{\lambda}} \frac{1}{\left(1-\phi^2 z\right)^2} \times e^{-\frac{pz}{\eta\zeta\Omega_{sn^\star} y_m\left(1-\phi^2 z\right)}}. \tag{10.19}$$

Inserting (10.19) into (10.18), the ergodic capacity can be expressed as

$$C_{\text{e}} = \sum_{m=1}^{V}\sum_{p=0}^{N}\binom{N}{p}\frac{(-1)^{p+1}\pi\sqrt{1-x_m^2}\left(\sqrt{y_m}\right)^{\alpha-4} p}{4\lambda^{\alpha}\phi^2 V\Gamma(\alpha)\eta\zeta\Omega_{sn^\star}} e^{-\frac{\sqrt{y_m}}{\lambda}} \times \int_0^{\frac{1}{\phi^2}} \frac{1}{\left(1-\phi^2 z\right)^2} e^{-\frac{pz}{\eta\zeta\Omega_{sn^\star} y_m\left(1-\phi^2 z\right)}} \mathrm{G}_{2,2}^{1,2}\left[z\middle|_{1,0}^{1,1}\right] \mathrm{d}z. \tag{10.20}$$

The evaluation of (10.20) is mathematically complex. Therefore, we employ the Gaussian–Chebyshev quadrature approximation method to derive a closed-form expression for the ergodic capacity, which is given as

$$C_{\mathrm{e}} \approx \sum_{u=1}^{U} \sum_{m=1}^{V} \sum_{p=0}^{N} \binom{N}{p} \frac{(-1)^{p+1} \pi p \sqrt{1-x_m^2} \sqrt{1-w_m^2}}{8\lambda^{\alpha} \phi^4 U V \Gamma(\alpha) \eta \zeta \Omega_{sn^\star}}$$
$$\times e^{-\frac{\sqrt{y_m}}{\lambda}} \frac{\left(\sqrt{y_m}\right)^{\alpha-4}}{\left(1-\phi^2 z_u\right)^2} e^{-\frac{p z_u}{\eta \zeta \Omega_{sn^\star} y_m \left(1-\phi^2 z_u\right)}} \mathrm{G}_{2,2}^{1,2}\left[z_u \middle|_{1,0}^{1,1}\right], \tag{10.21}$$

where $w_u = \cos\left(\frac{(2u-1)\pi}{2U}\right)$ and $z_u = \frac{1+w_u}{2\phi^2}$. From (10.21), it can be clearly observed that the ergodic capacity of the considered network depends on the system parameters, namely the number of reflecting elements (L), hardware imperfections (ϕ), and energy conversion efficiency (η).

10.3.4 Asymptotic ergodic capacity

At high SNR levels, the ergodic capacity can be approximated as [40]

$$C_{\mathrm{e}}^{\infty} \approx \mathbb{E}\{\log_2(\Lambda_b)\}$$
$$= \log_2\left(\int_0^{\infty} z f_{\Lambda_b}(z)\, \mathrm{d}z\right). \tag{10.22}$$

Inserting (10.19) into (10.22), the asymptotic ergodic capacity can be expressed as

$$C_{\mathrm{e}}^{\infty} \approx \log_2\left(\sum_{m=1}^{V} \sum_{p=0}^{N} \binom{N}{p} \frac{(-1)^{p+1} \pi \sqrt{1-x_m^2} \left(\sqrt{y_m}\right)^{\alpha-4}}{4\lambda^{\alpha} \phi^2 V \Gamma(\alpha) \eta \zeta \Omega_{sn^\star}}\right.$$
$$\left. \times\, p e^{-\frac{\sqrt{y_m}}{\lambda}} \int_0^{\frac{1}{\phi^2}} \frac{z}{\left(1-\phi^2 z\right)^2} e^{-\frac{p z}{\eta \zeta \Omega_{sn^\star} y_m \left(1-\phi^2 z\right)}} \mathrm{d}z\right). \tag{10.23}$$

The evaluation of (10.23) is mathematically challenging. Therefore, we employ the Gaussian–Chebyshev quadrature approximation method to derive the analytical expression for the asymptotic ergodic capacity, which is given as

$$C_{\mathrm{e}}^{\infty} \approx \log_2\left(\sum_{u=1}^{U} \sum_{m=1}^{V} \sum_{p=0}^{N} \binom{N}{p} \frac{(-1)^{p+1} \pi^2 p \sqrt{1-x_m^2} e^{-\frac{\sqrt{y_m}}{\lambda}}}{8\lambda^{\alpha} \phi^4 U V \Gamma(\alpha) \eta \zeta \Omega_{sn^\star}}\right.$$
$$\left. \times \left(\sqrt{y_m}\right)^{\alpha-4} \sqrt{1-w_m^2} \frac{z_u}{\left(1-\phi^2 z_u\right)^2} e^{-\frac{p z_u}{\eta \zeta \Omega_{sn^\star} y_m \left(1-\phi^2 z_u\right)}}\right). \tag{10.24}$$

From (10.24), it can be observed that the diversity order of the considered network is 1.5α, i.e., $1.5\left(\frac{L\pi^2}{16-\pi^2}\right)$, which depends on the number of reflecting elements.

10.3.5 Symbol error rate

Symbol error rate is an essential performance metric in wireless communication that measures the likelihood of a transmitted symbol being incorrectly received due to channel distortions, noise, or interference. It is crucial for assessing the reliability of modulation schemes and the overall efficiency of the system. The symbol error rate for various modulation schemes is given as [40]

$$\mathbb{P}_{\mathrm{e}} = \frac{a\sqrt{b}}{\sqrt{\pi}} \int_0^{\infty} \frac{e^{-bz}}{\sqrt{z}} \mathbb{P}_{\mathrm{out}}(z)\, \mathrm{d}z, \tag{10.25}$$

where $a = 1$ and $b = \sin^2(\pi/M)$ for an M-ary phase shift keying (PSK), $a = 2(1 - (1/\sqrt{M}))$ and $b = 3/(2(M-1))$ for a rectangular M-ary quadrature amplitude modulation (QAM), and M denotes the constellation size. Inserting (10.16) into (10.25), one can achieve

$$\mathbb{P}_{\mathrm{e}} = \frac{a\sqrt{b}}{\sqrt{\pi}} \sum_{m=1}^{V} \sum_{p=0}^{N} \binom{N}{p} \frac{(-1)^p \pi \sqrt{1-x_m^2}}{4\lambda^{\alpha}\phi^2 V \Gamma(\alpha)} \left(\sqrt{y_m}\right)^{\alpha-2} e^{-\frac{\sqrt{y_m}}{\lambda}} \times \int_0^{\frac{1}{\phi^2}} z^{-\frac{1}{2}} e^{-bz - \frac{pz}{\eta\zeta\Omega_{sn^\star}(1-\phi^2 z)y_m}} \mathrm{d}z. \tag{10.26}$$

The evaluation of (10.26) is mathematically complex. Therefore, we employ the Gaussian–Chebyshev quadrature approximation method to derive a closed-form expression for the SER, which is given as

$$\mathbb{P}_{\mathrm{e}} \approx \frac{a\sqrt{b}}{\sqrt{\pi}} \sum_{u=1}^{U} \sum_{m=1}^{V} \sum_{p=0}^{N} \binom{N}{p} \frac{(-1)^p \pi^2 \sqrt{1-x_m^2}}{8\lambda^{\alpha}\phi^4 V U \Gamma(\alpha)} \left(\sqrt{y_m}\right)^{\alpha-2} \times e^{-\frac{\sqrt{y_m}}{\lambda}} \sqrt{1-w_u^2} z_u^{-\frac{1}{2}} e^{-bz_u - \frac{pz_u}{\eta\zeta\Omega_{sn^\star}(1-\phi^2 z_u)y_m}}. \tag{10.27}$$

From (10.27), it can be clearly observed that the symbol error rate of the considered network depends on the system parameters, namely the number of reflecting elements (L), hardware imperfections (ϕ), energy conversion efficiency (η), and constellation size (M).

10.4 Results and discussion

This section presents the numerical and simulation results of the outage probability, asymptotic outage probability, ergodic capacity, asymptotic ergodic capacity, and symbol error rate for the considered network. A 3D coordinate system is chosen to illustrate the system model, where the power station, IoT device $M_{n^\star}$, the RIS, and the

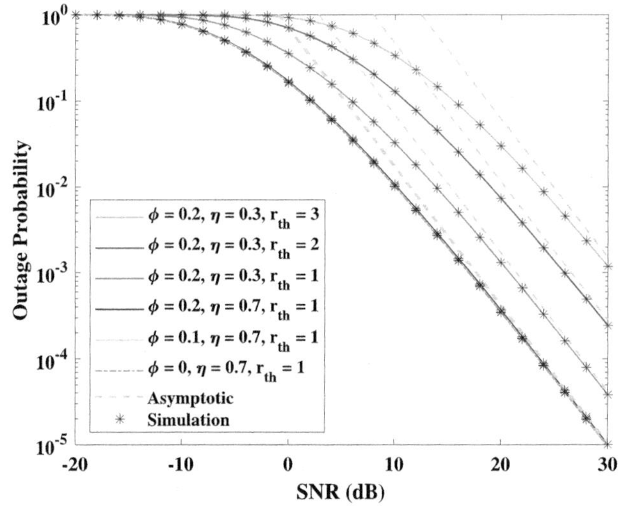

Figure 10.2:
Outage probability versus SNR curves with number of IoT devices $N = 4$ and the number of reflecting elements $L = 2$.

access point are situated at location $(-10, 0, 0)$, $(0, 0, 0)$, $(-2, 6, 0)$, and $(10, 0, 0)$, respectively [17]. A path-loss model is adopted, where $\Omega_{ij} = x_{ij}^{-v}$, with $v = 3$ as path-loss exponent. Here, x_{ij} represents the distance between the i and j nodes, where $i, j \in$ {power station, access point, RIS, IoT device}. Further, we set $U = V = 100$ to obtain the numerical results of outage probability, ergodic capacity, asymptotic ergodic capacity, and symbol error rate with acceptable accuracy and computational complexity.

Fig. 10.2 presents the outage probability versus SNR curves for the given parameters, with the number of IoT devices set to $N = 4$ and the number of reflecting elements set to $L = 2$. It is observed that the numerical and simulation results exhibit a perfect match, demonstrating the accuracy of the Gaussian–Chebyshev quadrature approximation method. Additionally, the asymptotic outage probability curves converge to the outage probability curves at high SNR levels. The outage performance improves as the threshold data rate r_{th} decreases. This is because a lower threshold data rate corresponds to a lower required SNR, reducing the likelihood of the SNR falling below the threshold. Consequently, the probability of outage occurrence decreases with lower threshold SNRs. To achieve an outage probability of 10^{-2}, one can obtain approximately 4.63 and 9.74 dB gains when r_{th} decreases from 3 to 2 and from 3 to 1 bps/Hz, respectively. Furthermore, it can be observed that the outage performance enhances with the increase in the energy conversion efficiency η due to the fact that the amount of harvested energy increases with η. To attain an outage probability of 10^{-2}, an approximate 3.82 dB gain is noted when η increases from 0.3 to 0.7. Furthermore, it is

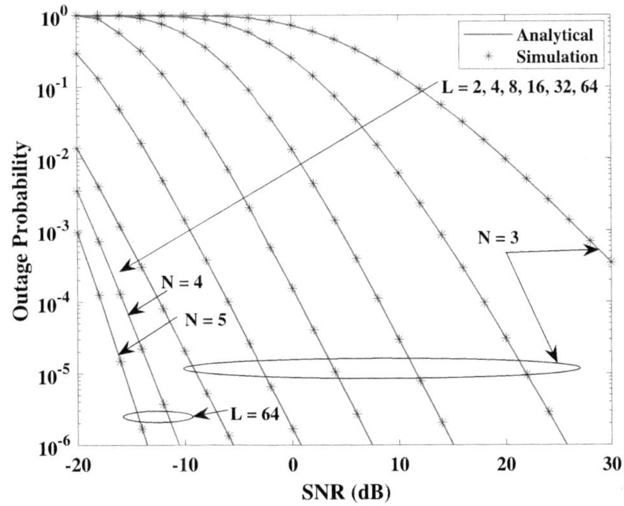

Figure 10.3:
Impact of the number of reflecting elements and the number of IoT devices on the outage performance.

observed that the outage performance deteriorates as the level of hardware impairments ϕ increases, due to the corresponding rise in distortion noise.

Fig. 10.3 illustrates the impact of reflecting elements (REs) and IoT devices on the outage performance of the considered network. It is observed that increasing the number of reflecting elements enhances outage performance, as the system effectively achieves fine-grained reflective beamforming [1]. To achieve an outage probability of 10^{-2}, approximate 10.87, 19.33, 26.55, 33.12, and 39.36 dB gains are noted when L increases from 2 to 4, 8, 16, 32, and 64, respectively. Furthermore, it is observed that the outage performance improves with an increase in number of IoT devices N, as the amount of harvested energy rises with N. It can be noted that the considered network achieves approximate 2.6 and 5.4 dB gains when N increases from 3 to 4 and 5, respectively, for the number of reflecting elements $L = 64$.

Fig. 10.4 illustrates the impact of the overall system ceiling, caused by hardware imperfections ϕ, on the outage probability of the considered network. It provides insight into the maximum achievable data rate before the system experiences an outage. It can be noted that the for a fixed ϕ, the overall system ceiling takes place at a specific threshold. This specifies the level of hardware impairment that the system can handle and the maximum data rate the system can achieve. Further, it is observed that the overall system ceiling takes place at approximately 3.85, 4.84, and 6.55 bps/Hz threshold data rates for $\phi = 0.1$, $\phi = 0.2$, and $\phi = 0.3$, respectively.

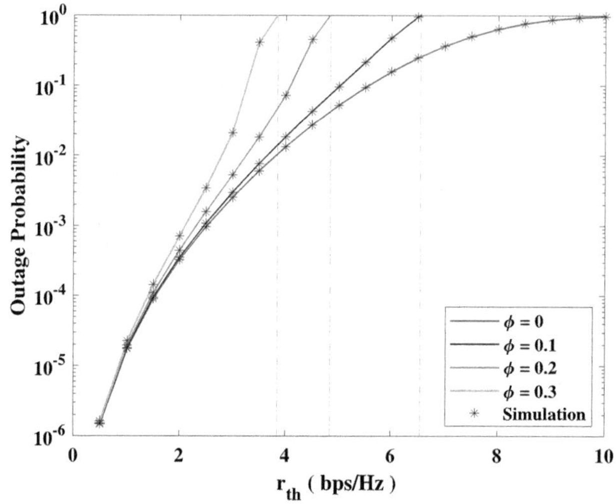

Figure 10.4:
Effect of overall system ceiling on the outage performance.

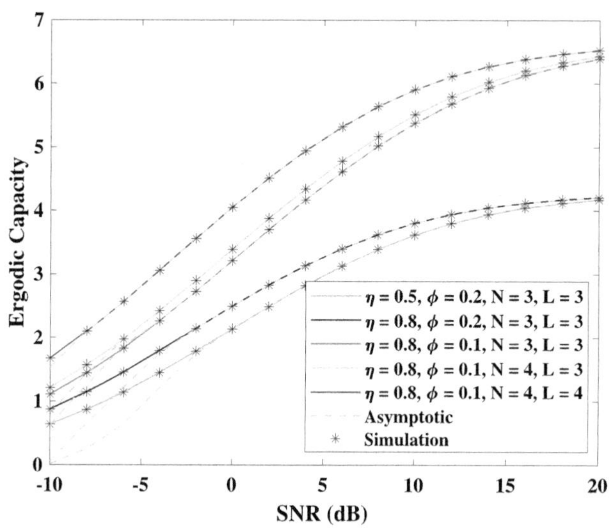

Figure 10.5:
Ergodic capacity versus SNR curves illustrating the impact of hardware impairments (ϕ), the number of reflecting elements (L), the number of IoT devices (N), and energy conversion efficiency (η).

Fig. 10.5 presents the ergodic capacity versus SNR curves for the considered network. We note that the derived analytical expression for ergodic capacity aligns wells the simulation results. In addition, the asymptotic ergodic capacity curves align with the ergodic capacity

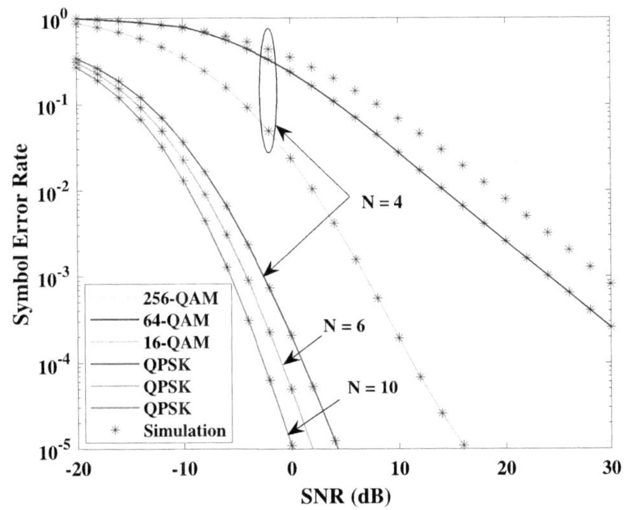

Figure 10.6:
Symbol error rate versus SNR curves with the number of reflecting element $L = 8$, energy conversion efficiency $\eta = 0.7$, and the level of hardware impairments $\phi = 0.1$.

curves at high SNR regime. It is observed that the ergodic capacity improves with η, as higher energy harvesting efficiency leads to an increase in harvested energy. Furthermore, the ergodic capacity increases with a decrease in the level of hardware impairments ϕ. To achieve an ergodic capacity of 4 bps/Hz, an approximate 6.26 dB gain is noted when ϕ decreases from 0.2 to 0.1. Further, it can be observed that the ergodic capacity increases with an increase in L and N.

Fig. 10.6 illustrates the symbol error rate versus SNR curves for the considered network. The numerical and simulation results align perfectly, illustrating the accuracy of the derived expression. It is observed that the symbol error rate improves as the constellation size M increases. It is noted that, to attain an symbol error rate of 10^{-2}, approximate 4.92 and 16.96 dB gains are attained when M decreases from 256 to 64 and 16 quadrature amplitude modulation (QAM) schemes, respectively. Furthermore, it can be observed that the symbol error rate improves with an increase in number of IoT devices N. To achieve an symbol error rate of 10^{-2} for quadrature phase shift keying (QPSK) modulation scheme, approximate 1.37 and 2.69 dB gains are achieved when N increases from 4 to 6 and 10, respectively.

Fig. 10.7 illustrates the impact of reflecting elements and energy conversion efficiency η on the symbol error rate performance for the considered network. It is observed that the symbol error rate improves as L increases, as the system effectively achieves fine-grained reflective beamforming at the destination node. It is noted that, to attain an symbol error rate of 10^{-2},

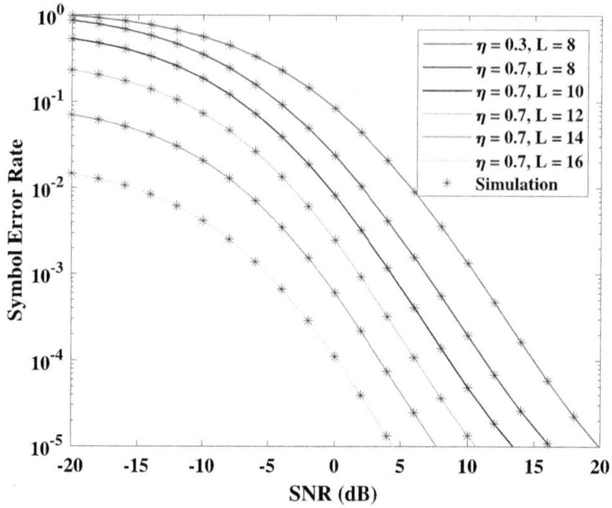

Figure 10.7:
Symbol error rate versus SNR curves for the 16 QAM scheme with the number of IoT devices $N = 4$ and the level of hardware impairments $\phi = 0.1$.

approximate 2.43, 5.43, 9.37, and 17.52 dB gains are noticed when L increases from 8 to 10, 12, 14, and 16, respectively. Further, the symbol error rate performance improvement is observed with an increase in energy conversion efficiency η.

10.5 Conclusion

This chapter provides a detailed performance analysis of an RIS-assisted wireless-powered IoT network with hardware impairments. The analytical expressions for outage probability, ergodic capacity, and symbol error rate are derived using the Gaussian–Chebyshev quadrature approximation method. Additionally, the asymptotic outage probability and asymptotic ergodic capacity expressions are obtained, along with the diversity order of the considered network. The results demonstrate that increasing the number of reflecting elements and IoT devices significantly enhances network performance. However, transceiver hardware impairments negatively impact system performance. Moreover, it has been observed that the overall system ceiling effect can impose limitations on the network's performance.

Future research on RIS-assisted IoT networks can aim to enhance efficiency, scalability, and resilience to meet the requirements of next-generation wireless systems. Advanced signal processing methods and artificial intelligence driven optimization techniques can be explored for dynamic RIS configuration, improving adaptability. The investigation of hybrid RIS (i.e., combines both the passive and active RIS) architectures can further boost signal amplification and network coverage. Additionally, addressing hardware impairments and developing effective compensation strategies will enhance system reliability. Moreover, integrating RIS with emerging technologies like nonorthogonal multiple access and rate splitting multiple access can unlock new possibilities for IoT applications. Lastly, experimental validation and large-scale real-world deployments will be crucial for assessing practical feasibility and optimizing RIS-assisted IoT networks for future B5G and 6G systems.

References

[1] Q. Wu, R. Zhang, Towards smart and reconfigurable environment: intelligent reflecting surface aided wireless network, IEEE Communications Magazine 58 (1) (Jan. 2020) 106–112.
[2] D. Kumar, C.K. Singh, O.L.A. López, V. Bhatia, M. Latva-aho, Performance analysis of active RIS-assisted downlink NOMA with transmit antenna selection, IEEE Transactions on Vehicular Technology 74 (5) (May 2025) 7774–7791.
[3] M.V. Katwe, A. Kaushik, L. Mohjazi, M. Abualhayja'a, D. Dardari, K. Singh, M.A. Imran, M.M. Butt, O.A. Dobre, An overview of intelligent meta-surfaces for 6G and beyond: opportunities, trends, and challenges, IEEE Communications Standards Magazine 8 (4) (Dec. 2024) 62–69.
[4] S.A. Ullah, N. Mazhar, A. Kaushik, A. Mahmood, M. Gidlund, S.A. Hassan, DRL-enhanced QoS-aware NOMA for ambient IoT: resource allocation optimization with RIS and RF energy harvesting diversity, IEEE Communications Standards Magazine 9 (3) (Sep. 2025) 49–56.
[5] H. Yang, F. Yang, X. Cao, S. Xu, J. Gao, X. Chen, M. Li, T. Li, A 1600-element dual-frequency electronically reconfigurable reflectarray at X/Ku-band, IEEE Transactions on Antennas and Propagation 65 (6) (Jun. 2017) 3024–3032.
[6] M. Najafi, V. Jamali, R. Schober, H.V. Poor, Physics-based modeling and scalable optimization of large intelligent reflecting surfaces, IEEE Transactions on Communications 69 (4) (Apr. 2020) 2673–2691.
[7] K. Zhi, C. Pan, H. Ren, K.K. Chai, M. Elkashlan, Active RIS versus passive RIS: which is superior with the same power budget?, IEEE Communications Letters 26 (5) (May 2022) 1150–1154.
[8] D. Kumar, P.K. Singya, O. Krejcar, K. Choi, V. Bhatia, Performance of IRS-aided FD two-way communication network with imperfect SIC, IEEE Transactions on Vehicular Technology 72 (4) (Apr. 2023) 5491–5496.
[9] Q. Wu, R. Zhang, Intelligent reflecting surface enhanced wireless network via joint active and passive beamforming, IEEE Transactions on Wireless Communications 18 (11) (Nov. 2019) 5394–5409.
[10] M. Zhong, Y. Yang, H. Yao, X. Fu, O.A. Dobre, O. Postolache, 5G and IoT: towards a new era of communications and measurements, IEEE Instrumentation & Measurement Magazine 22 (6) (Dec. 2019) 18–26.
[11] D. Kumar, P.K. Singya, V. Bhatia, Performance analysis of SWIPT enabled decode-and-forward based cooperative network, in: 2022 IEEE 11th Inter. Confer. Commun. Sys. Networ. Technol. (CSNT), Apr. 2022, pp. 476–481.
[12] N. Javaid, A. Sher, H. Nasir, N. Guizani, Intelligence in [iot-based 5g] networks: opportunities and challenges, IEEE Communications Magazine 56 (10) (Oct. 2018) 94–100.
[13] D. Kumar, P.K. Singya, V. Bhatia, Performance of SWIPT enabled full duplex IoT network with hardware impairments and imperfect SIC, in: IEEE Wireless Commun. Network. Confer. (WCNC), IEEE, May 2023, pp. 1–6.
[14] C.-X. Wang, X. You, X. Gao, X. Zhu, Z. Li, C. Zhang, H. Wang, Y. Huang, Y. Chen, H. Haas, J.S. Thompson, E.G. Larsson, M.D. Renzo, W. Tong, P. Zhu, X. Shen, H.V. Poor, L. Hanzo, On the road to 6G: visions, requirements, key technologies, and testbeds, IEEE Communications Surveys and Tutorials 25 (2) (2023) 905–974, 2nd Quart.

[15] M.A. Jamshed, A. Kaushik, M. Toka, W. Shin, M.Z. Shakir, S.P. Dash, D. Dardari, Synergizing airborne non-terrestrial networks and reconfigurable intelligent surfaces-aided 6G IoT, IEEE Internet of Things Magazine 7 (2) (Mar. 2024) 46–52.
[16] O.L. López, H. Alves, R.D. Souza, S. Montejo-Sánchez, E.M.G. Fernández, M. Latva-Aho, Massive wireless energy transfer: enabling sustainable IoT toward 6G era, IEEE Internet of Things Journal 8 (11) (Jun. 2021) 8816–8835.
[17] Z. Chu, Z. Zhu, F. Zhou, M. Zhang, N. Al-Dhahir, Intelligent reflecting surface assisted wireless powered sensor networks for Internet of things, IEEE Transactions on Communications 69 (7) (Jul. 2021) 4877–4889.
[18] Z. Chu, P. Xiao, D. Mi, W. Hao, M. Khalily, L.-L. Yang, A novel transmission policy for intelligent reflecting surface assisted wireless powered sensor networks, IEEE Journal of Selected Topics in Signal Processing 15 (5) (Aug. 2021) 1143–1158.
[19] N. Agrawal, A. Bansal, K. Singh, C.-P. Li, S. Mumtaz, Finite block length analysis of RIS-assisted UAV-based multiuser IoT communication system with non-linear EH, IEEE Transactions on Communications 70 (5) (May 2022) 3542–3557.
[20] Z. Chu, J. Zhong, P. Xiao, D. Mi, W. Hao, R. Tafazolli, A.P. Feresidis, RIS assisted wireless powered IoT networks with phase shift error and transceiver hardware impairment, IEEE Transactions on Communications 70 (7) (Jul. 2022) 4910–4924.
[21] J. Wu, S. Kim, B. Shim, Energy-efficient power control and beamforming for reconfigurable intelligent surface-aided uplink IoT networks, IEEE Transactions on Wireless Communications 21 (12) (Dec. 2022) 10162–10176.
[22] Z. Zhu, Z. Li, Z. Chu, Q. Wu, J. Liang, Y. Xiao, P. Liu, I. Lee, Intelligent reflecting surface-assisted wireless powered heterogeneous networks, IEEE Transactions on Wireless Communications 22 (12) (Dec. 2023) 9881–9892.
[23] Z. Zhu, Z. Li, Z. Chu, G. Sun, W. Hao, P. Liu, I. Lee, Resource allocation for intelligent reflecting surface assisted wireless powered IoT systems with power splitting, IEEE Transactions on Wireless Communications 21 (5) (May 2022) 2987–2998.
[24] Z. Shi, H. Wang, Y. Fu, X. Ye, G. Yang, S. Ma, Outage performance and AOI minimization of HARQ-IR-RIS aided IoT networks, IEEE Transactions on Communications 71 (3) (Mar. 2023) 1740–1754.
[25] Z. Chu, P. Xiao, D. Mi, W. Hao, Z. Lin, Q. Chen, R. Tafazolli, Wireless-powered intelligent radio environment with nonlinear energy harvesting, IEEE Internet of Things Journal 9 (18) (Sep. 2022) 18130–18141.
[26] D. Tyrovolas, P.-V. Mekikis, S.A. Tegos, P.D. Diamantoulakis, C.K. Liaskos, G.K. Karagiannidis, Energy-aware design of UAV-mounted RIS networks for IoT data collection, IEEE Transactions on Communications 71 (2) (Feb. 2023) 1168–1178.
[27] Q. Li, M. El-Hajjar, I. Hemadeh, D. Jagyasi, A. Shojaeifard, E. Basar, L. Hanzo, The reconfigurable intelligent surface-aided multi-node IoT downlink: beamforming design and performance analysis, IEEE Internet of Things Journal 10 (7) (Apr. 2023) 6400–6414.
[28] D. Kumar, C.K. Singh, O.L.A. López, V. Bhatia, M. Latva-Aho, Performance analysis of passive/active RIS aided wireless-powered IoT network with nonlinear energy harvesting, IEEE Transactions on Wireless Communications 24 (2) (Feb. 2025) 1132–1145.
[29] M. Abualhayja'a, M. Alquraan, K.A. Alblaihed, A. Kaushik, D. Nguyen, L. Mohjazi, Federated learning-empowered RIS-assisted UAV networks for IoT data collection and optimisation, Authorea Preprints (2025).
[30] R. Singh, A. Kaushik, A. Dolas, R.M. Bodile, W. Shin, Exploring reflecting phases in RIS-assisted indexed multiplexing for 6G IoT applications, in: IEEE Globecom Workshops (GC Wkshps), IEEE, Dec. 2024, pp. 1–6.
[31] D. Kumar, P.K. Singya, V. Bhatia, Performance of SWIPT enabled full duplex IoT network with hardware impairments and imperfect SIC, in: IEEE Wireless Commun. Network. Confer. (WCNC), IEEE, Mar. 2023, pp. 1–6.
[32] C. Kumaradasa, D. Kumar, N. Rajatheva, V. Bhatia, On performance of hybrid RIS-aided NOMA network, in: 2025 IEEE Wireless Commun. Networ. Confer. (WCNC), IEEE, Mar. 2025, pp. 1–6.

[33] E. Ahmed, I. Yaqoob, A. Gani, M. Imran, M. Guizani, Internet-of-things-based smart environments: state of the art, taxonomy, and open research challenges, IEEE Wireless Communications 23 (5) (Oct. 2016) 10–16.
[34] M.K. Shukla, H.H. Nguyen, O.J. Pandey, Multiuser full-duplex IoT networks with wireless-powered relaying: performance analysis and energy efficiency optimization, IEEE Transactions on Green Communications and Networking 4 (4) (Dec. 2020) 982–997.
[35] E. Bjornson, M. Matthaiou, M. Debbah, A new look at dual-hop relaying: performance limits with hardware impairments, IEEE Transactions on Communications 61 (11) (Nov. 2013) 4512–4525.
[36] D. Kumar, P.K. Singya, V. Bhatia, Impact of NLPA on SWIPT Enabled Two-Way AF Cooperative Network, IEEE Veh. Technol. Confer, vol. VTC2021-Spring, IEEE, Apr. 2021, pp. 1–5.
[37] W. Zhao, G. Wang, S. Atapattu, T.A. Tsiftsis, C. Tellambura, Is backscatter link stronger than direct link in reconfigurable intelligent surface-assisted system?, IEEE Communications Letters 24 (6) (Jun. 2020) 1342–1346.
[38] E. Süli, D.F. Mayers, An Introduction to Numerical Analysis, Cambridge Univ. Press, 2003.
[39] D. Kumar, P.K. Singya, V. Bhatia, ASER analysis of hybrid receiver based SWIPT two-way relay network, IEEE Transactions on Vehicular Technology 70 (10) (Oct. 2021) 10018–10030.
[40] Y. Deng, L. Wang, M. Elkashlan, K.J. Kim, T.Q. Duong, Generalized selection combining for cognitive relay networks over Nakagami-*m* fading, IEEE Transactions on Signal Processing 63 (8) (Apr. 2015) 1993–2006.

CHAPTER 11

Electromagnetic signal and information theory perspective on holographic integrated sensing and communications for 6G

Özlem Tuğfe Demir[a], Zehra Yigit[b], Fatih Cögen[b], Ahmet Faruk Coskun[b], Emre Arslan[b], and Mehmet Basaran[b]

[a]Bilkent University, Ankara, Turkiye [b]6GEN. LAB, Turkcell, Istanbul, Turkiye

11.1 Introduction

The rising demand for seamless, high-speed, and massively interconnected wireless networks, particularly in emerging sixth-generation (6G) systems, is enabling a wide range of applications—including the Internet of Things (IoT), smart cities, autonomous vehicles, and immersive communications—while simultaneously intensifying the scarcity of available spectrum resources. This spectrum congestion, primarily caused by limited resources such as bandwidth and overlapping requirements for communication, sensing, and other applications, has become one of the most pressing challenges in modern wireless systems [1]. Traditional methods of addressing this issue, such as spectrum sharing or improving spectral efficiency, are proving inadequate in the face of escalating demands. To this end, integrated sensing and communications (ISAC) emerges as a groundbreaking solution, offering a unified approach to simultaneously meet the needs of both communication and sensing functionalities to eliminate the redundancy of separate systems and enabling more efficient use of the radio frequency (RF) spectrum [2,3]. This dual-purpose approach not only alleviates spectrum scarcity but also introduces new operational efficiencies, such as reduced hardware costs and energy consumption. Using the same resources for both communication and sensing, ISAC unlocks the potential for advanced applications, including vehicular networks, smart cities, and beyond [4].

This evolution necessitates greater flexibility and enhanced sensing capabilities in 6G networks. A key technological enabler is metamaterial-based transceiver design, which leverages engineered materials with controllable electromagnetic (EM) responses to precisely

Electromagnetic Signal and Information Theory
https://doi.org/10.1016/B978-0-44-336354-2.00018-6
Copyright © 2027 Elsevier Inc. All rights are reserved, including those for text and data mining, AI training, and similar technologies.

manipulate wave propagation [5]. These materials can enable highly efficient beamforming, adaptive signal processing, and enhanced spectral utilization which are all critical for the seamless operation of ISAC systems [5].

Holographic multiple-input multiple-output (HMIMO) is the central to these advancements, providing a comprehensive framework in communication systems. HMIMO enables precise control of EM waves in terms of amplitude, phase, and direction, making it pivotal for addressing the high data rate, energy efficiency, and multiuser capacity requirements of 6G and beyond networks. It encompasses a suite of cutting-edge technologies, including reconfigurable holographic surfaces (RHS), dynamic metasurface antennas (DMA), and continuous-aperture phased MIMO (CAP-MIMO). Together, these technologies exemplify the adaptability of HMIMO systems, offering complementary solutions optimized for diverse use cases in the next-generation wireless systems [6]. By dynamically adjusting the characteristics of EM waves, these HMIMO technologies enable adaptive beamforming for improved communication coverage and sensing precision, supporting efficient operation of ISAC systems under dynamic network conditions [6], [7]. Moreover, with their low-cost, energy-efficient implementation, the HMIMO technologies enable the real-time beamforming and directionality, making them ideal for large-scale IoT deployments. Particularly, CAP-MIMO, a specialized extension of HMIMO, utilizes continuous surfaces with densely packed radiating elements or electromagnetically active materials to achieve unparalleled spatial resolution and precise beam steering. These innovations within the HMIMO paradigm showcase its transformative potential in supporting ISAC's dual functionality [6].

The subsequent discussion delves into the role of metamaterial-based transceiver designs, illustrating how they underpin the dual functionality of ISAC.

11.1.1 Spectrum congestion and ISAC as a native solution

Since the 1990s, communication technologies have introduced increasingly dense and heterogeneous wireless networks with each successive decade. Similarly, radar technologies—driven by electronic warfare measure–countermeasure dynamics and the need to detect increasingly sophisticated land, sea, and air platforms—have also focused on exploiting cost-effective portions of the RF spectrum [8]. As a result of this effect, known as spectrum congestion, systems belonging to the communication and radar domains, serving distinct functionalities without any information exchange or data flow between them, have become competitors before/during the spectrum allocation stage [8]. In addition, communication and radar systems that share neighboring regions of the RF spectrum as a result of the allocation process often operate as independent networks and are not synchronized, leading to persistent interference between them during operation. To prevent these in- or out-of-band emission-induced interference effects from jeopardizing the requirements of communication and radar systems, it is crucial to establish stringent in- and out-of-band emission limits, enabling their coexistence. Accordingly, while numerous coexistence-oriented studies [9–11] have been conducted for frequency bands currently used by 5G and former technologies, specifically the S-band (2–4 GHz), C-band (4–8 GHz), and L-band (1–2 GHz) as referenced in radar terminology, it is evident that similar preventative measures will be critically important for the upper-mid-band (golden band) frequencies identified for 6G technologies. These frequencies, designated as FR3 (7–15 GHz), overlap with the X-band (8–12 GHz), making interference mitigation strategies essential. To eliminate or minimize the adverse effects of spectrum congestion, several proposed approaches have been put forward in addition to the predominantly regulatory "coexistence" solutions. These include the development of advanced modulation techniques, integration with noncellular technologies, widespread deployment of small network models such as small cells, and the advancement of dynamic spectrum management tools. While these approaches are largely manageable from the perspective of cellular technologies, they still do not provide synchronization or joint planning with sensing systems (e.g., radars, passive sensing systems, and fixed observation systems). At this point, ISAC, one of the foundational pillars of 6G technology and a key differentiating usage scenario compared to previous generations, offers a natural solution to spectrum congestion. Beyond enabling the prediction and utilization of environment and user-specific information to improve the quality of existing communication services, the ISAC approach also represents an evolution toward the integration of communication and sensing networks. This evolution can be considered the ultimate solution to ensure a much more efficient use of the spectrum for both functionalities.

11.1.2 Advances in metamaterial-based transceiver and network equipment designs

Metamaterial-based transceiver designs have recently been representing innovative and promising solutions to address the high data rate, low latency, and energy efficiency requirements of 6G and beyond communication systems [12]. These designs leverage surfaces and systems capable of actively or passively manipulating EM waves, significantly enhancing performance metrics such as energy efficiency, bandwidth utilization, and user capacity. In this context, reconfigurable intelligent surfaces (RISs) and HMIMO technologies emerge as key advancements in metamaterial-based network equipment structures.

11.1.2.1 Reconfigurable Intelligent Surfaces (RISs)

RISs consist of large surfaces embedded with tunable elements that can dynamically shape and optimize the wireless propagation environment [5,13]. By controlling how EM waves are reflected, refracted, or absorbed, RIS enhances signal coverage, reduces interference, and improves the overall efficiency of wireless networks. RIS can be categorized into three main types: passive RIS [13,14], active RIS [15], and simultaneously transmitting and receiving RIS (STAR-RIS) [16].

Passive RIS. These surfaces feature low-cost and energy-efficient designs, where each element is typically controlled using electronic components such as PIN diodes or varactors. Passive RIS enhances signal strength by establishing virtual line-of-sight (LoS) links, particularly in coverage-challenged environments. Analysis of double-bounce channel conditions indicates that passive RIS can surpass traditional relay systems in spectral efficiency [13,14].

Active RIS. By incorporating additional RF amplifiers, active RIS elements provide signal amplification, a critical advantage for high-frequency regions such as mmWave and terahertz (THz) bands [15]. Active RIS compensates for signal degradation in wideband applications, significantly improving communication performance.

STAR-RIS. STAR-RIS is capable of reflecting and transmitting EM waves concurrently. Operating in energy splitting, mode switching, and time switching modes, STAR-RIS can serve multiple receivers simultaneously [16]. For example, STAR-RIS optimizes beam coverage to target specific regions while expanding overall coverage and supporting increased user density. A primary challenge in RIS technology lies in acquiring accurate channel state information (CSI). Traditional methods rely on pilot signals or iterative beam training techniques to estimate RIS-specific channels.

11.1.2.2 Holographic MIMO (HMIMO)

HMIMO serves as a comprehensive framework in communication systems, enabling the precise control of EM waves in terms of amplitude, phase, and direction [6]. It encompasses a range of innovative solutions in metamaterial-based antenna designs, uniting technologies such as RHS [17], DMA [18], and CAP-MIMO [19] under a unified structure. These technologies, as specialized subfields of HMIMO, address the high data rate, energy efficiency, and multiuser capacity requirements of 6G and beyond networks. While they share a common goal of EM wave manipulation, they are tailored to distinct application scenarios, making them complementary components within the HMIMO paradigm. Together, these technologies exemplify the adaptability of HMIMO, demonstrating its potential to be optimized for a wide range of use cases in next-generation wireless networks.

With their passive or semiactive characteristics, HMIMO aims to reconstruct EM wavefronts by creating holographic patterns that interact with incoming signals, and often alter propagation environments. On the other hand, by enabling real-time manipulation of EM waves through low-cost and flexible antenna structures, HMIMO offers significant energy-efficiency gains for large-scale scenarios such as IoT deployments. Therefore, with the help of its tunable subwavelength elements, HMIMO provides the ability to dynamically control wave propagation to achieve real-time beamforming and directionality in a compact design [6].

As a special version of HMIMO, CAP-MIMO uses a continuous surface with densely packed radiating elements or continuous electromagnetically active materials [19]. This enables precise control over EM waves, offering high spatial resolution and large degrees of freedom for shaping beams and wavefronts. CAP-MIMO integrates the principles of MIMO communication and aperture synthesis. Moreover, CAP-MIMO emphasizes the use of spatially continuous antenna surfaces and advanced phase control, making it particularly suited for precise beam steering and sensing applications. In the subsequent discussions, the ISAC capabilities of HMIMO systems built on CAP-MIMO techniques will be examined.

11.1.3 Utilization of HMIMO schemes for ISAC functionality

HMIMO schemes offer a groundbreaking solution for fulfilling ISAC functionality in future networks [20]. Traditional discrete antenna arrays face inherent limitations due to physical constraints, such as the half-wavelength spacing between elements, which restrict the maximum number of antennas and array gain [19]. HMIMO schemes, particularly CAP-MIMO, overcome these constraints by employing a continuous-aperture surface composed of subwavelength, tunable elements. This architecture enables precise manipulation of EM waves, significantly improving the performance of both communication and sensing functionalities within the same physical aperture.

One of the most notable advantages of HMIMO is its simultaneous improvement in sensing and communication performance [6]. Unlike conventional systems that allocate separate resources for communication and sensing, HMIMO supports joint resource utilization by enabling optimized beamforming for both functionalities [6]. Numerical simulations demonstrate that achieves a remarkable sensing signal-to-interference-plus-noise ratio (SINR) gain compared to discrete array counterparts under identical conditions [7]. Here, using Fourier-based continuous-to-discrete transformations, HMIMO designs optimize joint beamforming to maximize sensing accuracy for multiple targets while maintaining the SINR requirements for multiple communication users [7]. This leads to enhanced spectral efficiency, allowing HMIMO to efficiently manage the dual demands of high-resolution sensing and high-data-rate communication on shared waveforms.

The integrated and energy-efficient design of HMIMO is particularly suited to the stringent requirements of 6G, including spectrum scarcity and dense network topologies. While traditional discrete antenna arrays suffer from lower array gain and inefficient power radiation, the continuous-aperture configuration of CAP-MIMO enabled HMIMO systems ensures superior radiation efficiency [7]. The Fourier-based beamforming strategy employed by these systems not only overcomes the computational challenges of continuous patterns but also enables the system to adaptively adjust beam directions to address multiuser and multitarget scenarios. These HMIMO systems are particularly advantageous in complex environments where precise beam focusing and low-latency sensing are critical, such as autonomous vehicles, smart cities, and the next-generation industrial automation [7].

In conclusion, the HMIMO-based ISAC systems represent a transformative approach for meeting the multifaceted demands of 6G networks. By leveraging its continuous-aperture design and integrated beamforming capabilities, the CAP-MIMO enabled HMIMO offers superior performance in terms of sensing accuracy, communication capacity, and energy efficiency. These advantages position HMIMO as a key enabler for a wide range of applications, from high-density urban networks to mission-critical scenarios requiring ultra-reliable sensing and communication. Its ability to balance performance trade-offs in complex network environments underscores its potential as a cornerstone technology for 6G ISAC systems.

In this chapter, we introduce, for the first time, a tripolarization transmitter and receiver model in the context of holographic ISAC systems, enabling a more comprehensive representation of polarization effects. We derive the mathematical equations for the system model, capturing the interplay between sensing and communication functionalities under the tripolarization framework. Furthermore, we propose a novel concave–convex procedure (CCP)-based alternating optimization algorithm that maximizes the sensing SNR while ensuring that the communication SINR requirements are satisfied. We investigate HMIMO technology and demonstrate how it enhances ISAC performance over benchmark traditional discrete MIMO arrays. Together, these sections provide a comprehensive overview of the innovations driving the next generation of wireless communication systems.

Notation. Throughout this chapter, vectors and matrices are denoted by bold lowercase and uppercase letters, respectively (e.g., $\mathbf{a}$ and $\mathbf{A}$). The transpose, conjugate, and Hermitian (conjugate transpose) of a matrix $\mathbf{A}$ are respectively denoted as $\mathbf{A}^{\mathrm{T}}$, $\mathbf{A}^{*}$, and $\mathbf{A}^{\mathrm{H}}$. Further, $\mathbb{Z}$, $\mathbb{R}$, and $\mathbb{C}$ represent the integers, real numbers, and complex numbers, respectively, while $\{\cdot\}_{m=1}^{M}$ denotes a set indexed over the integer range from $m = 1$ to M. The expectation operator is denoted as $\mathbb{E}\{\cdot\}$, while identity matrix is represented by $\mathbf{I}$.

The rest of the chapter is organized as follows. Section 11.2 introduces the principles and transceiver structures of HMIMO schemes. In Section 11.3, the system model and the formulated optimization problem to maximize the sensing SNR under communication SINR constraints are introduced. Section 11.4 describes the proposed alternating optimization algorithm, detailing its steps to iteratively optimize the sensing and communication performances. In Section 11.5, numerical simulations to evaluate the performance of the proposed holographic ISAC system, highlighting its advantages over traditional discrete MIMO systems, are provided. Challenges and future research directions are outlined in Section 11.6. Finally, Section 11.7 concludes the chapter, summarizing key insights and potential future directions.

11.2 Principles and transceiver structures of holographic ISAC schemes

The HMIMO-enabled holographic ISAC systems with both communications and sensing capabilities employ several unique components on top of the conventional transmitter and receiver structures including basic wireless communications and radar processing units [7]. Here, the transmitter structure consists of a signal generation module embedded in the HMIMO surface, responsible for producing EM wave patterns to send signals to communication users and sensing targets, with HMIMO enhancing this process through its continuous aperture design for greater energy efficiency and accuracy.

On the receiver side, a module mounted to the HMIMO surface is responsible for processing the incoming signals that are reflected radar echoes from sensing targets and the conveyed uplink signals from communication users. Here, the beamforming algorithms are utilized to enhance the separation of signals arriving from different directions.

Similar to other metamaterial-based transceiver designs, HMIMO schemes employ control matrices, which are mathematical constructs used to adjust the phase and amplitude of surface elements, enabling the optimization of beam patterns over the continuous aperture, with Fourier-based modeling widely employed in their design.

Additionally, the unique components of holographic ISAC scheme, each contributing to their high-performance capabilities in communication and sensing, have been detailed in the following subsections.

11.2.1 Continuous aperture with subwavelength elements

As the main differentiator from other HMIMO designs, the continuous aperture is the surface composed of densely packed elements arranged at subwavelength intervals, replacing traditional discrete antenna arrays, and provides higher array gain by utilizing every point of the aperture to precisely manipulate EM waves in terms of amplitude, phase, and direction [6].

Here, subwavelength elements are individual components densely integrated across the surface, capable of modulating EM waves, and function to adjust their phase and amplitude

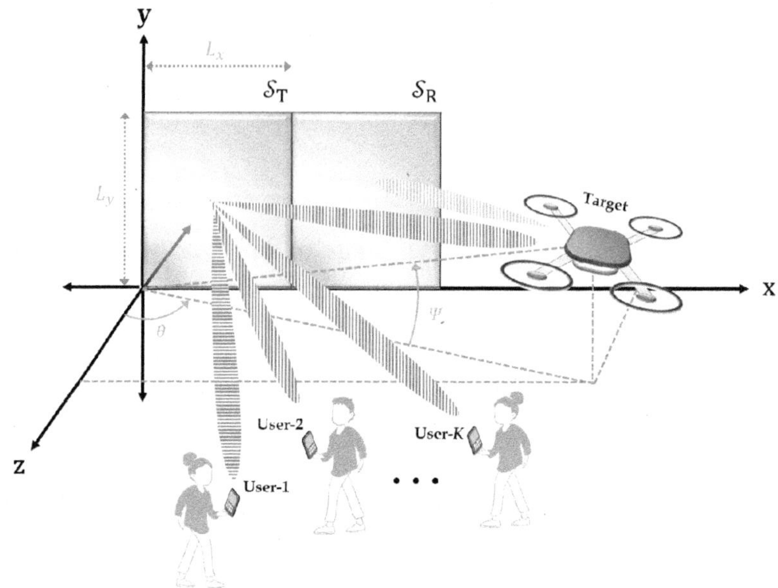

Figure 11.1:
A representative sketch for holographic ISAC scheme.

to reconfigure direction and optimize beamforming, often leveraging metamaterials or similar technologies [18,21,22].

Representative illustrations for both the transmitter and receiver functionalities are provided in Figs. 11.1 and 11.2, and denoted $\mathcal{S}_T$ and $\mathcal{S}_R$, respectively.

11.2.2 Beamforming unit

In holographic ISAC schemes, beamforming is performed using both digital beamforming at the radio unit of a base station (BS) and amplitude-controlled analog beamforming via the holographic surfaces [7]. By utilizing digital beamforming at the baseband signal level, the BS will be able direct communication data streams toward specific users. The BS adjusts the direction and power of the transmitted signals to ensure high SINR for communication tasks while minimizing interference with radar sensing signals [20,23]. On the other hand, with the help of analog beamforming performed by the holographic surface, the amplitude of each metamaterial element is adjusted to control the direction and shape of the radiated beam. By controlling the amplitude rather than the phase of the signal, HMIMO schemes can achieve more efficient beamforming with reduced power consumption [20,23].

This hybrid approach allows the system to simultaneously optimize both radar sensing and communication tasks. Here, the beamforming module consists of a hardware or software module that controls the amplitude, phase, and direction of EM waves, optimizing beam patterns for multiuser communication and multitarget sensing through mathematical approaches, such as Fourier transforms, to efficiently model the continuous aperture.

11.2.3 Continuous-to-discrete transformation module

The continuous-to-discrete transformation module in HMIMO systems converts the continuous aperture into a discrete model, enabling the use of traditional signal processing algorithms. This transformation is typically achieved through Fourier-based techniques, which represent the continuous surface as a sum of sinusoidal components that are then discretized for computational efficiency. This allows for optimization of beamforming and channel estimation using discrete signal processing methods [19]. By discretizing the continuous aperture, the module facilitates beam optimization and multiuser/multitarget scenarios, enabling techniques like multibeam formation, interference cancellation, and spatial multiplexing. The transformation also improves computational efficiency by simplifying complex integrals into manageable discrete calculations, making it feasible for real-time processing in large-scale systems.

The discrete model generated by the transformation reduces computational complexity and allows for efficient use of hardware accelerators like field-programmable gate arrays (FPGAs) and graphics processing units (GPUs). This ensures that HMIMO systems can handle high-performance tasks in dynamic environments, optimizing communication and sensing across multiple targets and users.

11.2.4 Algorithmic control unit

The algorithmic control unit comprises a software layer that governs optimization processes between the transmitter and receiver, managing tasks such as beamforming, signal power optimization, and multitarget optimization, with artificial intelligence (AI)-powered algorithms often incorporated in HMIMO-based systems [13].

This unit is also responsible for efficiently and optimally powering surface elements to ensure their operation with minimal energy consumption, and guaranteeing use-case-specific quality-of-service requirements [13].

These components collectively enable HMIMO systems to deliver high performance and flexibility for both communication and sensing applications.

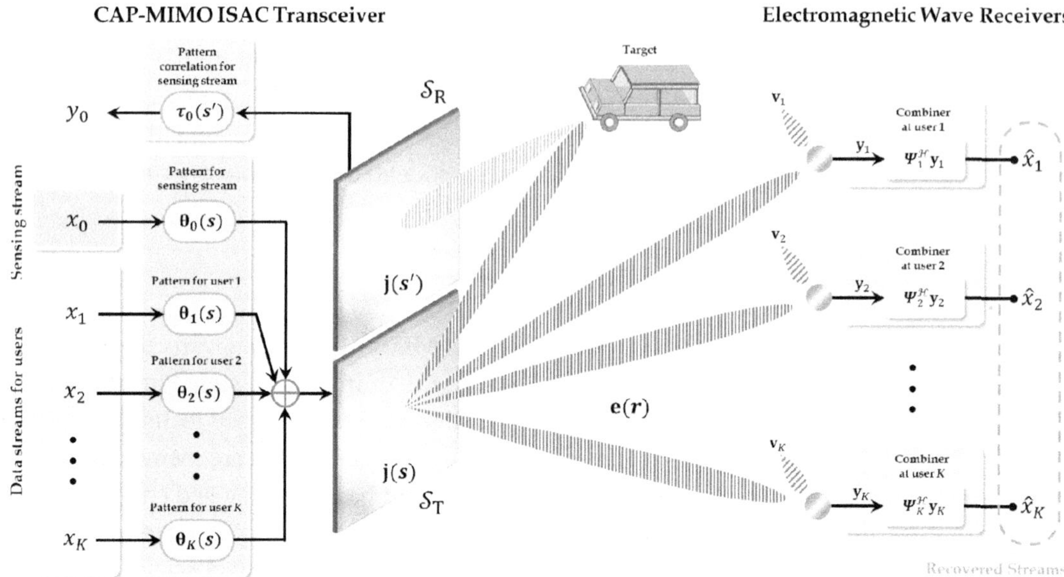

Figure 11.2:
Block diagram of ISAC-capable HMIMO scheme.

11.3 System model and problem formulation for holographic ISAC system

As illustrated in Fig. 11.2, we consider an ISAC system that serves for the wireless communications requirements of K user equipments (UEs), and the detection of a single target. The BS is equipped with separate HMIMO systems for downlink communications and sensing purposes denoted as $\mathcal{S}_\mathrm{T}$ and $\mathcal{S}_\mathrm{R}$, respectively. The transmitter HMIMO sends ISAC waveforms to simultaneously serve K UEs and detect the location of a potential target. To accomplish the detection task, the receiver HMIMO captures the EM waves reflected by the target, which result from a combination of various physical scattering phenomena.

We consider a transmitter HMIMO surface with aperture $\mathcal{S}_\mathrm{T}$ of area $A_\mathrm{T} = |\mathcal{S}_\mathrm{T}|$ working in a 3D homogeneous medium. Similarly, the receive HMIMO surface with aperture $\mathcal{S}_\mathrm{R}$ of area $A_\mathrm{R} = |\mathcal{S}_\mathrm{R}|$. We assume the ideal case that HMIMO surface has an almost continuous antenna aperture, which enables it to generate any continuous current distribution on its surface [19,24,25]. We consider a narrowband system and focus on the time-independent current density $\mathbf{j}(\mathbf{s}) \in \mathbb{C}^3$. In addition, we suppose that the UEs are tripolarization receivers. We let $\mathbf{x} \triangleq [x_0, x_1, \ldots, x_K]^\mathrm{T} \in \mathbb{C}^{K+1}$ denote the symbols transmitted to K UEs and the target, where x_0 denotes the symbols transmitted to the target, i.e., sensing stream. We assume that these symbols have unit power and are independent of each other. Then, the symbols $\mathbf{x}$ to be transmitted are modulated on $K+1$ different HMIMO patterns $\boldsymbol{\Theta}(\mathbf{s}) \triangleq \{\boldsymbol{\theta}_k(\mathbf{s})\}_{k=0}^{K}$. The combined current distribution $\mathbf{j}(\mathbf{s})$ on the transmit HMIMO surface aperture can be modeled as

$$\mathbf{j}(\mathbf{s}) = \sum_{k=0}^{K} \boldsymbol{\theta}_k(\mathbf{s})\, x_k, \quad \mathbf{s} \in \mathcal{S}_\mathrm{T}. \tag{11.1}$$

11.3.1 Electromagnetic channels

To model the radiated information-carrying EM waves in space, we define $\mathbf{e}(\mathbf{r}) \in \mathbb{C}^3$ as the electric field at point $\mathbf{r} \triangleq (r_x, r_y, r_z) \in \mathbb{R}^3$, which is induced by the current distribution $\mathbf{j}(\mathbf{s})$ on the transmit HMIMO surface aperture. Utilizing Green's method [19,26], the relationship between the current distribution $\mathbf{j}(\mathbf{s})$ at the transmitter and the electric field $\mathbf{e}(\mathbf{r})$ at a receiver point is given as

$$\mathbf{e}(\mathbf{r}) = j\kappa_0 Z_0 \int_{\mathcal{S}_\mathrm{T}} \mathbf{G}(\mathbf{r}, \mathbf{s})\mathbf{j}(\mathbf{s})\mathrm{d}\mathbf{s} \tag{11.2}$$

where the Green's function is $\mathbf{G}(\mathbf{r}, \mathbf{s}) \in \mathbb{C}^{3\times 3}$, playing the role of channel function. The Green's function is dependent on the wireless propagation environment. The κ_0 and Z_0 denote the wavenumber and wave impedance in free space, respectively. In unbounded and homogeneous mediums, the Green's function is given as [19]

$$\mathbf{G}(\mathbf{r}, \mathbf{s}) = \frac{e^{j\kappa_0 \|\mathbf{r}-\mathbf{s}\|}}{4\pi \|\mathbf{r}-\mathbf{s}\|} \left(\mathbf{I}_3 - \frac{(\mathbf{r}-\mathbf{s})(\mathbf{r}-\mathbf{s})^\mathrm{T}}{\|\mathbf{r}-\mathbf{s}\|^2} \right), \tag{11.3}$$

where we assume that all K UEs and potential target are located in the far-field region. Each UE is equipped with an ideal tripolarization antenna with area $A_\mathrm{iso} = \frac{\lambda^2}{4\pi}$, which satisfies $A_\mathrm{iso} \ll A_\mathrm{T}$ [27]. In this case, similar to a discrete antenna, each UE can be reasonably approximated by a point in 3D space.

We let $\mathbf{r}_k \in \mathbb{R}^3$ denote the 3D location of the kth UE. In line with [19], we consider that the kth UE with a tripolarization antenna can sense the three polarizations of the EM waves reaching point $\mathbf{r}_k$ and combine the three polarization components to decode the symbol x_k with a polarization combiner $\boldsymbol{\psi}_k \in \mathbb{C}^3$ (either in the analog or digital domain). Thus, according to (11.1) and (11.2), the EM wave received by the kth UE can be expressed as

$$\mathbf{y}_k = \mathbf{e}_k + \mathbf{v}_k = \underbrace{x_k j\kappa_0 Z_0 \int_{\mathcal{S}_\mathrm{T}} \mathbf{G}_k(\mathbf{s})\boldsymbol{\theta}_k(\mathbf{s})\,\mathrm{d}\mathbf{s}}_{\text{Desired signal to the } k\text{th UE}} + \underbrace{\sum_{i=0, i\neq k}^{K} x_i j\kappa_0 Z_0 \int_{\mathcal{S}_\mathrm{T}} \mathbf{G}_k(\mathbf{s})\boldsymbol{\theta}_i(\mathbf{s})\mathrm{d}\mathbf{s}}_{\text{Interferences from other UEs and sensing stream}} + \underbrace{\mathbf{v}_k}_{\text{Noise}}, \tag{11.4}$$

where $\mathbf{e_k} := \mathbf{e}(\mathbf{r_k})$, $\mathbf{G_k}(\mathbf{s}) := \mathbf{G}(\mathbf{r_k}, \mathbf{s})$, and $\mathbf{v}_k \in \mathbb{C}^3$ is the EM noise at the kth UE. For simplicity, we follow the isotropic propagation assumption used in [19,28], thus $\mathbf{v}_k$ for all $k \in \{1, \ldots, K\}$ can be modeled as mutually independent additive white Gaussian noise (AWGN) with zero mean and covariance matrix $\sigma^2\mathbf{I}_3$.

Next, the polarization combiner is applied and the signal $\boldsymbol{\psi}_k^{\mathrm{H}}\mathbf{y}_k$ is obtained with the corresponding SINR

$$\mathrm{SINR}_k\left(\boldsymbol{\Theta}(\mathbf{s})\right) = \frac{\left|\boldsymbol{\psi}_k^{\mathrm{H}}\boldsymbol{\alpha}_k\right|^2}{\boldsymbol{\psi}_k^{\mathrm{H}}\mathbf{J}_k\boldsymbol{\psi}_k}, \tag{11.5}$$

where $\boldsymbol{\alpha}_k$ and $\mathbf{J}_k$ are respectively given by

$$\boldsymbol{\alpha}_k = j\kappa_0 Z_0 \int_{\mathcal{S}_\mathrm{T}} \mathbf{G}_k(\mathbf{s})\boldsymbol{\theta}_k(\mathbf{s})\,\mathrm{d}\mathbf{s}, \tag{11.6}$$

$$\mathbf{J}_k = \sum_{i=0, i\neq k}^{K} \kappa_0^2 Z_0^2 \left(\int_{\mathcal{S}_\mathrm{T}} \mathbf{G}_k(\mathbf{s})\boldsymbol{\theta}_i(\mathbf{s})\mathrm{d}\mathbf{s}\right)\left(\int_{\mathcal{S}_\mathrm{T}} \mathbf{G}_k(\mathbf{s})\boldsymbol{\theta}_i(\mathbf{s})\mathrm{d}\mathbf{s}\right)^{\mathrm{H}} + \sigma^2\mathbf{I}_3. \tag{11.7}$$

The optimal combiner that maximizes the UE SINR is given as $\boldsymbol{\psi}_k^{\star} = \mathbf{J}_k{}^{-1}\boldsymbol{\alpha}_k$, which leads to the SINR

$$\mathrm{SINR}_k\left(\boldsymbol{\Theta}(\mathbf{s})\right) = \boldsymbol{\alpha}_k^{\mathrm{H}}\mathbf{J}_k^{-1}\boldsymbol{\alpha}_k. \tag{11.8}$$

To characterize the interactions between different polarization components of the reflection through the target, we will utilize the polarimetric radar cross-section (RCS) matrix $\boldsymbol{\Gamma} \in \mathbb{C}^{3\times 3}$. We let denote $\mathbf{e}_0$ is the incident electric field on the target given as

$$\mathbf{e}_0 = \sum_{i=0}^{K} x_i\, j\kappa_0 Z_0 \int_{\mathcal{S}_\mathrm{T}} \mathbf{G}_0(\mathbf{s})\boldsymbol{\theta}_i(\mathbf{s})\mathrm{d}\mathbf{s}. \tag{11.9}$$

Then, the reflected electric field through the polarimetric RCS matrix becomes

$$\bar{\mathbf{e}}_0 = \boldsymbol{\Gamma}\mathbf{e}_0. \tag{11.10}$$

For ease of presentation, we take the polarimetric RCS matrix as a known constant. Then, the received reflected echo signal from the target at receiver location $\mathbf{s}'$ is given as $\mathbf{G}_0^*(\mathbf{s}')\bar{\mathbf{e}}_0 + \mathbf{v}_0(\mathbf{s}')$, where $\mathbf{G}_0(\mathbf{s}') := \mathbf{G}(\mathbf{r}_0, \mathbf{s}')$ and $\mathbf{v}_0(\mathbf{s}') \in \mathbb{C}^3$ is the AWGN with each element having the variance σ^2. Denoting the HMIMO receiver pattern by $\boldsymbol{\tau}_0(\mathbf{s}') \in \mathbb{C}^3$, the received field is obtained as

$$y_0 = \int_{\mathcal{S}_\mathrm{R}} \boldsymbol{\tau}_0^{\mathrm{H}}(\mathbf{s}')\mathbf{G}_0^*(\mathbf{s}')\bar{\mathbf{e}}_0\mathrm{d}\mathbf{s}' + \int_{\mathcal{S}_\mathrm{R}} \boldsymbol{\tau}_0^{\mathrm{H}}(\mathbf{s}')\mathbf{v}_0(\mathbf{s}')\mathrm{d}\mathbf{s}'. \tag{11.11}$$

Therefore, the sensing SNR becomes

$$\gamma_0\left(\boldsymbol{\Theta}(\mathbf{s}), \boldsymbol{\tau}_0(\mathbf{s}')\right) = \frac{\mathbb{E}\left|\int_{\mathcal{S}_R} \boldsymbol{\tau}_0^H(\mathbf{s}')\mathbf{G}_0^*(\mathbf{s}')\bar{\mathbf{e}}_0 d\mathbf{s}'\right|^2}{\sigma^2 \int_{\mathcal{S}_R} \boldsymbol{\tau}_0^H(\mathbf{s}')\boldsymbol{\tau}_0(\mathbf{s}')d\mathbf{s}'}. \tag{11.12}$$

11.3.2 Problem formulation in continuous domain for improving sensing performance

As can be seen from (11.8) and (11.12), the UE SINR and sensing SNR are functions of the surface variables $\mathbf{s}$ and $\mathbf{s}'$, respectively. Here, in order to achieve a balance between sensing and communication performance within the proposed HMIMO ISAC system, we aim to maximize the sensing SNR while ensuring that the communication SINR requirements for K UEs are met. The resulting problem concerning the continuous pattern design is formulated as follows:

$$\max_{\boldsymbol{\Theta}(\mathbf{s}), \boldsymbol{\tau}_0(\mathbf{s}')} \quad \gamma_0\left(\boldsymbol{\Theta}(\mathbf{s}), \boldsymbol{\tau}_0(\mathbf{s}')\right) \tag{11.13a}$$

$$\text{such that} \quad \text{SINR}_k\left(\boldsymbol{\Theta}(\mathbf{s})\right) \geq \xi_k, \quad k = 1, \ldots, K, \tag{11.13b}$$

$$\sum_{k=0}^{K} \int_{\mathcal{S}_T} \|\boldsymbol{\theta}_k(\mathbf{s})\|^2 \, d(\mathbf{s}) \leq P_T, \tag{11.13c}$$

where ξ_k is the minimum required UE SINR, and P_T is the maximum transmit power that can be distributed across the surface $\mathcal{S}_T$.

The primary objective of the optimization problem (11.13) is to achieve a balance between the sensing SNR and the required UE SINR, as defined in (11.13b), while also satisfying the power budget constraint presented in (11.13c). However, since the infinite-dimensional integrals present in the both objective and constraints of problem (11.13), traditional beamforming techniques designed for discrete-array systems cannot be utilized. To tackle these infinite-dimensional functions in problem (11.13), a continuous-discrete transformation using the principles of the Fourier transform, will be presented in following subsection.

11.3.3 Continuous-discrete transformation

In this subsection, we present a continuous-to-discrete transformation aimed at addressing the complex optimization problem stated in (11.13). The primary approach involves converting patterns from the continuous domain into the discrete wavenumber domain through the application of the Fourier transform.

Here, if the continuous pattern function $\boldsymbol{\theta}_k(\mathbf{s})$ is absolutely integrable, it can be equivalently expressed using Fourier series expansion for $\mathbf{n} := (n_x, n_y, n_z) \in \mathbb{Z}^3$ as

$$\boldsymbol{\theta}_k(\mathbf{s}) = \sum_{\mathbf{n}}^{\infty} \mathbf{w}_{k,\mathbf{n}} \boldsymbol{\Omega}_{\mathbf{n}}(\mathbf{s}), \quad \mathbf{s} \in \mathcal{S}_{\mathrm{T}}, \tag{11.14}$$

where $\mathbf{n} := (n_x, n_y, n_z) \in \mathbb{Z}^3$, $\sum_{\mathbf{n}}^{\infty} := \sum_{n_x=-\infty}^{\infty} \sum_{n_y=-\infty}^{\infty} \sum_{n_z=-\infty}^{\infty}$, and $\mathbf{w}_{k,\mathbf{n}} \in \mathbb{C}^3$ is the projection length of the kth pattern in the discrete wavenumber domain, $\boldsymbol{\Omega}_{\mathbf{n}}(\mathbf{s}) \in \mathbb{C}$ is Fourier base function. The Fourier base function $\boldsymbol{\Omega}_{\mathbf{n}}(\mathbf{s})$ can be reexpressed as

$$\boldsymbol{\Omega}_{\mathbf{n}}(\mathbf{s}) = \frac{1}{\sqrt{A_{\mathrm{T}}}} e^{j2\pi \left(\frac{n_x}{L_x}\left(s_x - \frac{L_x}{2}\right) + \frac{n_y}{L_y}\left(s_y - \frac{L_y}{2}\right) + \frac{n_z}{L_z}\left(s_z - \frac{L_z}{2}\right) \right)}, \quad \mathbf{s} \in \mathcal{S}_{\mathrm{T}}, \tag{11.15}$$

where L_x, L_y, and L_z represent the maximum projection lengths of the volume $\mathcal{S}_{\mathrm{T}}$ along the x, y, and z axes, respectively. Additionally, the projection coefficients $\mathbf{w}_{k,\mathbf{n}}$ are given by

$$\mathbf{w}_{k,\mathbf{n}} = \frac{1}{\sqrt{A_{\mathrm{T}}}} \int_{\mathcal{S}_{\mathrm{T}}} \boldsymbol{\theta}_k(\mathbf{s}) \boldsymbol{\Omega}_{\mathbf{n}}^*(\mathbf{s})\, \mathrm{d}\mathbf{s}, \quad \mathbf{n} \in \mathbb{Z}^3, \tag{11.16}$$

where $\boldsymbol{\Omega}_{\mathbf{n}}^*(\mathbf{s})$ denotes the complex conjugate of the Fourier base function. Moreover, the Fourier bases $\boldsymbol{\Omega}_{\mathbf{n}}(\mathbf{s})$ satisfy the orthogonality conditions as follows:

$$\int_{\mathcal{S}_{\mathrm{T}}} \boldsymbol{\Omega}_{\mathbf{n}}(\mathbf{s}) \boldsymbol{\Omega}_{\tilde{\mathbf{n}}}^*(\mathbf{s}) \mathbf{ds} = \begin{cases} 1, & \tilde{\mathbf{n}} = \mathbf{n}, \\ 0, & \tilde{\mathbf{n}} \neq \mathbf{n}. \end{cases} \tag{11.17}$$

Therefore, by using the above discrete transformations, continuous pattern function $\boldsymbol{\theta}(\mathbf{s})$ in (11.13) can be equivalently replaced by their discrete projection lengths $\{\mathbf{w}_{k,\tilde{\mathbf{n}}}\}_{k=0}^{K}$. This substitution is accomplished by applying the following continuous-to-discrete transformation to the electric field.

By applying the Fourier series expansion of the pattern function in (11.14), the noise-free electric field at the kth UE (or target location when $k = 0$) can be equivalently rewritten as

$$\mathbf{e}_k = j\kappa_0 Z_0 \sum_{i=0}^{K} x_i \int_{\mathcal{S}_{\mathrm{T}}} \mathbf{G}_k(\mathbf{s}) \boldsymbol{\theta}_i(\mathbf{s}) \mathrm{d}\mathbf{s} = \sum_{i=0}^{K} \sum_{\mathbf{n}}^{\infty} \boldsymbol{\Delta}_{k,\mathbf{n}} \mathbf{w}_{i,\mathbf{n}} x_i, \quad k = 0, \ldots, K, \tag{11.18}$$

where

$$\boldsymbol{\Delta}_{k,\mathbf{n}} = j\kappa_0 Z_0 \int_{\mathcal{S}_{\mathrm{T}}} \mathbf{G}_k(\mathbf{s}) \boldsymbol{\Omega}_{\mathbf{n}}(\mathbf{s}) \mathrm{d}\mathbf{s}. \tag{11.19}$$

Here, $\boldsymbol{\Delta}_{k,\mathbf{n}} \in \mathbb{C}^{3\times 3}$ is the Fourier transform of the function $j\kappa_0 Z_0 \mathbf{G}_k(\mathbf{s})$ over the surface $\mathcal{S}_{\mathrm{T}}$ at wavenumber $\boldsymbol{\kappa}_{\mathbf{n}} = \left(2\pi \frac{n_x}{L_x}, 2\pi \frac{n_y}{L_y}, 2\pi \frac{n_z}{L_z}\right)$. Therefore, the Parseval's theorem implies an equivalence between the energy in the spatial domain $\mathcal{S}_{\mathrm{T}}$ and wavenumber domain $\boldsymbol{\kappa}_{\mathbf{n}}$ as follows:

$$\int_{\mathcal{S}_{\mathrm{T}}} \|\mathbf{G}_k(\mathbf{s})\|_{\mathrm{F}}^2 \, d\mathbf{s} = \sum_{\mathbf{n}}^{\infty} \left\| \boldsymbol{\Delta}_{k,\mathbf{n}} \right\|_{\mathrm{F}}^2, \tag{11.20}$$

where the left-hand side represents the integral of the Frobenius norm squares of continuous functions, while the right-hand side corresponds to the sum of the Frobenius norm squares of discrete matrices. Here, $\|\mathbf{G}_k(\mathbf{s})\|_{\mathrm{F}}^2$ and $\left\| \boldsymbol{\Delta}_{k,\mathbf{n}} \right\|_{\mathrm{F}}^2$ can be considered as channel gain in spatial and wavenumber domains, respectively. Moreover, according to Parseval's theorem, overall transmit power in (11.13c), can be rewritten as

$$\sum_{k=0}^{K} \int_{\mathcal{S}_{\mathrm{T}}} \|\boldsymbol{\theta}_k(\mathbf{s})\|^2 \, d\mathbf{s} = \sum_{k=0}^{K} \sum_{\mathbf{n}}^{\infty} \left\| \mathbf{w}_{k,\mathbf{n}} \right\|^2, \tag{11.21}$$

which ensures energy conservation within the system. It is clear from (11.20) and (11.21), continuous $\mathbf{G}_k(\mathbf{s})$ and $\boldsymbol{\theta}_k(\mathbf{s})$ functions can be replaced by their projection lengths $\left\{\boldsymbol{\Delta}_{k,\mathbf{n}}\right\}_{k=0}^{K}$ and $\left\{\mathbf{w}_{k,\mathbf{n}}\right\}_{k=0}^{K}$, respectively. However, since number of mathematical expansions are infinite as shown in (11.20) and (11.21), optimizing the projection lengths for HMIMO systems $\left\{\mathbf{w}_{k,\mathbf{n}}\right\}_{k=0}^{K}$ is impractical for typical computers. On the other hand, recently, it has been analytically proved that the channel gain $\left\| \boldsymbol{\Delta}_{k,\mathbf{n}} \right\|^2$ is predominantly concentrated in the low-wavenumber band, with negligible values in the high-wavenumber band [29]. As a result, we can approximate the Fourier expansion series $\sum_{\mathbf{n}}^{\infty} \boldsymbol{\Delta}_{k,\mathbf{n}} \mathbf{w}_{i,\mathbf{n}}$ by limiting it to a finite number of low-wavenumber with nonnegligible high-power terms within $[-\kappa_0, \kappa_0]$ band that satisfy $\left|2\pi \frac{n_x}{L_x}\right| \leq \kappa_0$, $\left|2\pi \frac{n_y}{L_y}\right| \leq \kappa_0$, and $\left|2\pi \frac{n_z}{L_z}\right| \leq \kappa_0$ while ignoring high-wavenumber and negligible terms [19]. Therefore, with $\mathbf{N} := (N_x, N_y, N_z) \in \mathbb{Z}^3$ being the retained expansion terms on 3D wavenumber domain such that $\left|2\pi \frac{n_x}{L_x}\right| \leq \kappa_0$, $\left|2\pi \frac{n_y}{L_y}\right| \leq \kappa_0$, and $\left|2\pi \frac{n_z}{L_z}\right| \leq \kappa_0$, for $n_x \in [-N_x, N_x]$, $n_y \in [-N_y, N_y]$, and $n_z \in [-N_z, N_z]$, the electric field in (11.18) can be approximated as

$$\mathbf{e}_k = j\kappa_0 Z_0 \sum_{i=0}^{K} x_i \int_{\mathcal{S}_{\mathrm{T}}} \mathbf{G}_k(\mathbf{s}) \boldsymbol{\theta}_i(\mathbf{s}) d\mathbf{s} \approx \sum_{i=0}^{K} \sum_{\mathbf{n}}^{\mathbf{N}} \boldsymbol{\Delta}_{k,\mathbf{n}} \mathbf{w}_{i,\mathbf{n}} x_i, \quad k = 0, \ldots, K, \tag{11.22}$$

where $\sum_{\mathbf{n}}^{\mathbf{N}}$ denotes $\sum_{-N_x}^{N_x} \sum_{-N_y}^{N_y} \sum_{-N_z}^{N_z}$. In a similar way, the transmit power (11.21) can be approximated as follows:

$$\sum_{k=0}^{K} \int_{\mathcal{S}_{\mathrm{T}}} \|\boldsymbol{\theta}_k(\mathbf{s})\|^2 \, d\mathbf{s} \approx \sum_{k=0}^{K} \sum_{\mathbf{n}}^{\mathbf{N}} \|\mathbf{w}_{k,\mathbf{n}}\|^2. \tag{11.23}$$

Using similar steps, for $\overline{\mathbf{n}} := (\overline{n}_x, \overline{n}_y, \overline{n}_z) \in \mathbb{Z}^3$, the HMIMO receiver pattern is approximated as

$$\boldsymbol{\tau}_0(\mathbf{s}') \approx \sum_{\overline{\mathbf{n}}}^{\overline{\mathbf{N}}} \overline{\mathbf{w}}_{0,\overline{\mathbf{n}}} \overline{\boldsymbol{\Omega}}_{\overline{\mathbf{n}}}(\mathbf{s}'), \quad \mathbf{s}' \in \mathcal{S}_{\mathrm{R}}, \tag{11.24}$$

where $\overline{\mathbf{w}}_{0,\overline{\mathbf{n}}} \in \mathbb{C}^3$ is the projection length of receiver pattern in the discrete wavenumber domain, $\overline{\mathbf{N}} := (\overline{N}_x, \overline{N}_y, \overline{N}_z) \in \mathbb{Z}^3$ being the retained expansion terms on the 3D wavenumber domain such that $\left|2\pi \frac{\overline{n}_x}{\overline{L}_x}\right| \le \kappa_0$, $\left|2\pi \frac{\overline{n}_y}{\overline{L}_y}\right| \le \kappa_0$, and $\left|2\pi \frac{\overline{n}_z}{\overline{L}_z}\right| \le \kappa_0$ for $\overline{n}_x \in [-\overline{N}_x, \overline{N}_x]$, $\overline{n}_y \in [-\overline{N}_y, \overline{N}_y]$, and $\overline{n}_z \in [-\overline{N}_z, \overline{N}_z]$. The $\overline{\boldsymbol{\Omega}}_{\overline{\mathbf{n}}}(\mathbf{s}') \in \mathbb{C}$ is the Fourier base function corresponding to the receiver. It is given as

$$\overline{\boldsymbol{\Omega}}_{\overline{\mathbf{n}}}(\mathbf{s}') = \frac{1}{\sqrt{A_{\mathrm{R}}}} e^{j2\pi\left(\frac{\overline{n}_x}{\overline{L}_x}\left(s_x - \frac{\overline{L}_x}{2}\right) + \frac{\overline{n}_y}{\overline{L}_y}\left(s_y - \frac{\overline{L}_y}{2}\right) + \frac{\overline{n}_z}{\overline{L}_z}\left(s_z - \frac{\overline{L}_z}{2}\right)\right)}, \quad \mathbf{s}' \in \mathcal{S}_{\mathrm{R}}, \tag{11.25}$$

where $\overline{L}_x$, $\overline{L}_y$, and $\overline{L}_z$ represent the maximum projection lengths of the volume $\mathcal{S}_{\mathrm{R}}$ along the x, y, and z axes, respectively. Additionally, the projection coefficients $\overline{\mathbf{w}}_{0,\overline{\mathbf{n}}}$ are given by

$$\overline{\mathbf{w}}_{0,\overline{\mathbf{n}}} = \frac{1}{\sqrt{A_{\mathrm{R}}}} \int_{\mathcal{S}_{\mathrm{R}}} \boldsymbol{\tau}_0(\mathbf{s}') \overline{\boldsymbol{\Omega}}^*_{\overline{\mathbf{n}}}(\mathbf{s}')\, \mathrm{d}\mathbf{s}', \quad \overline{\mathbf{n}} \in \mathbb{Z}^3. \tag{11.26}$$

Using these, the numerator and denominator of the sensing SNR in (11.12) can be reexpressed as

$$\begin{aligned}\mathbb{E}\left\{\left|\int_{\mathcal{S}_{\mathrm{R}}} \boldsymbol{\tau}_0^{\mathrm{H}}(\mathbf{s}') \mathbf{G}_0^*(\mathbf{s}') \overline{\mathbf{e}}_0 \mathrm{d}\mathbf{s}'\right|^2\right\} &\approx \mathbb{E}\left\{\left|\sum_{\overline{\mathbf{n}}}^{\overline{\mathbf{N}}} \overline{\mathbf{w}}_{0,\overline{\mathbf{n}}}^{\mathrm{H}} \overline{\boldsymbol{\Delta}}_{0,\overline{\mathbf{n}}}^{\mathrm{H}} \boldsymbol{\Gamma} \sum_{i=0}^{K} \sum_{\mathbf{n}}^{\mathbf{N}} \boldsymbol{\Delta}_{0,\mathbf{n}} \mathbf{w}_{i,\mathbf{n}} x_i\right|^2\right\} \\ &= \sum_{i=0}^{K} \left|\sum_{\overline{\mathbf{n}}}^{\overline{\mathbf{N}}} \overline{\mathbf{w}}_{0,\overline{\mathbf{n}}}^{\mathrm{H}} \overline{\boldsymbol{\Delta}}_{0,\overline{\mathbf{n}}}^{\mathrm{H}} \boldsymbol{\Gamma} \sum_{\mathbf{n}}^{\mathbf{N}} \boldsymbol{\Delta}_{0,\mathbf{n}} \mathbf{w}_{i,\mathbf{n}}\right|^2,\end{aligned} \tag{11.27}$$

$$\sigma^2 \int_{\mathcal{S}_{\mathrm{R}}} \boldsymbol{\tau}_0^{\mathrm{H}}(\mathbf{s}') \boldsymbol{\tau}_0(\mathbf{s}') \mathrm{d}\mathbf{s}' \approx \sigma^2 \sum_{\overline{\mathbf{n}}}^{\overline{\mathbf{N}}} \left\|\overline{\mathbf{w}}_{0,\overline{\mathbf{n}}}\right\|^2, \tag{11.28}$$

where

$$\overline{\boldsymbol{\Delta}}_{0,\overline{\mathbf{n}}} = \int_{\mathcal{S}_{\mathrm{R}}} \mathbf{G}_0(\mathbf{s}') \overline{\boldsymbol{\Omega}}_{\overline{\mathbf{n}}}(\mathbf{s}') \mathrm{d}\mathbf{s}'. \tag{11.29}$$

The finite-term approximation described above may lead to some performance loss. However, as the numbers of retained expansion terms, $\mathbf{N}$ and $\overline{\mathbf{N}}$, increase, the system approaches an ideal solution where the performance loss asymptotically approaches zero [19].

11.3.4 Problem formulation in discrete domain for improving sensing performance

First, to define the user SINR in finite-term approximation, the terms $\boldsymbol{\alpha}_k$ and $\mathbf{J_k}$ in (11.8) are rewritten as

$$\boldsymbol{\alpha}_k \approx \sum_{\mathbf{n}}^{\mathbf{N}} \boldsymbol{\Delta}_{k,\mathbf{n}} \mathbf{w}_{k,\mathbf{n}},$$

$$\mathbf{J}_k \approx \sum_{i=0,i\neq k}^{K} \left(\sum_{\mathbf{n}}^{\mathbf{N}} \boldsymbol{\Delta}_{k,\mathbf{n}} \mathbf{w}_{i,\mathbf{n}} \right) \left(\sum_{\mathbf{n}}^{\mathbf{N}} \boldsymbol{\Delta}_{k\mathbf{n}} \mathbf{w}_{i,\mathbf{n}} \right)^{\mathrm{H}} + \sigma^2 \mathbf{I}_3. \tag{11.30}$$

Then, by substituting the above finite-term approximations into (11.8), the SINR can be reexpressed in terms of the finite sums over the retained expansion terms **N**. Therefore, the original sensing SNR optimization problem in (11.13) can be reformulated in finite-term approximation as follows:

$$\max_{\{\overline{\mathbf{w}}_{0,\bar{\mathbf{n}}}\},\{\mathbf{w}_{i,\mathbf{n}}\}} \frac{\sum_{i=0}^{K} \left| \sum_{\bar{\mathbf{n}}}^{\overline{\mathbf{N}}} \overline{\mathbf{w}}_{0,\bar{\mathbf{n}}}^{\mathrm{H}} \overline{\boldsymbol{\Delta}}_{0,\bar{\mathbf{n}}}^{\mathrm{H}} \boldsymbol{\Gamma} \sum_{\mathbf{n}}^{\mathbf{N}} \boldsymbol{\Delta}_{0,\mathbf{n}} \mathbf{w}_{i,\mathbf{n}} \right|^2}{\sigma^2 \sum_{\bar{\mathbf{n}}}^{\overline{\mathbf{N}}} \left\| \overline{\mathbf{w}}_{0,\bar{\mathbf{n}}} \right\|^2} \tag{11.31a}$$

$$\text{such that} \quad \left(\sum_{\mathbf{n}}^{\mathbf{N}} \boldsymbol{\Delta}_{k,\mathbf{n}} \mathbf{w}_{k,\mathbf{n}} \right)^{\mathrm{H}} \left(\sum_{i=0,i\neq k}^{K} \left(\sum_{\mathbf{n}}^{\mathbf{N}} \boldsymbol{\Delta}_{k,\mathbf{n}} \mathbf{w}_{i,\mathbf{n}} \right) \left(\sum_{\tilde{\mathbf{n}}}^{\mathbf{N}} \boldsymbol{\Delta}_{k,\tilde{\mathbf{n}}} \mathbf{w}_{i,\tilde{\mathbf{n}}} \right)^{\mathrm{H}} + \sigma^2 \mathbf{I}_3 \right)^{-1}$$

$$\times \left(\sum_{\mathbf{n}}^{\mathbf{N}} \boldsymbol{\Delta}_{k,\mathbf{n}} \mathbf{w}_{k,\mathbf{n}} \right) \geq \xi_k, \quad k = 1, \ldots, K, \tag{11.31b}$$

$$\sum_{k=0}^{K} \sum_{\mathbf{n}}^{\mathbf{N}} \left\| \mathbf{w}_{k,\mathbf{n}} \right\|^2 \leq P_{\mathrm{T}}. \tag{11.31c}$$

11.4 Alternating optimization for sensing SNR maximization

Due to coupled optimization variables, for a tractable algorithm, we will employ alternating optimization where $\{\overline{\mathbf{w}}_{0,\bar{\mathbf{n}}}\}$ and $\{\mathbf{w}_{i,\mathbf{n}}\}$ are optimized alternately.

11.4.1 Optimization of $\{\overline{\mathbf{w}}_{0,\bar{\mathbf{n}}}\}$ for given $\{\mathbf{w}_{i,\mathbf{n}}\}$

When $\{\mathbf{w}_{i,\mathbf{n}}\}$ are fixed, the optimization problem in (11.31) becomes

$$\max_{\{\overline{\mathbf{w}}_{0,\bar{\mathbf{n}}}\}} \frac{\sum_{i=0}^{K} \left| \sum_{\bar{\mathbf{n}}}^{\overline{\mathbf{N}}} \overline{\mathbf{w}}_{0,\bar{\mathbf{n}}}^{\mathrm{H}} \overline{\boldsymbol{\Delta}}_{0,\bar{\mathbf{n}}}^{\mathrm{H}} \boldsymbol{\Gamma} \sum_{\mathbf{n}}^{\mathbf{N}} \boldsymbol{\Delta}_{0,\mathbf{n}} \mathbf{w}_{i,\mathbf{n}} \right|^2}{\sigma^2 \sum_{\bar{\mathbf{n}}}^{\overline{\mathbf{N}}} \left\| \overline{\mathbf{w}}_{0,\bar{\mathbf{n}}} \right\|^2}. \tag{11.32}$$

It can be expressed in a more tractable way by defining

$$\overline{\mathbf{w}} = \left[\overline{\mathbf{w}}_{0,(-\overline{N}_x,-\overline{N}_y,-\overline{N}_z)} \ \cdots \ \overline{\mathbf{w}}_{0,(\overline{N}_x,\overline{N}_y,\overline{N}_z)}\right]^{\mathrm{T}}, \tag{11.33}$$

$$\mathbf{a}_i = \left[\left(\boldsymbol{\Gamma}\sum_{\mathbf{n}}^{\mathbf{N}}\boldsymbol{\Delta}_{0,\mathbf{n}}\mathbf{w}_{i,\mathbf{n}}\right)^{\mathrm{H}}\overline{\boldsymbol{\Delta}}_{0,(-\overline{N}_x,-\overline{N}_y,-\overline{N}_z)} \ \cdots \ \left(\boldsymbol{\Gamma}\sum_{\mathbf{n}}^{\mathbf{N}}\boldsymbol{\Delta}_{0,\mathbf{n}}\mathbf{w}_{i,\mathbf{n}}\right)^{\mathrm{H}}\overline{\boldsymbol{\Delta}}_{0,(\overline{N}_x,\overline{N}_y,\overline{N}_z)}\right]^{\mathrm{T}}. \tag{11.34}$$

The objective function is then given as

$$\frac{\overline{\mathbf{w}}^{\mathrm{H}}\sum_{i=0}^{K}\mathbf{a}_i\mathbf{a}_i^{\mathrm{H}}\overline{\mathbf{w}}}{\sigma^2\left\|\overline{\mathbf{w}}\right\|^2}, \tag{11.35}$$

which is maximized by the eigenvector of the matrix $\sum_{i=0}^{K}\mathbf{a}_i\mathbf{a}_i^{\mathrm{H}}$ corresponding to the largest eigenvalue. Its norm can be set to any value, say to one.

11.4.2 Optimization of $\{\mathbf{w}_{i,\mathbf{n}}\}$ for given $\{\overline{\mathbf{w}}_{0,\overline{\mathbf{n}}}\}$

To obtain a more tractable form of the optimization problem, we define

$$\mathbf{w}_i = \left[\mathbf{w}_{i,(-N_x,-N_y,-N_z)} \ \cdots \ \mathbf{w}_{i,(N_x,N_y,N_z)}\right]^{\mathrm{T}}, \quad i = 0,\ldots,K, \tag{11.36}$$

$$\mathbf{b} = \left[\sum_{\overline{\mathbf{n}}}^{\overline{\mathbf{N}}}\left(\overline{\boldsymbol{\Delta}}_{0,\overline{\mathbf{n}}}\overline{\mathbf{w}}_{0,\overline{\mathbf{n}}}\right)^{\mathrm{H}}\boldsymbol{\Gamma}\boldsymbol{\Delta}_{0,(-N_x,-N_y,-N_z)} \ \cdots \ \sum_{\overline{\mathbf{n}}}^{\overline{\mathbf{N}}}\left(\overline{\boldsymbol{\Delta}}_{0,\overline{\mathbf{n}}}\overline{\mathbf{w}}_{0,\overline{\mathbf{n}}}\right)^{\mathrm{H}}\boldsymbol{\Gamma}\boldsymbol{\Delta}_{0,(N_x,N_y,N_z)}\right]^{\mathrm{T}}, \tag{11.37}$$

$$\mathbf{C}_i = \left[\boldsymbol{\Delta}_{i,(-N_x,-N_y,-N_z)} \ \cdots \ \boldsymbol{\Delta}_{i,(N_x,N_y,N_z)}\right]. \tag{11.38}$$

Using these for a fixed $\{\overline{\mathbf{w}}_{0,\overline{\mathbf{n}}}\}$, the optimization problem in (11.31) can be expressed as

$$\max_{\{\mathbf{w}_i\}} \sum_{i=0}^{K}\mathbf{w}_i^{\mathrm{H}}\mathbf{b}\mathbf{b}^{\mathrm{H}}\mathbf{w}_i \tag{11.39a}$$

$$\text{such that} \quad \mathbf{w}_k^{\mathrm{H}}\mathbf{C}_k^{\mathrm{H}}\left(\sum_{i=0,i\neq k}^{K}\mathbf{C}_i\mathbf{w}_i\mathbf{w}_i^{\mathrm{H}}\mathbf{C}_i^{\mathrm{H}} + \sigma^2\mathbf{I}_3\right)^{-1}\mathbf{C}_k\mathbf{w}_k \geq \xi_k, \quad k = 1,\ldots,K, \tag{11.39b}$$

$$\sum_{k=0}^{K}\left\|\mathbf{w}_k\right\|^2 \leq P_{\mathrm{T}}, \tag{11.39c}$$

where we have omitted the constant denominator term in the objective function. This optimization problem is not convex, and the closed-form solution is hard to obtain. To solve this problem effectively, we will employ CCP where the objective function and the left-hand side of the SINR constraints are replaced by their concave lower bounds. Then, at each iteration, a convex optimization problem is solved until convergence. The concave lower bound of the objective function is given as

$$\sum_{i=0}^{K} \mathbf{w}_i^{\mathrm{H}}\mathbf{b}\mathbf{b}^{\mathrm{H}}\mathbf{w}_i \geq \sum_{i=0}^{K}\left[2\Re\left(\mathbf{w}_i^{\mathrm{H}}\mathbf{b}\mathbf{b}^{\mathrm{H}}\mathbf{w}_i^{(r)}\right) - \left(\mathbf{w}_i^{(r)}\right)^{\mathrm{H}}\mathbf{b}\mathbf{b}^{\mathrm{H}}\mathbf{w}_i^{(r)}\right], \tag{11.40}$$

where $\mathbf{w}_i^{(r)}$ is the fixed value at iteration r, set from the value of $\mathbf{w}_i$ at the previous iteration. To obtain a concave lower bound to $\mathbf{w}_k^{\mathrm{H}}\mathbf{C}_k^{\mathrm{H}}\left(\sum_{i=0,i\neq k}^{K}\mathbf{C}_i\mathbf{w}_i\mathbf{w}_i^{\mathrm{H}}\mathbf{C}_i^{\mathrm{H}}+\sigma^2\mathbf{I}_3\right)^{-1}\mathbf{C}_k\mathbf{w}_k$, we resort to [30, Eq. (63b)]. To this end, we define

$$\boldsymbol{\alpha}_k^{(r)} = \mathbf{C}_k\mathbf{w}_k^{(r)}, \tag{11.41}$$

$$\mathbf{J}_k^{(r)} = \sum_{i=0,i\neq k}^{K}\mathbf{C}_i\mathbf{w}_i^{(r)}\left(\mathbf{w}_i^{(r)}\right)^{\mathrm{H}}\mathbf{C}_i^{\mathrm{H}}+\sigma^2\mathbf{I}_3. \tag{11.42}$$

Then we have

$$\begin{aligned}
&\mathbf{w}_k^{\mathrm{H}}\mathbf{C}_k^{\mathrm{H}}\left(\sum_{i=0,i\neq k}^{K}\mathbf{C}_i\mathbf{w}_i\mathbf{w}_i^{\mathrm{H}}\mathbf{C}_i^{\mathrm{H}}+\sigma^2\mathbf{I}_3\right)^{-1}\mathbf{C}_k\mathbf{w}_k\\
&\geq 2\Re\left(\left(\boldsymbol{\alpha}_k^{(r)}\right)^{\mathrm{H}}\left(\mathbf{J}_k^{(r)}\right)^{-1}\mathbf{C}_k\mathbf{w}_k\right)\\
&\quad-\left(\boldsymbol{\alpha}_k^{(r)}\right)^{\mathrm{H}}\left(\mathbf{J}_k^{(r)}\right)^{-1}\left(\sum_{i=0,i\neq k}^{K}\mathbf{C}_i\mathbf{w}_i\mathbf{w}_i^{\mathrm{H}}\mathbf{C}_i^{\mathrm{H}}+\sigma^2\mathbf{I}_3\right)\left(\mathbf{J}_k^{(r)}\right)^{-1}\boldsymbol{\alpha}_k^{(r)}\\
&= 2\Re\left(\left(\boldsymbol{\alpha}_k^{(r)}\right)^{\mathrm{H}}\left(\mathbf{J}_k^{(r)}\right)^{-1}\mathbf{C}_k\mathbf{w}_k\right)-\sum_{i=0,i\neq k}^{K}\mathbf{w}_i^{\mathrm{H}}\mathbf{C}_i^{\mathrm{H}}\left(\mathbf{J}_k^{(r)}\right)^{-1}\boldsymbol{\alpha}_k^{(r)}\left(\boldsymbol{\alpha}_k^{(r)}\right)^{\mathrm{H}}\left(\mathbf{J}_k^{(r)}\right)^{-1}\mathbf{C}_i\mathbf{w}_i\\
&\quad-\sigma^2\left(\boldsymbol{\alpha}_k^{(r)}\right)^{\mathrm{H}}\left(\mathbf{J}_k^{(r)}\right)^{-2}\boldsymbol{\alpha}_k^{(r)}.
\end{aligned} \tag{11.43}$$

At the rth iteration of the CCP, we solve the following convex problem:

$$\max_{\{\mathbf{w}_i\}} \sum_{i=0}^{K}\Re\left(\mathbf{w}_i^{\mathrm{H}}\mathbf{b}\mathbf{b}^{\mathrm{H}}\mathbf{w}_i^{(r)}\right) \tag{11.44a}$$

$$\text{such that}\quad 2\Re\left(\left(\boldsymbol{\alpha}_k^{(r)}\right)^{\mathrm{H}}\left(\mathbf{J}_k^{(r)}\right)^{-1}\mathbf{C}_k\mathbf{w}_k\right)-\sum_{i=0,i\neq k}^{K}\mathbf{w}_i^{\mathrm{H}}\mathbf{C}_i^{\mathrm{H}}\left(\mathbf{J}_k^{(r)}\right)^{-1}\boldsymbol{\alpha}_k^{(r)}\left(\boldsymbol{\alpha}_k^{(r)}\right)^{\mathrm{H}}\left(\mathbf{J}_k^{(r)}\right)^{-1}\mathbf{C}_i\mathbf{w}_i$$

$$-\sigma^2\left(\boldsymbol{\alpha}_k^{(r)}\right)^{\mathrm{H}}\left(\mathbf{J}_k^{(r)}\right)^{-2}\boldsymbol{\alpha}_k^{(r)}\geq\xi_k,\quad k=1,\ldots,K, \tag{11.44b}$$

$$\sum_{k=0}^{K}\|\mathbf{w}_k\|^2\leq P_{\mathrm{T}}, \tag{11.44c}$$

where we have omitted the constant terms in the objective function. The overall alternating procedure is summarized in Algorithm 1.

Algorithm 1 Alternating Optimization Algorithm.

Initialization: Randomly select $\mathbf{w}_i$ for $i=0,\ldots,K$. Set convergence thresholds ϵ_1, ϵ_2.
while Sensing SNR improvement larger than ϵ_1 **do**
  Compute $\mathbf{a}_i$, for $i=0,\ldots,K$ according to (11.34).
  Set $\overline{\mathbf{w}}$ to the unit-norm eigenvector of the matrix $\sum_{i=0}^{K}\mathbf{a}_i\mathbf{a}_i^{\mathrm{H}}$ corresponding to the largest eigenvalue.
  Set inner iteration counter to $r=0$. Compute $\mathbf{b}$ according to (11.37).
  Randomly select $\mathbf{w}_i^{(0)}$, for $i=0,\ldots,K$.
  while Sensing SNR improvement larger than ϵ_2 **do**
    Compute $\boldsymbol{\alpha}_k^{(r)}$ and $\boldsymbol{J}_k^{(r)}$ according to (11.41)–(11.42).
    Solve the convex optimization problem in (11.44). Set $\mathbf{w}_i^{(r+1)}$ to be the solution.
    $r\leftarrow r+1$
  end while
end while
Output: $\overline{\mathbf{w}}$, $\mathbf{w}_i$, for $i=0,\ldots,K$.

Here, for I_s being the sampling number of the continuous HMIMO surface ($\int_{\mathcal{S}_{\mathrm{T}}}\mathrm{d}\mathbf{s}$) serving K communication UEs and a single sensing target, the computational complexity of the proposed CCP-based algorithm for each iteration can be expressed as $\mathcal{O}((I_s(K+1)+K+1)^{3.5})$ [19,31].

11.5 Simulation and performance evaluation

In this section, we evaluate the performance of the proposed algorithm within the context of the considered holographic ISAC system and quantify its performance improvements over benchmark traditional discrete MIMO systems. The traditional discrete MIMO system is implemented following the approach in [19], where the continuous aperture is divided into

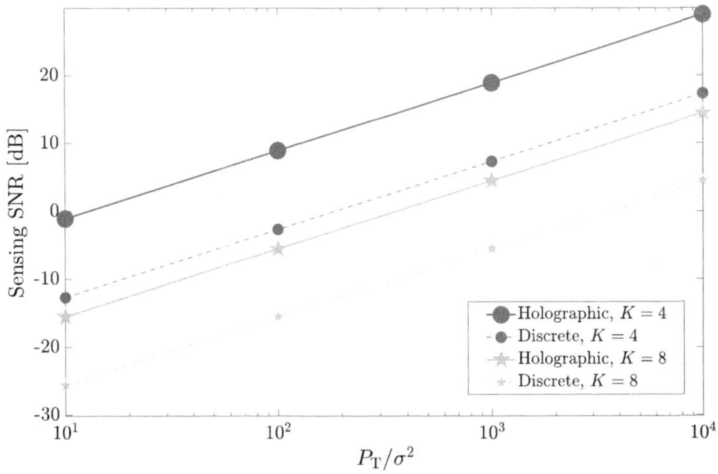

Figure 11.3:
Sensing SNR versus P_{T}/σ^2 for holographic and discrete MIMO.

multiple patches with a spacing of half a wavelength. Each patch's pattern is modeled as a rectangular function, with its amplitude and phase optimized similarly to discrete antennas using the same algorithmic steps.

Unless otherwise stated, the system parameters are as follows: the transmit power to noise power ratio is $P_{\mathrm{T}}/\sigma^2 = 100$, and the transmitter and receiver surfaces each have an area of $A_{\mathrm{T}} = A_{\mathrm{R}} = 0.5 \times 0.5\ \mathrm{m}^2$. The carrier frequency is $f_c = 2.4$ GHz, and the free-space wave impedance is $Z_0 = 376.73\ \Omega$. The SINR requirement for each UE is set to $\xi_k = 10$. The parameters are set as $N_x = N_y = \overline{N}_x = \overline{N}_y = 7$ and $N_z = \overline{N}_z = 0$. The polarimetric RCS matrix is defined as $\mathbf{\Gamma} = \mathbf{I}_3$. In the figures, each data point represents the average of 10 random setups. In each random setup, UEs and the target are placed randomly within a 10 m neighborhood of the holographic transceiver's center along the x-axis and at a distance of [2, 30] m along the z-axis.

In Fig. 11.3, we vary P_{T}/σ^2 and plot the sensing SNR for both $K = 4$ and $K = 8$ UEs, comparing the performance of holographic MIMO with traditional discrete MIMO. Notably, the holographic MIMO system demonstrates a significant sensing SNR advantage, with a power gain of approximately 10 dB over benchmark discrete MIMO system, consistent with the observations in [32]. This highlights the superior efficiency of holographic MIMO in leveraging continuous apertures. Furthermore, as the number of UEs increases from $K = 4$ to $K = 8$, the sensing SNR deteriorates significantly due to the increased interference and less sensing resources. Finally, the sensing SNR exhibits an almost linear relationship with P_{T}/σ^2 when viewed on a logarithmic scale.

In Fig. 11.4, we vary the area of the transmitter while keeping the receiver area fixed at $A_R = 0.5^2$ m^2. The results show a consistent pattern across all cases: increasing the transmitter area significantly enhances the sensing SNR. Once again, the advantages of HMIMO over traditional discrete MIMO are evident. With holographic beamforming in HMIMO, it becomes possible to achieve the same sensing SNR with a much smaller physical transmitter area, underscoring its efficiency in utilizing continuous apertures.

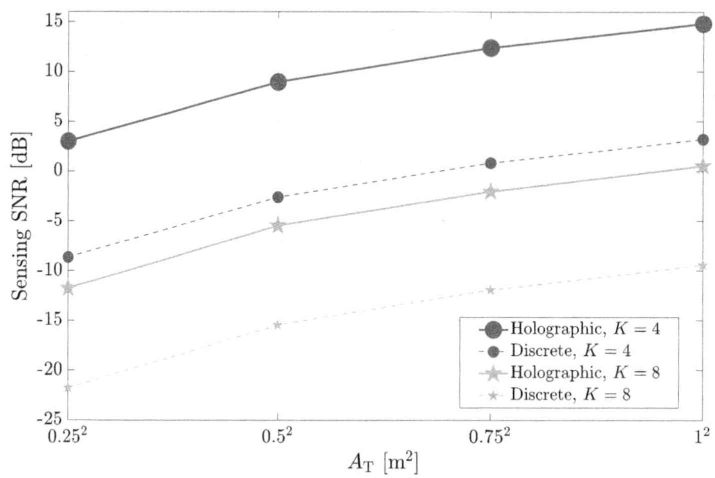

Figure 11.4:
Sensing SNR versus transmitter area A_T for holographic and discrete MIMO.

In Fig. 11.5, we fix the transmitter area at $A_T = 0.5^2$ m^2 and vary the receiver area. The resulting trends are remarkably similar to those in the previous figure, indicating that the effect of increasing the transmitter or receiver area on sensing SNR exhibits comparable characteristics. This highlights the symmetric impact of aperture size on sensing performance.

11.6 Challenges and future research directions

While HMIMO-enabled holographic ISAC systems present exciting opportunities for integrating sensing and communication, their implementation faces significant challenges that must be taken into consideration and, unless addressed, may act as a bottleneck. Combating these challenges and exploring these research directions will be key to realizing the full potential of this technology.

One critical challenge lies in the scalability and complexity of hardware design of HMIMO systems, where the continuous aperture design of introduces strict requirements for manufacturing precision. Moreover, developing scalable and cost-effective fabrication

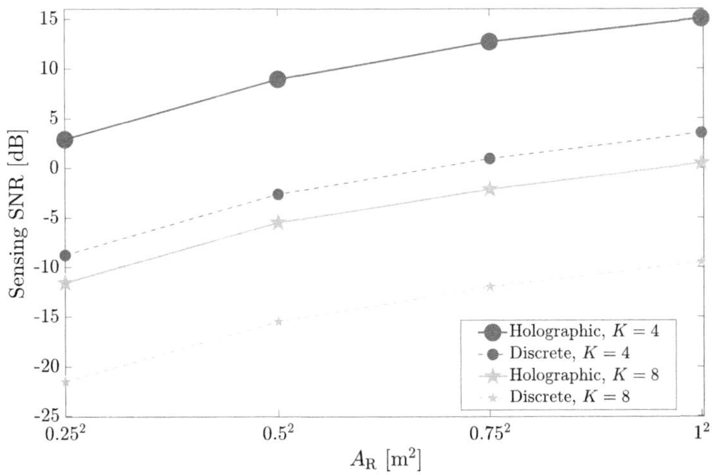

Figure 11.5:
Sensing SNR versus receiver area A_R for holographic and discrete MIMO.

techniques for HMIMO, capable of maintaining high spatial resolution and EM control, remains an open area for research. Furthermore, the deployment of large-scale HMIMO systems is constrained by the computational complexity associated with controlling continuous aperture surfaces. This will require sophisticated and real-time optimization algorithms, possibly leveraging machine learning (ML) or artificial intelligence (AI)-based approaches which will be necessary to dynamically adapt to environmental changes.

As in most ISAC systems, the unavoidable trade-off between sensing and communication is another critical issue for HMIMO systems, particularly in dynamic, resource-constrained environments. Efficient resource allocation strategies that balance the requirements of both functionalities are important. Existing solutions largely depend on traditional optimization methods, but in CAP-MIMO-enabled HMIMO systems, continuous aperture introduces new challenges in real-time computation. Research into adaptive resource allocation frameworks, such as reinforcement learning-based strategies or game-theoretic models, could optimize the dynamic allocation of spatial and spectral resources. These methods would enable HMIMO systems to prioritize radar or communication tasks as needed while maintaining high performance for both.

Interference management is also a barrier to the seamless operation of HMIMO systems. The coexistence of multiple UEs and sensing targets within a shared spectral and spatial environment increases the likelihood of interference. This issue is further compounded in high-density environments, where multiple HMIMO systems or other wireless technologies may compete for the same resources. Advanced beamforming techniques, such as dynamic polarization control and frequency-selective beamforming, could help mitigate UE-to-UE and UE-to-target interference. Additionally, very important research into joint waveform designs must be conducted so that interference between sensing and communication streams must be minimized which can strengthen system robustness.

Additionally, the channel modeling required for HMIMO systems presents unique challenges. Unlike discrete antenna systems, the continuous nature of CAP-MIMO-enabled HMIMO introduces complexities in accurately modeling EM wave interactions over the HMIMO and with the environment. Existing channel models often rely on simplified assumptions that may not capture real-world conditions. Future research should focus on developing robust and realistic channel models that account for multipath propagation, non-line-of-sight conditions, polarization effects and much more. These models would enable better prediction and optimization of HMIMO system performance in diverse scenarios.

Energy efficiency is another critical consideration for HMIMO systems, especially as these systems scale in size and complexity. The dense arrangement of metamaterial elements and the continuous nature of the HMIMO design introduce significant power requirements. While the elimination of traditional phase shifters reduces energy consumption, further solutions are needed to ensure sustainable operation. In addition, low-power metamaterial designs and the integration of energy harvesting or self-sustaining mechanisms could provide viable solutions. Moreover, optimization algorithms should incorporate energy efficiency metrics to balance power consumption with performance requirements dynamically.

Overall, the challenges associated with HMIMO systems include hardware scalability, dynamic resource allocation, interference management, channel modeling, energy efficiency, and more. Addressing these challenges requires a multidisciplinary approach, combining advances in material science, computational techniques, and theoretical modeling. By tackling these obstacles, HMIMO can become a cornerstone of next-generation holographic ISAC systems, unlocking unprecedented opportunities of 6G ISAC vision.

11.7 Conclusion

As the demand for more efficient wireless communication and advanced sensing capabilities continues to rise, the need for innovative solutions to optimize the use of limited spectral resources becomes increasingly urgent. Holographic ISAC emerges as a powerful technology designed to address these challenges by combining both communication and radar sensing tasks into a single, unified system. This integration significantly improves spectral efficiency, reduces hardware complexity, and allows for dynamic allocation of resources between communication and sensing functions.

The cornerstone of holographic ISAC is envisioned as utilizing HMIMO designs with continuous apertures, which offer unparalleled control over EM waves. By using densely packed metamaterial elements, HMIMO schemes enable precise beamforming with lower power consumption and greater spatial resolution compared to traditional phased-array systems. This makes it a highly attractive solution for next-generation networks, particularly as we move toward 6G, where both high-speed communication and accurate sensing will be crucial in applications such as autonomous vehicles, smart cities, and industrial IoT systems.

Throughout this chapter, we have explored the fundamental principles of holographic ISAC, discussed the role of HMIMO designs in enabling advanced beamforming, and examined the system's channel model and performance metrics. We have formulated the tripolarized ISAC system model, capturing the interactions between sensing and communication functions. Moreover, we have developed a CCP-based alternating optimization algorithm to enhance sensing SNR while meeting the required communication SINR levels. Furthermore, the sensing performance of HMIMO-based holographic ISAC system has been investigated and its superior performance over benchmark traditional discrete MIMO arrays has been demonstrated.

In summary, holographic ISAC represents a transformative approach to 6G ISAC, offering significant improvements in spectral efficiency, energy consumption, and system performance. As the world becomes increasingly reliant on high-performance networks, this technology will play a pivotal role in ensuring that both communication and sensing tasks can coexist seamlessly, enabling the next generation of wireless innovations.

Acknowledgments

The work by Ö. T. Demir was supported by 2232-B International Fellowship for Early Stage Researchers Programme funded by the Scientific and Technological Research Council of Türkiye.

The work by Turkcell team has been supported by The Scientific and Technological Research Council of Turkiye (TUBITAK) 1515 Frontier R&D Laboratories Support Program for Turkcell 6GEN. LAB under Project 5229902.

References

[1] R. Liu, M. Li, H. Luo, Q. Liu, A.L. Swindlehurst, Integrated sensing and communication with reconfigurable intelligent surfaces: opportunities, applications, and future directions, IEEE Wireless Communications 30 (1) (2023) 50–57.

[2] F. Liu, Y. Cui, C. Masouros, J. Xu, T.X. Han, Y.C. Eldar, S. Buzzi, Integrated sensing and communications: toward dual-functional wireless networks for 6G and beyond, IEEE Journal on Selected Areas in Communications 40 (6) (2022) 1728–1767.

[3] A. Kaushik, R. Singh, S. Dayarathna, R. Senanayake, M. Di Renzo, M. Dajer, H. Ji, Y. Kim, V. Sciancalepore, A. Zappone, et al., Toward integrated sensing and communications for 6G: key enabling

technologies, standardization, and challenges, IEEE Communications Standards Magazine 8 (2) (2024) 52–59.
[4] A. Kaushik, Integrated Sensing and Communications for Future Wireless Networks: Principles, Advances and Key Enabling Technologies, 1st edition, Elsevier, Jan. 2025.
[5] M.V. Katwe, A. Kaushik, L. Mohjazi, M. Abualhayja'a, D. Dardari, K. Singh, M.A. Imran, M.M. Butt, O.A. Dobre, An overview of intelligent meta-surfaces for 6G and beyond: opportunities, trends, and challenges, IEEE Communications Standards Magazine 8 (4) (2024) 62–69.
[6] S. Zhang, B. Di, A. Kaushik, Y.C. Eldar, Holographic-pattern based multi-user beam training in RHS-aided hybrid near-field and far-field communications, IEEE Transactions on Wireless Communications (2025).
[7] Y. Liu, M.D. Renzo, X. Zhang, Holographic intelligence surface-assisted ISAC, arXiv preprint, arXiv: 2402.01234, 2024.
[8] H. Griffiths, L. Cohen, S. Watts, E. Mokole, C. Baker, M. Wicks, S. Blunt, Radar spectrum engineering and management: technical and regulatory issues, Proceedings of the IEEE 103 (1) (Jan. 2015) 85–102.
[9] Characteristics of radiolocation radars, and characteristics and protection criteria for sharing studies for aeronautical radionavigation and meteorological radars in the radiodetermination service operating in the frequency band 2700–2900 MHz, Tech. Rep., ITU Recommendation M.1464-1, July 2003.
[10] S. Heuel, A. Roessler, Coexistence of s-band radar and 4G mobile networks, in: 2014 15th International Radar Symposium (IRS), 2014.
[11] L. Zheng, M. Lops, Y.C. Eldar, X. Wang, Radar and communication coexistence: an overview: a review of recent methods, IEEE Signal Processing Magazine 36 (5) (Sept. 2019) 85–99.
[12] C. You, Y. Cai, Y. Liu, M.D. Renzo, T.M. Duman, A. Yener, A.L. Swindlehurst, Next generation advanced transceiver technologies for 6G and beyond, arXiv:2403.16458v3, July 2024.
[13] M. Di Renzo, A. Zappone, M. Debbah, M.-S. Alouini, C. Yuen, J. De Rosny, S. Tretyakov, Smart radio environments empowered by reconfigurable intelligent surfaces: how it works, state of research, and the road ahead, IEEE Journal on Selected Areas in Communications 38 (11) (2020) 2450–2525.
[14] E. Basar, M. Di Renzo, J. De Rosny, M. Debbah, M.-S. Alouini, R. Zhang, Wireless communications through reconfigurable intelligent surfaces, IEEE Access 7 (2019) 116753–116773.
[15] Z. Zhang, L. Dai, X. Chen, C. Liu, F. Yang, R. Schober, H.V. Poor, Active RIS vs. passive RIS: which will prevail in 6G?, IEEE Transactions on Communications 71 (3) (2022) 1707–1725.
[16] X. Mu, Y. Liu, L. Guo, J. Lin, R. Schober, Simultaneously transmitting and reflecting (STAR) RIS aided wireless communications, IEEE Transactions on Wireless Communications 21 (5) (2021) 3083–3098.
[17] R. Deng, B. Di, H. Zhang, D. Niyato, Z. Han, H.V. Poor, Reconfigurable holographic surfaces for future wireless communications, IEEE Wireless Communications 28 (6) (Dec. 2021) 126–131.
[18] N. Shlezinger, G.C. Alexandropoulos, M.F. Imani, Y.C. Eldar, D.R. Smith, Dynamic metasurface antennas for 6G extreme massive MIMO communications, IEEE Wireless Communications 28 (2) (Apr. 2021) 106–113.
[19] Z. Zhang, L. Dai, Pattern-division multiplexing for multi-user continuous-aperture MIMO, IEEE Journal on Selected Areas in Communications 41 (8) (2023) 2350–2366.
[20] X. Zhang, Y. Liu, H. Wang, Holographic integrated sensing and communication, IEEE Journal on Selected Areas in Communications 40 (8) (2022) 1574–1589.
[21] D. González-Ovejero, G. Minatti, G. Chattopadhyay, S. Maci, Multibeam by metasurface antennas, IEEE Transactions on Antennas and Propagation 65 (6) (Feb. 2017) 2923–2930.
[22] C.L. Holloway, E.F. Kuester, J.A. Gordon, J. O'Hara, J. Booth, D.R. Smith, An overview of the theory and applications of metasurfaces: the two-dimensional equivalents of metamaterials, IEEE Antennas & Propagation Magazine 54 (2) (Apr. 2012) 10–35.
[23] A. Magbool, M.D. Renzo, C. Yuen, A survey on ISAC with intelligent metasurfaces, IEEE Communications Surveys and Tutorials (2024), In press.
[24] A. Pizzo, T.L. Marzetta, L. Sanguinetti, Spatially-stationary model for holographic MIMO small-scale fading, IEEE Journal on Selected Areas in Communications 38 (9) (2020) 1964–1979.

[25] D. Dardari, Communicating with large intelligent surfaces: fundamental limits and models, IEEE Journal on Selected Areas in Communications 38 (11) (2020) 2526–2537.
[26] F.K. Gruber, E.A. Marengo, New aspects of electromagnetic information theory for wireless and antenna systems, IEEE Transactions on Antennas and Propagation 56 (11) (2008) 3470–3484.
[27] S. Hu, F. Rusek, O. Edfors, Beyond massive MIMO: the potential of data transmission with large intelligent surfaces, IEEE Transactions on Signal Processing 66 (10) (2018) 2746–2758.
[28] L. Sanguinetti, A.A. D'Amico, M. Debbah, Wavenumber-division multiplexing in line-of-sight holographic MIMO communications, IEEE Transactions on Wireless Communications (Sep. 2022).
[29] Z. Wan, J. Zhu, Z. Zhang, L. Dai, C.-B. Chae, Mutual information for electromagnetic information theory based on random fields, IEEE Transactions on Communications 71 (4) (Feb. 2023) 1982–1996.
[30] Y. Liu, Q. Shi, Q. Wu, J. Zhao, M. Li, Joint node activation, beamforming and phase-shifting control in IoT sensor network assisted by reconfigurable intelligent surface, IEEE Transactions on Wireless Communications 21 (11) (2022) 9325–9340, https://doi.org/10.1109/TWC.2022.3175740.
[31] L.-N. Tran, M.F. Hanif, M. Juntti, A conic quadratic programming approach to physical layer multicasting for large-scale antenna arrays, IEEE Signal Processing Letters 21 (1) (2013) 114–117.
[32] L. Zhao, X. Zhang, M.D. Renzo, Performance analysis of holographic MIMO-based ISAC, arXiv preprint, arXiv:2401.05789, 2024.

CHAPTER 12

ESIT-aided integrated sensing and communication (ISAC): enhancing efficiency and performance

Vidya Bhasker Shukla and Vimal Bhatia

Center for Wireless Communications, University of Oulu, Oulu, Finland

12.1 Fundamentals of ISAC

Future wireless systems, including beyond 5G and sixth-generation (6G) networks, are anticipated to deliver diverse high-precision sensing capabilities, such as indoor localization for robot navigation, Wi-Fi sensing for smart homes, and radar sensing for autonomous vehicles. Traditionally, sensing and communication systems are designed independently and operate in separate frequency bands. However, with the widespread adoption of millimeter-wave and massive MIMO technologies, communication signals in the next-generation wireless systems offer high resolution in both time and angular domains, enabling high-accuracy sensing using these signals. Consequently, there is a growing interest in jointly designing sensing and communication systems to share the same frequency band and hardware, thereby enhancing spectrum efficiency and reducing hardware costs. This has spurred research into integrated sensing and communication (ISAC) [1].

ISAC is an emerging paradigm in wireless systems that unifies sensing (e.g., radar-like environmental perception) and communication (e.g., data transmission) within a single framework. By sharing spectrum, hardware, and signal processing resources, ISAC enhances efficiency and performance, making it a cornerstone for applications like 6G, autonomous vehicles [3], and smart cities. Basic block diagram of ISAC system is given in Fig. 12.1.

A general system model for ISAC can be given as [4]

$$\mathbf{Y}_c = \mathbf{H}_c\mathbf{S} + \mathbf{Z}_c, \tag{12.1}$$

$$\mathbf{Y}_s = \mathbf{H}_s\mathbf{S} + \mathbf{Z}_s, \tag{12.2}$$

where $\mathbf{H}_c$ and $\mathbf{H}_s$ are channel matrices for communication and sensing, respectively; $\mathbf{Y}_c$ and $\mathbf{Y}_s$ are received communication and sensing signals, respectively; $\mathbf{S}$ denotes the transmitted

Electromagnetic Signal and Information Theory
https://doi.org/10.1016/B978-0-44-336354-2.00019-8
Copyright © 2027 Elsevier Inc. All rights are reserved, including those for text and data mining, AI training, and similar technologies.

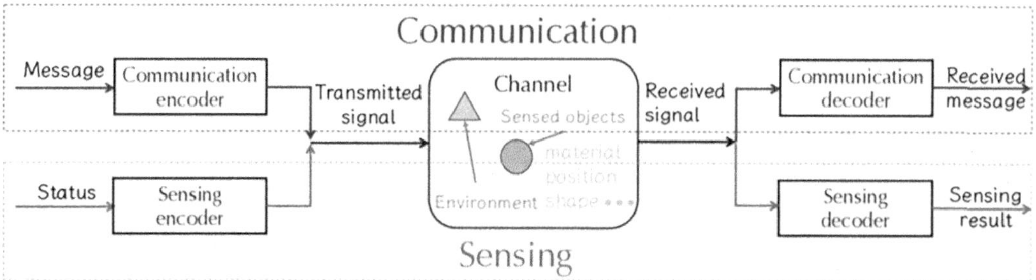

Figure 12.1:
General block diagram of ISAC system [2].

dual-functional waveform emitted from an ISAC transmitter for performing both sensing and communication tasks; $\mathbf{Z}_c$ and $\mathbf{Z}_s$ denote the received communication and sensing noise, modeled as independently and identically distributed (i.i.d.) circularly symmetric complex Gaussian random variables with zero mean, namely $\text{vec}(\mathbf{Z}_c) \sim \mathcal{CN}(\mathbf{0}, \sigma_c^2\mathbf{I})$. Similarly, we also model the sensing noise $\text{vec}(\mathbf{Z}_s) \sim \mathcal{CN}(\mathbf{0}, \sigma_s^2\mathbf{I})$.

Instead of operating sensing and communication functions independently, ISAC uses shared resources (e.g., spectrum, antennas, waveforms) to perform both tasks simultaneously or synergistically. This dual-purpose approach is motivated by:

- Spectrum scarcity – growing demand for wireless services requires efficient use of limited frequency bands.
- Hardware constraints – shared transceivers reduce cost, power, and size.
- Application needs – modern systems (e.g., autonomous vehicles) require both sensing (for navigation) and communication (for coordination).

ISAC systems are built on several foundational principles, including:

Shared Resources.

- Spectrum. Sensing and communication operate in the same frequency band, e.g., L-band (1–2 GHz), S-band (2–4 GHz), C-band (4–8 GHz), and mmWave band (30–300 GHz), reducing bandwidth fragmentation.
- Hardware. Common antennas, transceivers, and processors handle both tasks, minimizing redundancy.
- Waveforms. Dual-function waveforms serve both sensing (e.g., detecting targets) and communication (e.g., carrying data).

Joint signal processing. Signals are processed to extract both sensing parameters (e.g., range, velocity) and communication data (e.g., bits). This requires sophisticated algorithms to balance competing objectives.

Trade-off management. Sensing and communication have different requirements (e.g., high resolution for sensing vs. high data rate for communication). ISAC optimizes these trade-offs through adaptive design.

System Components.

An ISAC system typically includes [5]:

1. Transmitter, which generates dual-function waveforms that encode communication data and enable sensing. This includes orthogonal frequency-division multiplexing (OFDM) or chirp-based signals. Configurable parameters (e.g., power, beam direction) are adjusted based on sensing and communication needs.

2. Receiver, which processes received signals to extract communication data (via demodulation) and sensing information (via parameter estimation, e.g., time-of-flight for range). It may also include multiple antennas for spatial diversity (e.g., MIMO).

3. Signal processor, which performs joint processing, such as channel estimation, equalization, and target detection.

4. Control unit, which coordinates sensing and communication tasks, allocating resources dynamically based on environmental feedback.

5. Sensing module, which uses radar, mmWave, or other sensors to collect environmental data, informing both sensing and communication strategies.

Key Techniques in ISAC.

ISAC systems rely on advanced techniques to achieve dual functionality:

1. Dual-function waveform design. Waveforms like OFDM, frequency-modulated continuous wave (FMCW) [5], or direct-sequence spread spectrum (DSSS) are designed to balance sensing resolution and communication rate. OFDM waveforms, common in 5G, can carry data in subcarriers while enabling range-Doppler estimation for sensing.

ESIT optimizes waveforms using electromagnetic propagation models and information-theoretic metrics (e.g., mutual information).

2. Channel estimation

3. Beamforming. This directs energy toward communication users and sensing targets using multiantenna arrays (e.g., MIMO). Adaptive beamforming, informed by ESIT, improves both signal-to-noise ratio (SNR) and target detection. For example, in autonomous vehicles, beams focus on nearby obstacles for sensing and other vehicles for V2V communication.

4. Optimization. A basic optimization problem can be stated as follows [6]:

$$\max \rho I\,(S; \mathbf{Y}_c \mid \mathbf{H}_c) + (1-\rho) I\,(\mathbf{H}_s; \mathbf{Y}_s \mid \mathbf{S}) \tag{12.3}$$

$$\text{such that other specific practical constraints are satisfied,} \tag{12.4}$$

where the weighting factor $\rho \in [0, 1]$ controls the priorities assigned to sensing and communication functionalities.

5. Interference management. Sensing and communication signals can interfere with each other. ESIT-aided techniques (e.g., interference cancellation, dynamic spectrum allocation) mitigate this using physics aware signal processing.

6. Design challenges. ISAC systems face several challenges, including:

- Trade-offs. Sensing requires high-resolution waveforms (e.g., wide bandwidth, low ambiguity), while communication prioritizes high data rates and low latency. Balancing these is nontrivial.
- Interference. Self-interference between sensing and communication signals degrades performance. Advanced signal processing (e.g., successive interference cancellation) is needed.
- Complexity. Joint processing and optimization increase computational demands. Low-complexity algorithms are critical for real-time applications.
- Channel dynamics. Fast-changing channels (e.g., in vehicular scenarios) require robust estimation techniques.
- Hardware limitations. Shared hardware must support diverse requirements (e.g., high power for sensing, low latency for communication).

ISAC performance is evaluated using dual metrics:

Communication metrics such as:

- Data rate (bits/s/Hz).
- Bit error rate (BER).
- Outage probability.

Sensing Metrics such as:

- Range resolution ($\Delta R = \frac{c}{2B}$, where c is the speed of light, B is bandwidth).
- Velocity resolution.
- Detection probability.

ISAC is poised to impact various domains:

Autonomous vehicles – simultaneous obstacle detection (sensing) and vehicle-to-vehicle (V2V) communication.

6G Networks – mmWave and THz bands enable high-resolution sensing and ultra-high-speed communication.

Smart cities – integrated traffic monitoring and broadband connectivity.

Healthcare – noninvasive monitoring (e.g., vital signs) combined with data transmission.

Defense – joint radar and communication for situational awareness.

ESIT-aided ISAC focuses on advanced signal processing for ISAC and enhances ISAC by providing physics-aware signal processing and information-theoretic optimization.

Conclusion

ISAC systems integrate sensing and communication using shared resources, dual-function waveforms, and joint signal processing to achieve spectral and hardware efficiency. Key techniques include waveform design, channel estimation, beamforming, and optimization, all enhanced by ESIT's physics-aware approach. Despite challenges like trade-offs and interference, ISAC is transformative for 6G, autonomous systems, and beyond. Understanding these fundamentals equips advance research or applications in ISAC, particularly with focus on ESIT and optimization.

12.2 ESIT

Electromagnetic signal and information theory (ESIT) addresses the limitations of purely abstract information theory by integrating Maxwell's equations and other electromagnetic principles directly into the communication model. It reconsiders the relationship between abstract "bits" and their physical representation in the electromagnetic field. ESIT is an interdisciplinary field that integrates electromagnetic theory, signal processing, and information theory. ESIT focuses on studying and developing physically consistent communication systems for transmitting and processing information in networks. In essence, ESIT is physics-informed information theory and signal processing tailored for communication systems. ESIT-informed perspectives is compared with traditional view and associated challenges are discussed in Table 12.1

12.2.1 The function of ESIT in information theory, communication systems, and signal processing

ESIT could significantly contribute to information and communication theories, wireless communications, and signal processing as follows [7]:

- Precisely modeling and comprehending the function of evanescent waves and near fields could enable optimal generation and shaping of electromagnetic waves.
- Precisely modeling and understanding electromagnetic mutual coupling could enable the evaluation and quantification of similarities and differences between continuous and discrete antenna-array models.

Table 12.1: ESIT-informed Perspectives.

Aspect	Traditional View	ESIT-informed View	Associated Challenge
Signal Representation	Information is represented as a sequence of bits or symbols, while the underlying physical characteristics—such as the waveform or electromagnetic field structure—are not explicitly captured.	The signal is represented as an electromagnetic field characterized by its physical parameters.	A major challenge in electromagnetic (EM) modeling lies in the lack of a unified, field-domain representation capable of capturing continuous EM field behavior beyond discrete symbol abstractions. Accurately modeling the intricate spatial, temporal, and polarization characteristics of EM waveforms remains computationally intensive, while conventional orthogonal basis sets often fail to represent localized spatial field variations effectively.
Channel Modeling	Propagation physics are not explicitly modeled; instead, channels are typically treated as black-box systems characterized by stochastic models such as AWGN or Rayleigh fading.	These models are derived from Maxwell's equations or their paraxial approximations, incorporating effects such as reflection, scattering, diffraction, and surface-imposed boundary conditions.	Conventional statistical fading models are inadequate for representing dynamic propagation behaviors and near-field effects, and they often neglect boundary interactions. Additionally, certain surfaces exhibit memory effects, wherein the channel response depends on prior interactions, further complicating accurate modeling.
Information Capacity	The channel capacity is typically expressed using Shannon's formula, $$C = B\log_2(1+\text{SNR}),$$ which represents an average measure of performance derived from statistical channel assumptions.	Encompasses spatial entropy, modal capacity, and degrees of freedom, derived from field correlation, localization, and coherence area.	A universally accepted metric or capacity formulation to characterize their information-theoretic limits is still lacking. Maximizing capacity within the field domain constitutes a complex, high-dimensional, and nonconvex optimization problem, which is computationally prohibitive in most scenarios. Furthermore, the extent to which a reconfigurable surface can reliably encode or convey information remains uncertain.

continued on next page

Table 12.1: (*continued*)

Aspect	Traditional View	ESIT-informed View	Associated Challenge
Optimization Metrics	Emphasizes maximizing data throughput, minimizing bit error rate, or reducing energy consumption per bit through advanced signal processing techniques.	Focuses on optimizing electromagnetic field distribution, enhancing spatial diversity, tuning impedance configurations, and concentrating energy density.	In ESIT-enabled systems, achieving an optimal balance among energy efficiency, latency, capacity, and hardware complexity is considerably more challenging than in conventional architectures. Traditional performance metrics, such as spectral efficiency, fail to adequately represent electromagnetic field-level characteristics, underscoring the necessity for ESIT-specific cost functions. Furthermore, the joint optimization of surface configuration, user scheduling, and waveform design further increases system complexity.
Design Philosophy	The components of a communication system—transmitter, channel, and receiver—are typically decoupled and designed independently.	Components are jointly designed using unified models that enforce physical constraints like causality and energy conservation across layers.	Integrating electromagnetic theory with digital communications, hardware design, and artificial intelligence necessitates profound cross-disciplinary expertise. A key challenge lies in the lack of standardized frameworks or protocols tailored for EM-assisted communication systems. Achieving system-level codesign—spanning antennas, intelligent surfaces, and communication protocols—requires holistic, yet practically implementable, design methodologies.

- Employing electromagnetically consistent models could enable the creation of innovative and effective signal processing techniques, such as channel estimation, in the presence of mutual coupling and fully-connected control circuits.
- Using electromagnetically consistent models could facilitate the development of new optimization algorithms that consider the effects of unwanted scattering and external interference during the design phase.

12.2.2 Why ESIT is emerging?

For decades, communication system design relied on a layered abstraction model, where the physical aspects of EM wave propagation were largely separated from the logical design of information encoding and decoding. However, newer communication technologies have created a need for a more integrated approach:

- **Engineered surfaces.** Programmable metasurfaces and reconfigurable intelligent surfaces (RISs) are designed to dynamically reshape electromagnetic wave propagation, influencing the physical channel itself [8].
- **Beyond 5G and 6G.** As wireless technology advances, traditional communication models are becoming insufficient for understanding the performance of future systems. ESIT is essential for accurately characterizing the fundamental limitations and potentials of next-generation wireless communications [9].
- **Near-field communications.** With engineered surfaces, the near-field interactions of EM waves become more critical. ESIT provides a framework for designing and analyzing communication schemes that explicitly account for these electromagnetic effects [10].

12.3 Technical framework of ESIT-aided ISAC

The technical framework of ESIT-aided ISAC involves leveraging ESIT to enhance the performance of ISAC systems, which integrate sensing and communication functionalities into a unified platform. ESIT provides a theoretical foundation for analyzing and optimizing electromagnetic signal propagation, scattering, and information transfer. It encompasses signal design, channel modeling, and information-theoretic limits, which are critical for balancing sensing and communication performance in ISAC systems. In ESIT-aided ISAC, ESIT principles are applied to optimize waveform design, resource allocation, and signal processing to achieve dual functionality (sensing and communication) with minimal interference and maximal efficiency.

We can construct a technical framework by combining ESIT principles with established ISAC architectures, drawing on general ISAC concepts and related technologies. Below is a detailed explanation of the framework, addressing key components, enabling techniques, and considerations for ESIT role in ISAC.

12.4 Key ESIT techniques in ISAC

ESIT provides a robust framework for enhancing ISAC systems by integrating electromagnetic principles with information-theoretic approaches. This synergy optimizes the dual functionality of sensing and communication in ISAC, particularly for the next-generation 6G networks. Below, it is explained how ESIT enhances ISAC systems, drawing on the paraphrased statements you provided and expanding on their implications:

1. Optimal waveform design through electromagnetic modeling.

Evanescent waves and near fields. ESIT's accurate modeling of evanescent waves and near fields enables precise generation and shaping of electromagnetic waves. In ISAC systems, this enhances sensing by improving spatial resolution (e.g., for target localization) and communication by optimizing signal propagation (e.g., focused beamforming). For example, evanescent waves can be leveraged to enhance near-field sensing for applications like gesture recognition while ensuring reliable data transmission.

Information-theoretic insight. Information theory guides the design of waveforms that maximize mutual information for communication while minimizing sensing errors. ESIT combines these to create dual-purpose waveforms (e.g., OFDM-based) that balance sensing accuracy and channel capacity.

Impact on ISAC. This results in improved radar-like sensing (e.g., higher range resolution) and communication efficiency (e.g., higher data rates), especially in complex environments like urban or non-line-of-sight scenarios.

2. Improved antenna array design via mutual coupling analysis.

   ESIT focus on modeling electromagnetic mutual coupling allows for evaluating differences between continuous and discrete antenna-array models. In ISAC, this is critical for designing shared antenna arrays that support both sensing and communication. Understanding mutual coupling ensures accurate beamforming for sensing (e.g., target tracking) and communication (e.g., multiuser MIMO).

   Information-theoretic optimization. Information theory quantifies the impact of mutual coupling on channel capacity and sensing performance, enabling optimal array configurations that maximize system throughput and minimize localization errors.

   Impact on ISAC. Optimized antenna designs reduce hardware complexity and cost while enhancing ISAC performance, such as in mmWave or sub-THz systems where mutual coupling is significant.

3. Robust Optimization Algorithms for Interference Management.

   ESIT adoption of electromagnetically consistent models supports the design of optimization algorithms that account for undesired scattering and external interference at the design stage. In ISAC, this ensures robust performance in dynamic environments with clutter or cochannel interference.

   Information-theoretic framework. Information theory provides metrics (e.g., signal-to-interference-plus-noise ratio, SINR) to optimize resource allocation (e.g., spectrum, power) for both sensing and communication, mitigating interference effects. These algorithms improve coexistence of sensing and communication, enabling reliable operation in dense 6G networks. For example, they can minimize sensing errors caused by reflections while maintaining high communication throughput.

4. Advanced Signal Processing Schemes.

Mutual coupling and control circuits. ESIT facilitates efficient signal processing schemes, such as channel estimation in the presence of mutual coupling and fully-connected control circuits. In ISAC, accurate channel estimation enhances sensing-assisted communication, where real-time sensing data (e.g., user location) informs beamforming or resource allocation.

Information-theoretic techniques. Information theory optimizes signal processing by maximizing the mutual information between transmitted and received signals, improving both channel estimation accuracy and sensing performance. This leads to lower pilot overhead, higher sensing accuracy (e.g., for Doppler estimation), and improved communication reliability, especially in high-mobility scenarios like autonomous vehicles.

5. Novel encoding and decoding schemes.

   ESIT use of integral operators in communications and signal processing enables the identification of optimal encoding and decoding schemes. In ISAC, these schemes allocate resources (e.g., time, frequency, or spatial domains) to balance sensing and communication objectives.

   Information-theoretic design. Information theory ensures these schemes maximize system capacity while meeting sensing requirements (e.g., Cramér–Rao bounds for localization accuracy). For example, novel coding strategies can prioritize sensing in critical scenarios (e.g., collision avoidance) while maintaining communication links. This enhances ISAC's ability to support advanced 6G applications, such as extended reality (XR) or smart factories, by achieving high data rates and precise environmental mapping simultaneously.

6. Integration with emerging technologies.

   Reconfigurable intelligent surfaces (RIS). ESIT electromagnetic modeling, combined with information-theoretic optimization, enhances RIS-enabled ISAC systems. RIS manipulates the propagation environment to improve signal quality, while ESIT ensures optimal RIS configurations for sensing and communication. RIS-assisted ISAC systems achieve higher sum-rates and lower localization errors by focusing energy on sensed targets, as seen in sensing-assisted communication frameworks. This is particularly effective in mmWave or sub-THz bands.

7. Energy efficiency and sustainability.

   ESIT optimization of electromagnetic resources (e.g., power allocation) and information-theoretic metrics (e.g., energy efficiency) supports energy-efficient ISAC designs, a key requirement for 6G sustainability. By minimizing power consumption while maintaining sensing and communication performance, ESIT enables scalable ISAC deployments in energy-constrained environments, such as IoT networks.

8. Addressing ISAC trade-offs.

ESIT electromagnetic modeling addresses performance trade-offs between sensing and communication, such as degradation in non-line-of-sight conditions. Information theory provides a framework to balance these objectives by optimizing resource sharing. This ensures robust performance in challenging scenarios, enhancing applications like autonomous driving (reliable sensing) and high-speed connectivity (low-latency communication).

Practical implications for 6G ISAC systems. ESIT-enhanced ISAC supports transformative 6G applications, including autonomous vehicles (improved localization and V2X communication), smart cities (real-time monitoring and connectivity), and XR (high-fidelity environmental reconstruction).

Standardization. ESIT aligns with ETSI ISG ISAC efforts to develop channel models and KPIs for 6G. Its electromagnetic and information-theoretic insights fill gaps in existing models (e.g., 3GPP), particularly for scattering and mutual coupling in sensing scenarios.

Scalability. By optimizing hardware, algorithms, and signal processing, ESIT enables scalable ISAC deployments across diverse frequency bands (e.g., mmWave, sub-THz) and environments.

12.4.1 Key ESIT technologies and their roles in ISAC balance

Below are the major ESIT-inspired technologies that jointly optimize field-level energy, spatial structure, and information content to achieve balanced ISAC performance.

Reconfigurable Intelligent Surface (RIS)

An RIS is a passive or semipassive surface with programmable elements that manipulate EM waves (phase, amplitude, polarization) to control propagation in real time, as summarized in Table 12.2.

ESIT Perspective

Under the ESIT framework, an RIS is viewed as an EM field transformer, not just a channel modifier. Its configuration defines a field distribution function $F(r, f)$ that directly impacts both communication capacity and sensing precision.

Thus ESIT-based field modeling ensures both are optimized through the same EM configuration.

Table 12.2: ESIT for RIS.

Function	RIS contribution
Communication	Enhances received SNR and spatial multiplexing by directing EM energy toward users, improving channel quality and data rate.
Sensing	Shapes reflected fields to increase angular and range resolution of sensed targets.
Balance Mechanism	ESIT-optimized RIS phase profiles minimize interference between information-bearing and sensing echoes while maintaining field orthogonality.
Mathematical View	$$\max_{\Phi} \lambda_C\, C(\Phi) + \lambda_S\, S(\Phi),$$ where $C(\Phi)$ is the communication capacity, $S(\Phi)$ is the sensing accuracy, and Φ denotes the RIS reflection matrix.

Holographic MIMO (H-MIMO)

H-MIMO employs continuous-aperture electromagnetic transceivers that synthesize and sample EM fields with subwavelength granularity.

ESIT Perspective

H-MIMO treats the aperture as a holographic field generator where each point contributes continuously to the overall field—enabling field-domain information processing and wavefront reconstruction, as shown in Table 12.3.

Table 12.3: ESIT for HMIMO.

Function	H-MIMO Contribution
Communication	Realizes ultra-high spatial DoFs, near-field focusing, and interference suppression.
Sensing	Achieves fine spatial resolution for imaging and localization by reconstructing the full EM field.
Balance Mechanism	ESIT formulates both channel estimation and target detection as field reconstruction problems; by optimizing field patterns, the same holographic beam serves both data and sensing functions.
Mathematical View	$$I_{\text{field}} = I_{\text{comm}} + I_{\text{sense}} + I_{\text{cross}},$$ where I_{cross} (field-domain correlation term) is maximized in ESIT-H-MIMO to promote information-sharing between sensing and communication.

Fluid Antenna System (FAS)

FAS allows dynamic reconfiguration of antenna positions or shapes in space or time, forming a "fluid" antenna that adapts to the propagation environment.

ESIT Perspective

FAS acts as a spatially adaptive EM aperture, redistributing field intensity to match channel topology or sensing needs—enabling field-domain mobility, as outlined in Table 12.4.

Table 12.4: ESIT for FAS.

Function	Fluid Antenna System (FAS)
Communication	Maximizes instantaneous SNR or channel gain by repositioning the antenna or fluid elements.
Sensing	Samples EM fields across multiple spatial positions, improving spatial diversity and resolution.
Balance Mechanism	ESIT-driven control dynamically adjusts antenna configuration using mutual information gradient.
Mathematical View	$$\nabla_{\text{pos}}\left[I_{\text{comm}} + I_{\text{sense}}\right],$$ ensuring balanced adaptation between data throughput and sensing fidelity.

Thus each ESIT technology (RIS, H-MIMO, FAS) manipulates different aspects of $\mathbf{E}(\mathbf{r}, t)$ such as:

- RIS – spatial phase and amplitude control
- H-MIMO – continuous aperture shaping
- FAS – dynamic aperture adaptation

Through field-domain optimization guided by ESIT, these techniques achieve mutually reinforcing sensing and communication performance instead of the classical trade-off.

12.5 ESIT-aided ISAC trade-offs across physical domains

In traditional ISAC, trade-offs occur at the **signal level**, e.g., allocating power or bandwidth between radar and communication or designing waveforms that compromise between range resolution and data rate.

In contrast, in the ESIT framework, the trade-offs are **field-domain trade-offs** between how EM field resources are structured and how efficiently information and sensing observables can be extracted from them.

Mathematically, ESIT optimizes

$$\max_{\mathbf{E}(\mathbf{r},t)} \alpha I_{\text{comm}}(\mathbf{E}) + (1-\alpha) I_{\text{sense}}(\mathbf{E}),$$

subject to **Maxwell's equations** and physical constraints. Here $\mathbf{E}(\mathbf{r}, t)$ is the EM field distribution as a function of space $\mathbf{r}$ and time t; $\mathbf{r}$ denotes the spatial coordinate vector (e.g., position in 2D or 3D space), t is the time variable representing temporal evolution of the EM field, α is the weighting coefficient ($0 \leq \alpha \leq 1$), balancing communication and sensing objectives, $I_{\text{comm}}(\mathbf{E})$ denotes the information metric for communication (e.g., mutual information or channel capacity), $I_{\text{sense}}(\mathbf{E})$ is the information metric for sensing (e.g., detection accuracy or sensing information rate), and $(1 - \alpha)$ is the complementary weighting factor for the sensing objective.

Thus, different ESIT variants (or properties) alter this optimization landscape and lead to different trade-offs between sensing and communication.

Spatial-domain trade-offs (field distribution/aperture control)

Relevant ESIT technologies: holographic MIMO (H-MIMO), reconfigurable intelligent surface (RIS), fluid antenna systems (FAS).

Effect:

- Communication prefers directive beams for high SNR and interference suppression.
- Sensing prefers diverse and broad beams for better spatial coverage and resolution.

ESIT determines how field aperture energy is distributed:

$$\int_A |E(\mathbf{r})|^2 \, d\mathbf{r} = P_{\text{tot}},$$

where $E(\mathbf{r})$ is the electric field distribution (magnitude and phase) as a function of the spatial coordinate $\mathbf{r}$, representing the electromagnetic field intensity across region A; $|E(\mathbf{r})|^2$ is the instantaneous power density proportional to the local field energy at position $\mathbf{r}$; while A is the spatial domain or aperture region over which the field is defined and integrated (e.g., antenna surface, reconfigurable surface, or sensing aperture); $d\mathbf{r}$ is the differential area or volume element depending on the dimensionality of A, representing an infinitesimal integration element; P_{tot} is the total transmitted or radiated power of the electromagnetic field over region A, typically used as a power constraint in optimization problems. Changes in aperture discretization or field continuity (RIS vs. H-MIMO) shift the balance between field focusing (communication) and field diversity (sensing).

Trade-off: Directive vs. diversified field shaping—higher directivity improves data throughput but reduces sensing coverage and resolution.

Spectral-domain trade-offs (frequency–wavenumber structure)

Relevant ESIT properties: wideband or multicarrier field modeling; spatio-spectral holography.

Effect:

- Communication systems exploit frequency selectivity to increase throughput (via multiplexing).
- Sensing systems exploit bandwidth for finer range resolution.

ESIT variants determine how spectral resources are entangled with spatial field properties via

$$E(\mathbf{r}, f) = \sum_k E_k(\mathbf{r})\,\delta(f - f_k),$$

where $E(\mathbf{r}, f)$ is the EM field distribution as a joint function of spatial position $\mathbf{r}$ and frequency f, representing the frequency-domain field response at location $\mathbf{r}$; $E_k(\mathbf{r})$ is the complex amplitude (magnitude and phase) of the EM field component corresponding to discrete frequency f_k at spatial coordinate $\mathbf{r}$; $\delta(f - f_k)$ denoes the Dirac delta function, selecting the field contribution at frequency f_k, used to express discrete spectral components in the continuous frequency domain; f is the continuous frequency variable representing the spectral domain; f_k is the discrete frequency component (or subcarrier) indexed by k; while k is the index variable representing the kth frequency component in the summation.

Thus, coupling between space and frequency directly impacts both range and channel coherence.

Trade-off:

Spectral efficiency vs. range resolution—larger bandwidth aids sensing but increases field dispersion (beam squinting), which can degrade communication unless compensated by ESIT-aware field control.

Temporal-domain trade-offs (waveform dynamics/pulse structure)

Relevant ESIT variants: time-varying H-MIMO or RIS.

Effect:

- Sensing requires time-domain diversity and correlation for Doppler or velocity estimation.
- Communication requires temporally coherent waveforms for symbol synchronization and decoding.

ESIT defines how temporal field coherence is balanced with temporal diversity via

$$R_E(\tau) = \mathbb{E}[E(t)E^*(t+\tau)],$$

where $R_E(\tau)$ is autocorrelation function of the electromagnetic field (or signal), $E(t)$ is a function of the time lag τ, quantifying the similarity between $E(t)$ and its delayed version $E(t + \tau)$; $\mathbb{E}[\cdot]$ is expectation operator, representing the statistical average over time or ensemble realizations; $E(t)$ is the complex time-domain electric field (or signal) at time t; $E^*(t+\tau)$ is the complex conjugate of the electric field at a delayed time $(t+\tau)$; t denotes the time variable representing the temporal evolution of the field; τ is the time lag (delay variable) used to measure temporal correlation of the field. Shorter correlation (wider spectrum) benefits sensing, while longer correlation benefits communication.

Trade-off:

Temporal coherence vs. Doppler diversity—a tighter time correlation (for stable communication) reduces Doppler estimation precision (for sensing).

Polarization-domain trade-offs

Relevant ESIT property: polarization-multiplexed field modeling.

Effect:

- Communication uses dual-polarization channels for higher capacity.
- Sensing uses polarization diversity to infer target material or orientation.

The ESIT variant defines how polarimetric information is encoded or preserved in the field.

Trade-off:

Polarization multiplexing vs. polarimetric sensing fidelity—encoding more data via polarization may distort polarization signatures critical for sensing.

Material-domain trade-offs (electromagnetic environment control)

Relevant ESIT technologies: metasurfaces, intelligent materials, and RIS-assisted environments.

Effect:

- The EM environment can be tuned to favor either communication (high directivity) or sensing (strong reflection diversity).

- ESIT characterizes how material-induced field transformations modify both I_{comm} and I_{sense}.

Trade-off:

Environment adaptability vs. stability—frequent reconfiguration (beneficial for sensing) may disrupt steady-state communication links.

Information-theoretic trade-offs (mutual vs. Fisher information)

In ESIT, communication and sensing are unified under the field-domain information measures:

$$I_{\text{comm}} = I(x; y), \qquad I_{\text{sense}} = I(\theta),$$

and the total field information is

$$I_{\text{total}} = I_{\text{comm}} + I_{\text{sense}} + I_{\text{cross}},$$

where I_{cross} quantifies shared information between sensing and communication tasks.

Trade-off:

Information sharing vs. information orthogonality—maximizing shared field information (I_{cross}) aids synergy but may increase task interference if objectives are misaligned.

Conclusion

ESIT enhances ISAC systems by combining electromagnetic modeling (e.g., evanescent waves, mutual coupling) with information-theoretic optimization (e.g., mutual information, resource allocation). This synergy improves waveform design, antenna performance, interference management, signal processing, and encoding/decoding, addressing ISAC's performance trade-offs and energy efficiency challenges. By integrating with technologies like RIS and supporting 6G use cases, ESIT paves the way for robust, high-performance ISAC systems.

12.6 Challenges and future directions

The primary challenges in integrated sensing and communication (ISAC) systems can be grouped into four key areas: theoretical foundations, physical layer design, network and cross-layer design, and application-level issues. Theoretically, questions arise about the information-theoretic boundaries of ISAC, the extent to which sensory data can reveal channel information, and methods to measure integration and coordination benefits. At the physical layer, challenges include managing clock synchronization and phase offsets in bistatic and distributed setups, assessing how close current designs are to the Pareto-optimal frontier, and leveraging superresolution techniques to enhance sensing performance. Network and cross-layer design challenges focus on defining the cellular architecture for future sensing networks and developing tailored resource management schemes and protocols for ISAC networks. At the application level, ensuring security and privacy in ISAC applications and achieving multiobject multitask (MOMT) recognition using wireless signals are critical hurdles. With the help of ESIT many of above challenges can be mitigated.

References

[1] S. Lu, F. Liu, Y. Li, K. Zhang, H. Huang, J. Zou, X. Li, Y. Dong, F. Dong, J. Zhu, et al., Integrated sensing and communications: Recent advances and ten open challenges, IEEE Internet of Things Journal 11 (11) (2024) 19094–19120.

[2] F. Shang, et al., Towards the limits: Sensing capability measurement for ISAC through channel encoder, arXiv preprint, arXiv:2405.09497, 2024.

[3] K. Zheng, Q. Zheng, H. Yang, L. Zhao, L. Hou, P. Chatzimisios, Reliable and efficient autonomous driving: The need for heterogeneous vehicular networks, IEEE Communications Magazine 53 (12) (2015) 72–79.

[4] Y. Xiong, F. Liu, Y. Cui, W. Yuan, T.X. Han, G. Caire, On the fundamental tradeoff of integrated sensing and communications under gaussian channels, IEEE Transactions on Information Theory 69 (9) (2023) 5723–5751.

[5] F. Liu, C. Masouros, A.P. Petropulu, H. Griffiths, L. Hanzo, Joint radar and communication design: Applications, state-of-the-art, and the road ahead, IEEE Transactions on Communications 68 (6) (2020) 3834–3862.

[6] C. Ouyang, Y. Liu, H. Yang, On the performance of uplink ISAC systems, IEEE Communications Letters 26 (8) (2022) 1769–1773.

[7] M. Di Renzo, M.D. Migliore, Electromagnetic signal and information theory, IEEE BITS the Information Theory Magazine (2024).

[8] R. Singh, J. An, A. Kaushik, C. Yuen, A. Zhang, O.A. Dobre, Redefining electromagnetic signal and information theory with programmable surfaces: Open critical questions and future horizon, Authorea Preprints.

[9] A. Kaushik, R. Singh, M. Li, H. Luo, S. Dayarathna, R. Senanayake, X. An, R.A. Stirling-Gallacher, W. Shin, M. Di Renzo, Integrated sensing and communications for IoT: Synergies with key 6G technology enablers, IEEE Internet of Things Magazine 7 (5) (2024) 136–143.

[10] J. Shao, Y. Wang, L. Chen, Near-surface characterization using high-speed train seismic data recorded by a distributed acoustic sensing array, IEEE Transactions on Geoscience and Remote Sensing 60 (2022) 1–11 .

Index

9780443363542